Annual School of Young Scientists 2020: "Synthesis, structure and properties of high entropy materials"

IOP Conference Series: Materials Science and Engineering
Volume 1014

Belgorod, Russia
14 – 16 October 2020

ISBN: 978-1-7138-4548-5
ISSN: 1757-8981

Printed from e-media with permission by:

Curran Associates, Inc.
57 Morehouse Lane
Red Hook, NY 12571

Some format issues inherent in the e-media version may also appear in this print version.

Institute of Physics
Dirac House, Temple Back
Bristol BS1 6BE UK

Phone: 44 1 17 929 7481
Fax: 44 1 17 920 0979

techtracking@iop.org

Additional copies of this publication are available from:

Curran Associates, Inc.
57 Morehouse Lane
Red Hook, NY 12571 USA
Phone: 845-758-0400
Fax: 845-758-2633
Email: curran@proceedings.com
Web: www.proceedings.com

TABLE OF CONTENTS

Preface

The development of new materials drives the progress of human society, which experienced the stone age, the iron age, the steam engine age, and the information age. Metal and alloys were considered as state of art materials for structural applications for centuries. For example, during the XX century progress in metallurgy has enabled the humanity to build cars, planes, and space rockets, i.e. to create modern transportation industry. However, currently metallic alloys face severe competition from other materials like plastics, ceramics, composites, etc. To widen the application range of metals and alloys, we need to obtain alloy materials with superior combinations of different properties for daily life, national defense, and outer space exploration.

In recent years, “high-entropy alloys (HEAs)” or “complex concentrated alloys (CCAs)”, consisting of equal or near equal atomic multi-elements, have attracted a large number of scholars from different fields. HEAs/CCAs belong to central, weakly explored, areas of multicomponent phase diagrams, and thus can have unique structures and properties. For example, it was found that they can have superior mechanical properties in a wide temperature range (from cryogenic to extremely high temperatures). Recently, it has been shown that they also demonstrate promising functional properties. Moreover, extension of HEAs/CCAs concept to other, non-metallic materials has brought us new classes of materials like high entropy ceramics, high entropy intermetallics, etc. These new materials can also have unique structural and functional properties.

Therefore, International Conference and Annual School of Young Scientists: “Synthesis, structure and properties of high entropy materials” is organized and held at Belgorod State University. The main aim of the school is to gather the young scientists interested in HEAs and related materials and give them the platform to listen to established international experts in the field and share their own research. The first School was successfully held on October 9-11, 2019 with 17 invited speakers and more than 60 attendees. Despite the Coronavirus pandemic, the next meeting was held on October 14-16, 2020. The Conference and School has attracted 11 invited speakers and more than 80 young scientists. The topics of interest included:

High- and medium-entropy alloys, complex concentrated alloys, methods of their preparation, structure, mechanical and functional properties, phase stability and structural phase transformations, deformation mechanisms, diffusion and ordering processes.

High-entropy and complex concentrated coatings, methods of their preparation, structure, and properties.

High-entropy ceramics, methods of their preparation, structure, and properties.

Research and development of new complex concentrated materials for technology and medicine, including nanostructured materials, microstructural design of multicomponent materials, prospects for practical application.

Advanced methods of obtaining and processing metallic and non-metallic materials for structural and functional applications, including additive technologies, new casting, powder metallurgy, welding, surface treatment technologies.

Other promising directions in the development of new metallic and non-metallic materials.

Computational modeling of complex concentrated alloys, coatings, and ceramics in different environments.

The selected works presented at the International Conference and Annual School of Young Scientists 2020: “Synthesis, structure and properties of high entropy materials” are published in this special issue. The School was supported by Russian Science Foundation Grant № 19-79-30066.

Organizing committee:

Gennady Salishchev
Anderey Rempel
Nikolay Repnikov
Yuri Chumlyakov
Nikolay Kashaev
Vladimir Sanin
Yulia Ivanisenko

Irina Kireeva
Dmitry Luzgin
Evgeniy Trofimov
Elena Astafurova
Dmitry Moskovskikh
Sergey Zherebtsov
Nikita Stepanov
Alexey Lipnitsky
Marina Tikhonova

IOP Conf. Series: Materials Science and Engineering 1014 (2021) 011002 doi:10.1088/1757-899X/1014/1/011002

Peer review declaration

All papers published in this volume of *IOP Conference Series: Materials Science and Engineering* have been peer reviewed through processes administered by the Editors. Reviews were conducted by expert referees to the professional and scientific standards expected of a proceedings journal published by IOP Publishing.

- **Type of peer review:** Single-blind
- **Describe criteria used by Reviewers when accepting/declining papers.**

- Scientific novelty.
- Technical correctness.
- Quality of figures and tables.
- Proper literature survey.
- Interpretation of presented data.
- Language.

Was there the opportunity to resubmit articles after revisions?

Yes, there was.

- **Conference submission management system:** http://shea.bsu.edu.ru/shea/2020/pub/
- **Number of submissions received:** 71
- **Number of submissions sent for review:** 71
- **Number of submissions accepted:** 65
- **Acceptance Rate (Number of Submissions Accepted / Number of Submissions Received X 100):** 91 %
- **Average number of reviews per paper:** 2
- **Total number of reviewers involved:** 43
- **Any additional info on review process:**

- Each reviewer confirmed no conflict of interest with the authors.
- Each reviewer was assigned in accordance with his scientific background.

- **Contact person for queries:** tikhonova@bsu.edu.ru

IOP Conf. Series: Materials Science and Engineering 1014 (2021) 011002 doi:10.1088/1757-899X/1014/1/011002

Tikhonova Marina
PhD, Head of Department of Materials Science and Nanotechnology,
Senior Research Associate
Laboratory of Mechanical Properties
of Nanoscale Materials and Superalloys,
Belgorod State National Research University

Simulation of heat transfer processes in a diesel engine equipped with pistons coated by using Galvanic-plasma Method

M S Al-Bdeiri*, V V Krasilnikov, S V Sergeyev

Belgorod State National Research University, Pobeda, 85, Belgorod, 308015, Russia

*Corresponding author: engmahmood86@gmail.com

Abstract. The methodology for analyzing the thermodynamics and heat transfer of the combustion chambers of a diesel engine is presented. The work is devoted to the modified quasi-steady method (MQM) for the analysis of a diesel engine piston coated with an aluminum alloy. The oxide-coated piston using a galvanic-plasma modification (GPM) depending on the thickness of the coating, including comparison with the results for uncoated piston temperature, to achieve higher characteristics of the Cummins KTA-50 diesel engine. In thermodynamic modeling of a diesel engine, instantaneous gas temperatures and convective heat exchanges are first predicted. The time-dependent boundary conditions are then applied to the gas-blown surfaces of two-dimensional, transitional finite element models of the components of the combustion chamber. Further, the predictions on the finite elements of the instantaneous heat flux passing through each surface of the component are used to determine when the engine goes into quasi-stationary operation. The results show that our path in methodology can identify the complex transition paths of the heat process in the engine combustion chamber and significantly improve heat conduction and convection heat models when modeling a diesel engine.
Keywords: modified quasi-steady method (MQM), thermodynamics, a galvanic-plasma modification (GPM), heat transfer, diesel engine piston, aluminium alloy coating.

1. Introduction

Heat transfer from gas to the wall of a diesel engine critically affects engine performance, fuel economy, and exhaust emissions [1]. In addition, it affects the design of the elements of the combustion chamber, their cooling system, and lubrication. The heat transfer in cylinders, affecting the expected increase in average effective pressure and exhaust energy, led researchers to develop many thermodynamic simulations of the engine to determine the decrease in cooling or insulation of the components of the combustion chamber using ceramics and other materials with low thermal conductivity[2, 3]. The accuracy of engine performance and heat removal predictions is critically dependent on the modeling the heat transfer mechanisms in the piston and their relationship with components structural for the combustion chamber. Multidimensional, transitional models of wall thermal conductivity in complex geometries were developed on the basis of the finite element method (FEM) and were used together with the found cycle time and boundary conditions for gases to study the cyclic thermal behavior of individual components. This approach significantly contributed to our understanding of the problems associated with engine design, such as thermal failure of ceramic coatings caused by stress near sharp angles, heat transfer from the exhaust gas to the cooling oil, and thermal breakdown of the lubricant [4, 5]. However, from the point of view of computational efficiency, FEM simulations of the cyclic heat equation were highly discretized.

2. Engine configuration and boundary condition

The engine is a 16-cylinder, two-stage turbocharged engine with direct injection and high pressure. Figure 1 shows the geometry (to scale) and material composition of a cylinder-piston group system (CPG).

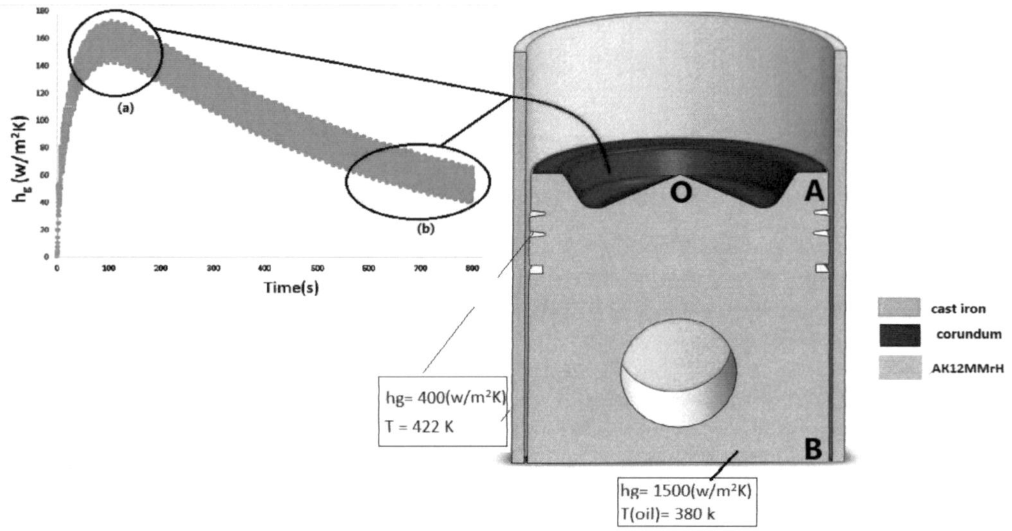

Figure 1. Determination of the geometry and materials of (CPG), as well as their thermal environment. Boundary conditions (including MQM profiles of average gas temperature and convective heat transfer coefficient) are used for FEM analysis

Figure 1 defines the boundary conditions used to perform transient thermal analysis of the components of the combustion chamber. The instantaneous convective heat transfer coefficient of the gas is obtained by the MQM analysis. The instantaneous extended convective coefficient is calculated in the FEM based on the radiation temperature of the gas and the local component of the surface temperature We can see in Figure 1 zone (a) the maximum change in temperature at the start of the engine and this continues till 100 seconds, Hence the transient temperature turns gradually going to be stable as shown in the zone (b). A reduced cyclic convective coefficient (without radiation) is applied to the surfaces of the gap between the piston and the sleeve above the main ring. Corundum (Al_2O_3) [6]with a thermal conductivity of k = 35 W / (m * K) is applied to the upper insert of an aluminum piston with a thickness of 0.1 mm and 0.2 mm, respectively[7]. Our study focuses on cyclic thermal transients when the engine reaches a quasi-stationary mode of operation at a constant speed of 1,500 rpm [8].

3. Results and discussion

Figure 2 shows the predicted instantaneous temperatures at the outer edge of the ceramic-coated piston through many thermodynamic cycles of the engine as shown in Figure 1. The effect of realistic piston geometry on the heat flux can be seen in Figure 2a. The increasing surface temperature near the center of the piston (r = 0) is observed. When you move away from the center of the piston (r ~ 30 mm) the temperature drops as the cooling effects of the piston interface increase the radial heat transfer. The surface temperature rises sharply near the outer edge of the bowl (r ≅ 55 mm). As expected, the surface temperature of the aluminum substrate is significantly lower than that of the uncoated piston surface due to the low thermal conductivity of the coating. The temperature curves obtained for different bottom thicknesses are similar to each other and approximately equidistant. Lowering the temperature on the metal surface makes a positive contribution to the strength of the piston material. Changes in maximum temperatures with coating thickness for substrate coating surfaces are shown.

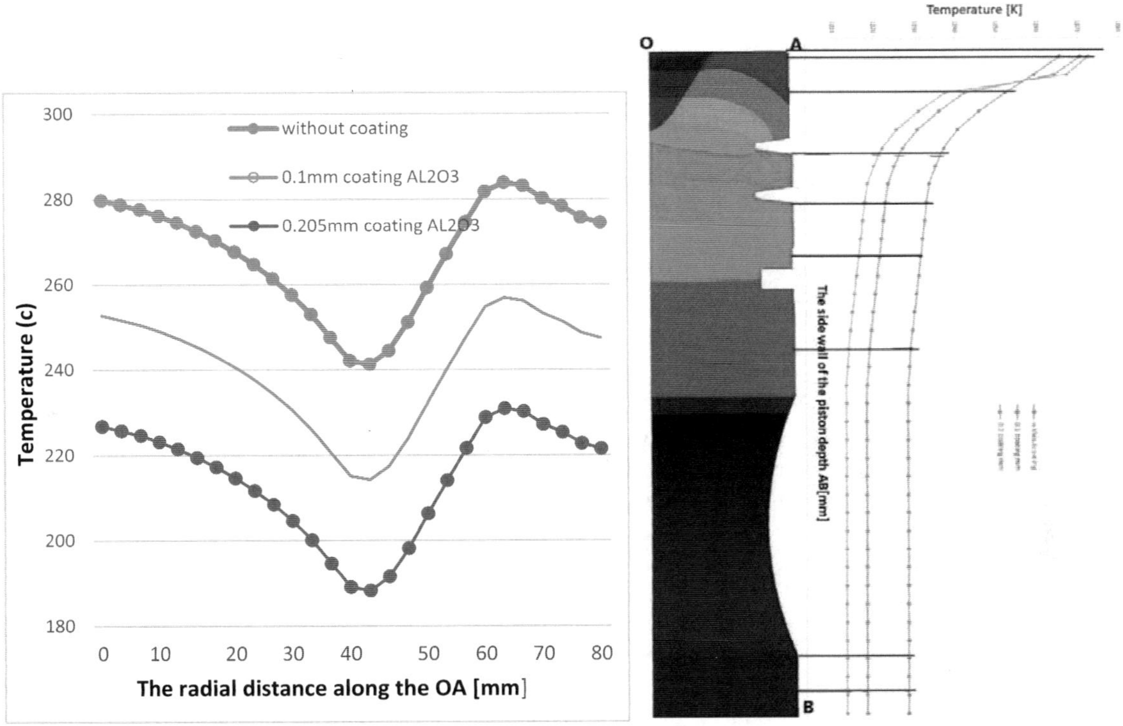

a) b)

Figure 2. It is shown the distribution of the temperature of the piston surface. **a)** The radial distance represents a radius of piston OA and **b)** Distribution of transient temperatures along the sidewall of the piston AB

Figure 2b) shows the distribution of the transient temperature along the sidewall of the ceramic-coated piston. As expected, because of the heat transfer process in coated piston through a three-layer wall. The first layer is an oxidized layer, the second is the base material of the piston, and the third is the layer of cooling engine oil, this would lead the transient temperatures gradually decreases towards the skirt of the piston.

In the process of thermodynamic modeling of a diesel engine, the instantaneous gas temperatures and convective heat transfers are determined, and the predictions on the finite elements of the instantaneous heat flow are used to determine the moment when the engine goes into quasi-stationary operation. This technique allows us to identify complex trajectories of heat flow transition in the combustion chamber and significantly improve the thermal conductivity of diesel engine heat flow models. This makes it possible to plan thermal areas in components, providing designers with tools to track the heat flow path.

4. Conclusion

The methodology of conjugate thermodynamic and heat transfer analyzes of a diesel engine is presented. The methodology is based on the modified quasi-steady method (MQM) and finite element methods (FEM). It was shown that this methodology can reveal the complex transitional paths of the heat flux in the combustion chamber of the engine and reveal the details of the distribution of heat losses in various cooling media. Using two-dimensional resistor networks, it is possible to more accurately analyze the heat generation and the predictions of engine performance with geometrically similar components. Despite the fact that the methodology of coupled thermodynamics and heat transfer presented here was demonstrated using the MQM and FEM programs developed by our group, it is certainly general and can be adapted to any programs that have capabilities similar to those used in this analysis . In addition, the methodology can be easily expanded to determine the values of resistors associated with heat flow paths obtained from the analysis of three-dimensional FEM. However, it is believed that such additional complexity is not justified, unless the boundary conditions for gas used in the FEM analysis are not for space, but for time are formulated. The latter approach will require the use of a program based on fluid mechanics (in contrast to the MQM analysis in combination with three-dimensional models of thermostructural components).

5. References

[1] Havale S A and Wankhade S 2017 Design, Thermal Analysis and Optimization of a Piston using ANSYS International Research Journal of Engineering and Technology **4** 8-15

[2] Shariyat M, Sola J F and Jazayeri S 2016 Experimentally validated combustion and piston fatigue life evaluation procedures for the bi-fuel engines, using an integral-type fatigue criterion Latin American Journal of Solids and Structures **13** 1030-53

[3] Kakaee A H, Gharloghi J, Foroughifar A and Khanlari A 2015 Thermo-mechanical analysis of an SI engine piston using different boundary condition treatments Journal of Central South University **22** 3817-29

[4] Munyao E M, He J G, Zhiyuan Y and Yi Z X 2014 Simulation of Thermal-Mechanical Strength for Marine Engine Piston Using FEA Journal of Engineering Research and Applications **4** 319-23

[5] Zhang X, Zhang Y, Chang L, Jiang Z, Yao Z and Liu X 2012 Effects of frequency on growth process of plasma electrolytic oxidation coating Materials Chemistry and Physics **132** 909-15

[6] Nabi M N and Rasul M G 2019 One-Dimensional Thermodynamic Model Development for Engine Performance, Combustion and Emissions Analysis Using Diesel and Two Paraffin Fuels Energy Procedia **156** 259-65

[7] Sergeev S, Albieri M S, Yatsenko V and Natalya D 2019 Theoretical and Practical Study of Possibility to Decrease Thermal Stress in Pistons of Internal Combustion Diesel Engine by Using Galvanic Plasma Modification International Journal of Advanced Science and Technology **28** 550-62

[8] Baker D M and Assanis D N 1994 A methodology for coupled thermodynamic and heat transfer analysis of a diesel engine Applied mathematical modelling **18** 590-601

IOP Conf. Series: Materials Science and Engineering 1014 (2021) 012002 doi:10.1088/1757-899X/1014/1/012002

Spark plasma sintering of Si_3N_4-YAG nanocomposite

P V Andreev[1*], E Ye Rostokina[1], L S Alekseeva[2], A A Murashov[2], M S Boldin[2] and S S Balabanov[1]

[1]G.G. Devyatykh Institute of Chemistry of High-Purity Substances of the Russian Academy of Sciences, Nizhny Novgorod, 603951, Russia
[2]Physical-Technical Research Institute, Lobachevsky State University of Nizhny Novgorod, Nizhny Novgorod, 603950, Russia

[*]Corresponding author: andreev@phys.unn.ru

Abstract. Spark plasma sintering (SPS) of mixtures based on silicon nitride nanopowders (Si_3N_4) was considered. A method for creating a mixture based on nanosized Si_3N_4 and a sintering additive of yttrium-aluminum garnet (YAG, $Al_5Y_3O_{12}$) was developed. The precursor of the sintering additive was formed by the method of spray drying using a mixture of colloidal solutions of $Al_5(NO_3)_3(OH)_{12}$ and $Y(OH)_2(CH_3COO)$. The produced mixture samples were studied by X-ray diffraction analysis and SEM. The kinetics of the SPS of the prepared mixtures was studied. The effect of preliminary annealing (up to 1000 °C) of the mixture on the formation of the YAG sintering additive have been considered, as well as the method of reactive synthesis of the sintering additive in the SPS process. It was found out that sintering of the annealed mixture does not allow forming high density ceramics. As a result of reactive sintering, ceramics based on Si_2N_2O were formed.

1. Introduction

Ceramics based on silicon nitride Si_3N_4 due to its physical and mechanical properties, such as high strength and hardness, bending strength, corrosion resistance, which in addition to refractoriness provides a number of advantages as a material for both cutting and machining tools, and for critical engine components in the automobile and aircraft industries [1]–[8].

Sintering of pure Si_3N_4 is difficult because of its covalent nature, which causes the slowing down of diffusion processes, particularly the processes of viscous flow, affecting the mass transfer and densification during sintering. For this reason, in the main modern methods of sintering ceramics Si_3N_4, activating additives are used, which at high temperatures form a liquid phase at the grain boundaries of the main component and contribute to the consolidation of ceramics.

Recently, ceramics with an ultrafine grain structure, including those based on Si_3N_4, have become increasingly popular for researchers [9]. In some cases, the reduction in grain size has a positive effect on the mechanical characteristics of ceramics. Among the sintering methods, in this case, preference is given to the method of spark plasma sintering (SPS). SPS allows compacting mixtures in the high-speed modes of heating, which does not lead to significant growth of crystalline grains during sintering and allows you to form an ultra-fine grained structure in ceramics.

The aim of the study was to investigate the possibility of fabrication of ceramics based on an ultrafine grained Si_3N_4-YAG mixture using the SPS method.

2. Materials and Methods

The objects of study were powder mixtures based on Si_3N_4 produced by "abcr GmbH" (Germany) with a particle size of up to 15 nm.

The application of the sintering additive by spray drying is based on a mixture of $Al_5(NO_3)_3(OH)_{12}$ and $Y(OH)_2(CH_3COO)$ colloid solutions with Si_3N_4 powder. The water suspension contained 52.5% wt. of Si_3N_4-YAG in conversion to solid phase. To stabilize the suspension, 1.5% wt. ammonium polyacrylate was added. After mixing in a planetary mill, the suspension was dried by spray drying at 150 °C.

The obtained mixture was annealed in four stages at temperatures of 300, 500, 800, 1000 °C for 2-8 hours with dispersion in an agate mortar between stages. The purpose of the annealing was to complete the YAG synthesis with the formation of sintering additive layer on the surface of Si_3N_4 particles.

The SPS was performed on DR. SINTER model SPS-625 equipment. Sintering was performed at a pressure of P = 70 MPa, the heating rate was V = 100 °C/min. To study the reactive synthesis of the sintering additive SPS was also carried out for the initial mixtures without preliminary annealing.

The crystal structure of the samples was studied by X-ray diffraction analysis using a Shimadzu XRD-7000 X-ray diffractometer. The microstructure of the powders was investigated by scanning electron microscopy using a JEOL JSM-6490 instrument. The density of sintered samples was measured by hydrostatic weighing in distilled water using Sartorius CPA scales.

3. Results and discussion

SEM images of mixtures before and after sintering are shown in figure 1. The images show spherical particles forming agglomerates of Si_3N_4 particles and sintering additives.

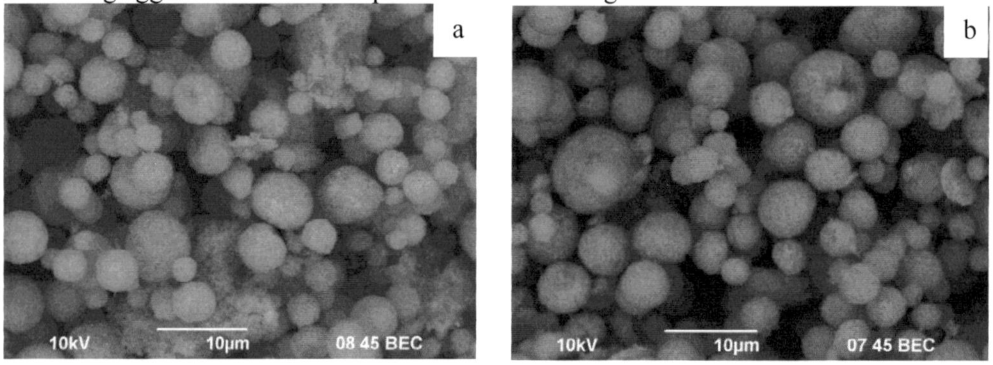

Figure 1. SEM images of the mixture based on Si_3N_4 nanopowder and YAG precursor before annealing (a) and after annealing (b).

The temperature dependences of shrinkage for sintering of the analyzed mixtures are presented in figure 2. The value of the α-Si_3N_4-YAG composite density (91.5-8.5% wt.) is accepted as ρ_0. The sintering kinetics of the studied samples differs significantly. The sample, which is not preliminary annealed, has a characteristic step on the shrinkage curve at about 1370 °C, typical for reactive sintering. The intensive shrinkage stage for this sample is completed at temperatures about 1700 °C. The use of preliminarily annealing for the mixture leads to a shift in the intensive shrinkage stage to lower temperatures by ~ 300 °C. Further heating above T = 1270 °C causes melting and decomposition of the sample.

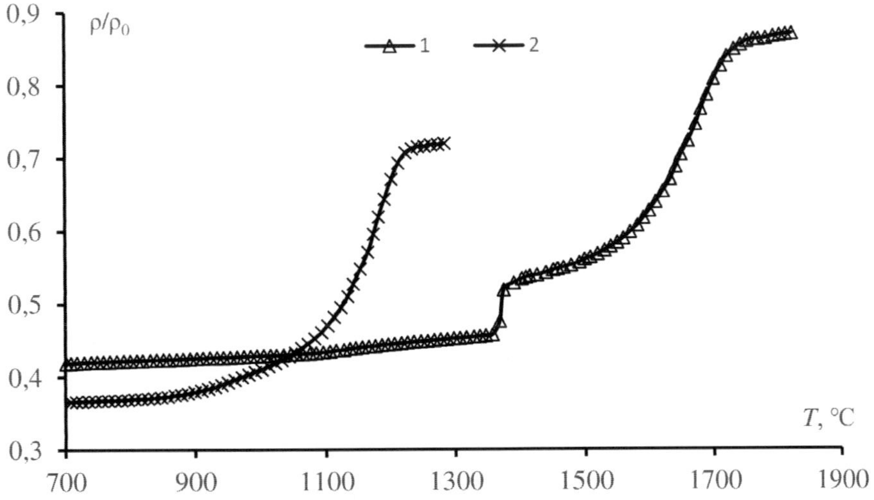

Figure 2. Densification versus temperature for reactive sintering (1) and sintering of annealed mixture (2).

X-ray diffraction patterns of the considered mixtures and ceramics samples prepared of them by the method of SPS are presented in figure 3.

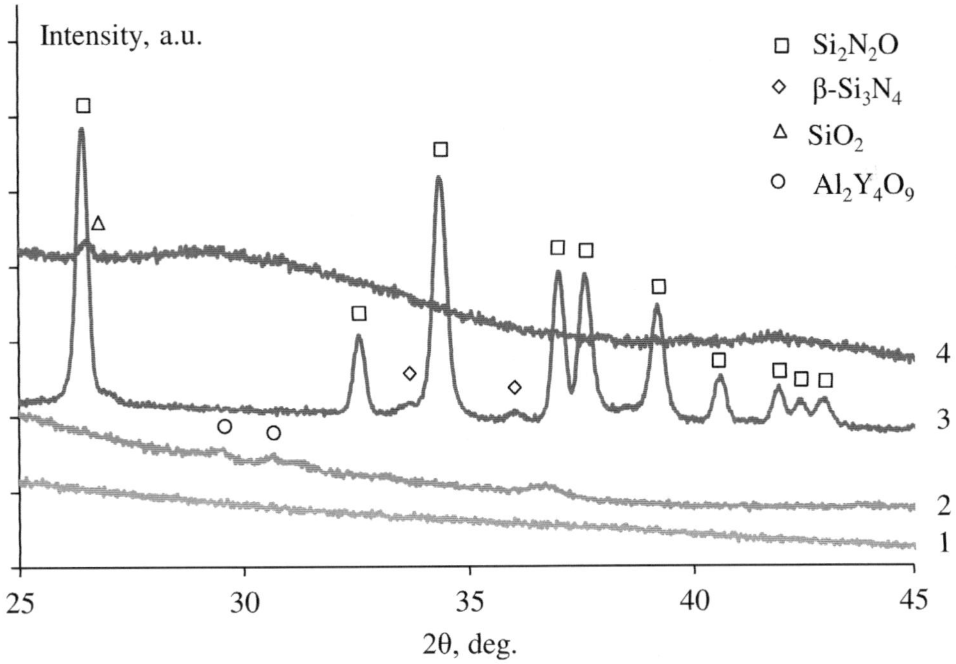

Figure 3. Diphractograms of mixtures before (1) and after (2) annealing, and also made of them by the method of SPS ceramics (3 and 4 accordingly)..

As shown in the X-ray diffraction patterns in figure 3, we can conclude that the process of annealing of the originally amorphous mixture does not lead to the formation of a significant volume of rather coarse crystallites. Most likely, particles of mixed yttrium and aluminum oxides of various stoichiometry are formed during annealing, as is proven by a small number of low-intensity diffraction peaks. It is interesting that at the following SPS, intensive peak growth did not occur, but crystallites of one of SiO_2

modifications were found. As mentioned, the synthesis was stopped at about 1100 °C. Experiments have shown that heating the mixture above 1200 °C leads to intense melting of the sample accompanied by a sharp increase in pressure and subsequent break of the mold. This is probably due to the process of intense formation of the SiO_2 crystalline phase, which has a lower melting point than Si_3N_4.

During the reactive sintering of the initial mixture without preliminarily annealing, intensive growth of Si_2N_2O crystallites occurred, which may be associated with a high content of adsorbed oxygen in the mixture. It is also should be noted that a small fraction of β-Si_3N_4 was formed.

4. Conclusion

Therefore, oxygen adsorbed on the surface of Si_3N_4 particles has an effect on the sintering of the nanocomposites. In this case, the mixture preliminarily annealed in air leads to the formation of a mixture, the sintering of which is impossible - the sample melts intensively at a temperature of about 1200 °C. Reactive sintering allows obtaining a ceramic sample, but the Si_2N_2O phase is formed, as well as a small proportion of β-Si_3N_4. The method we have considered can be used for synthesis of ceramics based on Si_2N_2O with β-Si_3N_4 particles as a reinforcing additive.

Acknowledgments
The reported study was funded by RFBR, project number 19-33-60084.

References

[1] Knoch H and Heinrich J 1980 *Materwiss. Werksttech.* **11** 361
[2] Ziegler G, Heinrich J, and Wötting G 1987 *J. Mater. Sci.* **22** 3041
[3] Pyzik A J and Carroll D F 1994 *Annu. Rev. Mater. Sci.* **24** 189
[4] Riley F L 2000 *J. Am. Ceram. Soc.* **83** 245
[5] Richerson D W 2006 *Proceedings of the ASME Turbo Expo (Barcelona)* vol 2 (ASME) pp 241–254
[6] Hampshire S 2007 *J. Achiev. Mater. Manuf. Eng.* **24** 43
[7] Bocanegra-Bernal M H and Matovic B 2010 *Mater. Sci. Eng. A* **527** 1314
[8] Klemm H *J. Am. Ceram. Soc.* **93** 1501
[9] Perevislov S N, Lukyanova O A, Lysenkov A S, Kim K A, and Vysotin A B *IOP Conf. Ser. Mater. Sci. Eng.* **848** 012068

Chemical and phase composition of the novel superelastic Ti-Zr-Nb alloy with a high Zr content

D V Barilyuk[1*], A I Bazlov[1], S D Prokoshkin[1] and A S Konopatsky[1]

[1]National University of Science and Technology (NUST) (MISiS), 4 Leninsky, Moscow 119049, Russia

[*]Corresponding author: danybary@mail.ru

Abstract. The novel superelastic Ti-44Zr-10Nb (at.%) alloy was obtained by the vacuum arc remelting method and subjected to homogenization (950 °C, 60 min) and thermomechanical treatment (TMT), which consisted of cold rolling (CR) with a 25% thickness reduction rate and consequent post-deformation annealing (PDA) at 550 °C for 30 min. The analysis of the chemical composition was conducted by the EDS method. It was demonstrated that the obtained chemical composition of the alloy was in good correspondence with the nominal chemical composition. The phase composition of the alloy after TMT was investigated by the XRD technique. The lattice parameters of β and α'' phases were measured using XRD data to calculate the crystallographic recovery strain limit. The maximum transformation strain of 7.3% was obtained in the $[011]_\beta$ direction.

1. Introduction

The development of completely biocompatible metallic materials for bone tissue replacement is one of the most promising areas of modern materials science, which aims to solve applied medical problems [1]. The Ti-Zr-Nb alloys are promising materials with enhanced biomechanical compatibility with bone tissue due to their pronounced superelastic behavior. The superelastic properties of the material can be estimated by evaluating the crystallographic transformation strain, the value of which reaches 8% for the Ti-Zr-Nb alloys with an increased Zr content [2]. Another important issue that should be taken into account is obtaining the optimal phase composition beneficial for the superelasticity of the material at room temperature (RT) [3]. The chemical composition of the material to a great extent defines its phase composition. Our current work is devoted to the manufacturing of a superelastic Ti-Zr-Nb alloy with novel chemical composition and the investigation of its phase composition and crystallographic transformation strain after TMT.

2. Materials and methods

An ingot of the Ti-44Zr-10Nb alloy was manufactured by the vacuum arc remelting (VAR) technique. The mass of the ingot was 150 g. To prevent an increase in the level of impurities, a protective Ar atmosphere was used. Five consequent remeltings were performed with the ingot turned upside down. Impurities were analyzed using the high-temperature gas extraction technique. The results of the analysis are listed in Table 1.

Table 1. Impurities content in the Ti-44Zr-10Nb alloy.

O_2 (wt.%)	N (wt.%)	C (wt.%)	S (wt.%)	H_2 (wt.%)
0.018±0.002	<0.001	0.033±0.002	<0.001	0.004±0.0004

The ingot was subjected to homogenization annealing at 950 °C for 1 hour. Small (10x10x2 mm^2) plates were prepared for metallographic studies after homogenization. For that, polishing with abrasive paper (up to P2000) and polishing fabric was applied. For the etching, the following solution was used: 10 vol% hydrofluoric acid, 30 vol% nitric acid, and 60 vol% water. The chemical composition of the alloy was investigated by Energy Dispersive Spectroscopy (EDS) using an X-Max detector (Oxford instruments) and a JEOL 5600F scanning electron microscope. Additional samples 5x5x1.4 mm^2 were subjected to the following TMT scheme: cold rolling (CR) with a 25% thickness reduction rate and post-deformation annealing (PDA) at 550 °C for 30 min (quenching in water). The phase composition was investigated by the X-ray diffraction (XRD) method using a DRON-2 diffractometer operated at 52 kV and 32 mA with Cu-kα radiation (λ=0.1540 nm) at room temperature (RT). The XRD profile was acquired in a 2θ range from 30° to 70° with a step size of 0.05° and an exposure time of 5 sec. The surface of the samples for XRD was subjected to polishing with an abrasive paper (P800) and short time etching to remove defective surface layers. The crystallographic transformation strain calculation was based on the phenomenological theory of martensitic transformations (PTMT). According to PTMT, the arbitrary vector x of the β phase transforms to the vector x' of α'' during the β→α'' martensitic transformation as follows [4]:

$$x' = T.x \tag{1}$$

Where T is the lattice distortion matrix for each martensite correspondence variant (CV) expressed in the coordinates of the parent phase. There are 6 CVs of martensite in the BCC (β) to the orthorhombic (α'') lattice transformation.

To calculate the crystallographic transformation strain ε^i induced by the lattice distortion in any crystallographic direction, we can use the following expression [4]:

$$\varepsilon^i = \frac{|x'| - |x|}{|x|} \tag{2}$$

3. Results and discussion

Figure 1 shows a micrograph of the Ti-44Zr-10Nb alloy after homogenization at 950 °C for 1 hour.

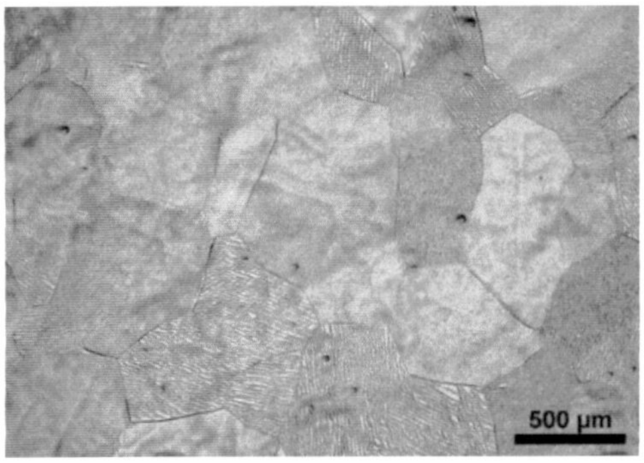

Figure 1. Optical image of Ti-44Zr-10Nb alloy microstructure.

It follows from Figure 1 that no typical defects of the cast structure such as dendrites can be seen after homogenization [5]. The grain structure of the materials is represented by large equiaxed

grains of 427±90 μm. The defects free structure with equiaxed grains is beneficial for the consequent TMT, including CR since in this case the mechanical stress is distributed more uniformly.
Figure 2 collects the results of EDS studies: an electron image, the corresponding EDS profile, and elemental maps of the investigated alloy. It can be seen from Figure 2b that the measured and nominal chemical compositions are in good correspondence. Elemental maps (Figure 2c) demonstrate that all elements are distributed uniformly.

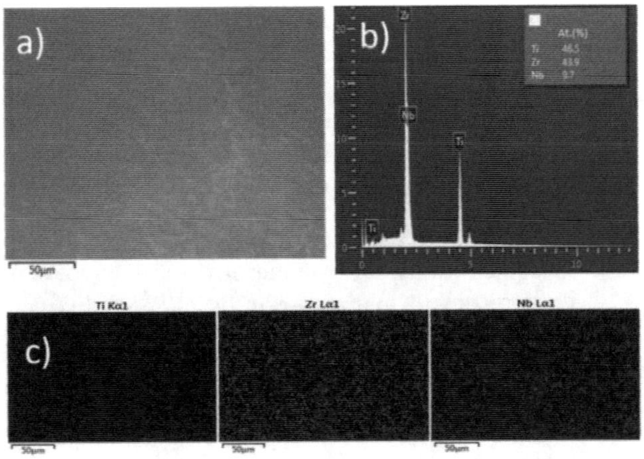

Figure 2. Chemical composition of theTi-44Zr-10Nb alloy; (a) electron image, (b) EDS spectra, (c) elemental maps.

Figure 3 presents two XRD profiles of the investigated alloy after TMT and TMT with additional plastic deformation of 10%. As shown, the main and only phase in the alloy after TMT is β. It means that the temperature of the martensitic transformation is lower than RT, which contributes to a distinct superelastic behavior. The absence of the α'' phase makes it impossible to calculate the transformation strain (since it depends on the lattice parameters of each phase). An additional CR was used to induce martensitic transformation and stabilize a certain amount of the α''- phase at RT[6]. The corresponding XRD profile contains a noticeable amount of the α'' phase. The calculation of the lattice parameters and, therefore, the maximum transformation strain of the alloy was conducted based on XRD data.

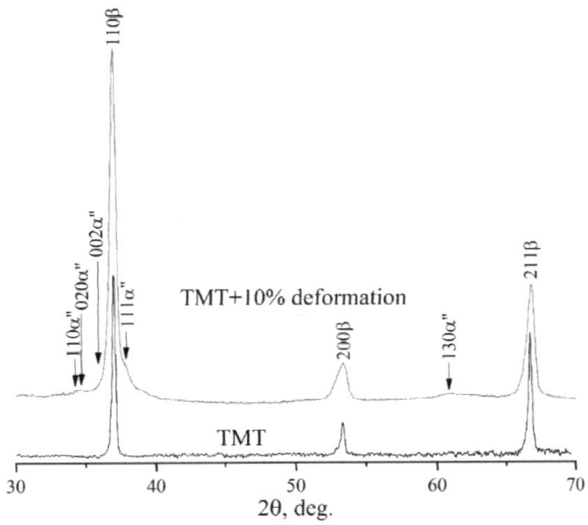

Figure 3. XRD profiles for Ti-44Zr-10Nb after TMT.

According to XRD data, the lattice parameters of the Ti-44Zr-10Nb alloy are a_0=0.3431 nm, a=0.3137 nm, b=0.5206 nm, and c=0.5016 nm. The results of calculating the transformation strain along three representative directions $[001]_\beta$, $[011]_\beta$, and $[-111]_\beta$ are shown in Table 2.

Table 2. Transformation strain of the Ti-44Zr-10Nb alloy along three representative directions.

	$\varepsilon^{[001]}$	$\varepsilon^{[011]}$	$\varepsilon^{[-111]}$
CV1	0.0501	0.0729	0.0228
CV2	0.0501	0.0338	-0.0045
CV3	0.0501	-0.0136	-0.0045
CV4	0.0501	-0.0136	0.0228
CV5	-0.0857	-0.0136	-0.0045
CV6	-0.0857	-0.0136	0.0228

As we can see, the most favorable variant of martensite to form upon application of tensile stress is CV1, because it shows only positive values along all orientations. On the other hand, the best option for compression is CV5. The maximum transformation strain of 7.3% is obtained for the $[011]_\beta$ direction of CV1, it should be noted that this estimation is given for the single crystal approximation. Materials with high transformation strain tend to demonstrate enhanced functional properties. [7].

4. Conclusions

In this study, a novel Ti-44Zr-10Nb alloy was obtained by the vacuum arc remelting method. The EDS analysis showed that all elements of the alloy were distributed uniformly and that the obtained chemical composition was in good correspondence with the nominal chemical composition. The XRD analysis revealed that the phase composition of the alloy after TMT consists of the β phase only. From the XRD data obtained for the sample after additional deformation of 10%, the lattice parameters of the β and α'' phases were calculated. The maximum transformation strain was evaluated as 7.3%. These results suggest that Ti-44Zr-10Nb is a promising material for biomedical applications.

Acknowledgments
The present work has been carried out under the financial support of the Russian Science Foundation (project №20-63-47063).

References

[1] Machado L G and Savi M A 2003 *Braz. J. Med. Biol. Res.* **36** 683-91
[2] Kim H Y, Fu J, Tobe H, Kim J I and Miyazaki S 2015 *Shape Memory and Superelasticity* **1** 107-16
[3] López Pavón L, Lopez Cuellar E, Villalpando Hernandez S, E.Moreno-Cortez I, Kim H Y and Miyazaki S 2019 *Journal of Alloys and Compounds* **782** 893-8
[4] Kim H Y and Miyazaki S 2018 *Ni-Free Ti-Based Shape Memory Alloys* (Oxford: Butterworth-Heinemann) pp 57-9
[5] Konopatskii A S, Zhukova Yu S, Dubinskii S M, Korobkova A A, Filonov M R and Prokoshkin S D 2016 *Metallurgist* **60** 223-8
[6] Konopatsky A S, Dubinskiy S M, Zhukova Yu S, Sheremetyev V, Brailovski V, Prokoshkin S D and Filonov M R 2017 *Materials science and Engineering* A **702** 301-11
[7] Li S, Choi M-S, Nam T-H 2020 *Materials science and Engineering* A **782** 139278

IOP Conf. Series: Materials Science and Engineering 1014 (2021) 012004 doi:10.1088/1757-899X/1014/1/012004

A new combined SPD technique to improve mechanical properties and electrical conductivity of long-sized billets

E V Bobruk[1,2*], A V Botkin[2], M Yu Murashkin[1,2] and G I Raab[2]

[1]Saint Petersburg State University, 28 Universitetsky pr., Peterhof, Saint Petersburg, 198504, Russia
[2]Ufa State Aviation Technical University, 12 K. Marx, Ufa, 450008, Russia

[*] Corresponding author: e-bobruk@yandex.ru

Abstract. As a result of computer simulation carried out using the DEFORM 3D software package, the influence of the shape of the extrusion die-set, combined with equal-channel angular pressing with parallel channels (ECAP-PC), on the stress-strain state of cylindrical billets of an aluminum alloy was established. The obtained data made it possible to manufacture an experimental die-set for extrusion, combined with ECAP-PC, and to produce in it a cylindrical long-sized UFG billet from the 6063 alloy. It was established that one cycle of direct extrusion combined with ECAP-PC in the matrix with a non-axisymmetric deforming section for direct extrusion leads to a significant microstructure refinement and an improvement of mechanical properties of the alloy.

1. Introduction

Equal channel angular pressing (ECAP) has been the most common method of severe plastic deformation (SPD), which enables producing billets in the form of bars with an ultrafine-grained (UFG) structure from metallic materials including those based on aluminum for structural and electrical purposes [1]. Simple shear is realized, as the billet passes through the conjugation zone of channels with equal cross-sections, made in special equipment, while the diameter retains. This tool shape allows repeatedly deforming the billet and achieving ultra-high strain values ($e \geq 4$), which is a prerequisite for the UFG structure formation and, accordingly, for enhancement of mechanical and service properties of the billet material.

The development of the ECAP method is ECAP-PC [2], which, in comparison with ECAP, provides a decrease in processing cycles due to shear occurring in two deformation zones corresponding to two consecutive zones of intersection of channels in the tool. But even such a number of processing cycles of ECAP-PC is difficult to be used in batch industrial production.

Direct extrusion combined with ECAP-PC of a cylindrical billet is a promising development of the ECAP-PC method, focused primarily on improving the productivity of processing UFG material by three or more times. A distinctive feature of direct extrusion combined with ECAP-PC is the feasibility of shear in three deformation zones, in two mutually perpendicular planes - during direct extrusion in the die with a non-axisymmetric deforming section and ECAP-PC in two intersecting tool channels. In this case, the deformation of the billet according to the scheme described above can reach $e \sim 3.5$ and more, which makes it possible to provide a significant refinement of the microstructure of the processed material already in one processing cycle and, accordingly, enhance the mechanical properties. In this work, experimental evaluation of the obtained simulation results of the processing by direct extrusion combined with ECAP-PC of a cylindrical billet was carried out using the example of a commercial heat-hardened aluminum alloy 6063 of the Al-Mg-Si system.

2. Experimental

In the course of research, computer simulation of direct extrusion combined with ECAP-PC using the DEFORM 3D software package and a full-scale experiment were carried out. In computer simulation, the following conditions and assumptions were accepted: the tool – an absolutely rigid body (3D models of the tool were previously created in "KOMPAS 3D 8V"); temperature - constant equal to 100 °C; the thermal effect of deformation was neglected due to the low strain rate; billet of the 6063 alloy – a plastic body; the diagram of deformation of the alloy $\sigma_i = 219\ln(\varepsilon_i + 1) + 68$ constructed upon the results of mechanical tests of the samples, was introduced when preparing the database in the form of a table function; the friction coefficient was introduced when preparing the database in the form of table function (f_σ) ($\tau = f_\sigma \sigma_S / \sqrt{3}$) of the modulus of compressive, normal contact stress $|\sigma_н|$; the number of finite elements - 80,000; deformation velocity - 5mm/s; time step 0.1s; the number of simulation steps is 100...400.

Before experimental direct pressing combined with ECAP-PC, the initial cast billets were annealed in a SNOL electric furnace at a temperature of 530 °C.

The direct extrusion combined with ECAP-PC of the billets was carried out in the die-set mounted on a hydraulic press with a nominal force of 600 kN. The press was equipped with a measuring unit including an analogue digital converter and a computer with the "IMADET 5.0" processing program. To study the strained state changing in the billet during deformation, the channel for direct extrusion was designed as axisymmetric and non-axisymmetric (figure 1).

Electron microscopic studies of the materials under study were carried out on a JEOL JEM 2100 transmission electron microscope. Static tensile tests of cylindrical specimens with a gauge length of 15 mm and a gauge diameter of 3 mm were carried out in an Instron 8862 machine at room temperature. Specific electrical conductivity (μ) of the samples was determined using an eddy current meter for electrical conductivity of non-ferrous metals and alloys with a relative error of ± 2%.

3. Results and discussion

3.1. Simulation results

In the simulation, the relative length of the original cylindrical billet and the die shape (L/d) were varied. The deformation zone of the billet was investigated during direct extrusion in the die with axisymmetric and non-axisymmetric deforming sections. The deformation pattern during extrusion was estimated by the Lode coefficient (μ_ε), which was calculated for 80 points uniformly taken over three sections of the deformation zone.

The Lode coefficient values for the points of the deformation zone of a billet deformed in the die with a non-axisymmetric deforming section for extrusion satisfy the inequality: $-0.65 \leq \mu_\varepsilon \leq 0.38$. The inequality for the die with an axisymmetric deforming section for extrusion is $-0.9 \leq \mu_\varepsilon \leq -0.25$. The average values of the Lode coefficient are -0.14 and -0.57 respectively. The deformation schemes of the points of the billet deformed in the die with a non-axisymmetric deforming section for extrusion are close to shear. In the case the die has an axisymmetric deforming region, the scheme is close to stretching. From the obtained modeling data, it can be noted that a more uniform strain distribution in the billet deformed in the die with a non-axisymmetric deforming section for direct extrusion is due to a relatively greater strain heterogeneity in the section of a part of the billet deformed only by direct extrusion.

The study of the dependence of the deforming force on the punch displacement during pressing of a cylindrical billet 30 mm in diameter and 40 mm in length showed that the deforming force in all cases of simulation reaches its maximum value at the moment of completion of the plastic deformation centers formation. The maximum values of the deforming forces do not differ significantly - by 5%, and are 462 kN (axisymmetric deforming section,) 436 kN (non-axisymmetric deforming section). The lower value of the deforming force during pressing in the die with a non-axisymmetric deforming section for direct extrusion is possibly due to the influence of the deformation pattern close to shear realized during pressing.

Figure 2 shows a graphical dependence of the maximum extrusion force combined with ECAP-PC in the die with a non-axisymmetric deforming section for direct extrusion on the relative length of the original cylindrical billet. It can be seen that the dependences of the force and stress on the relative length of the initial cylindrical sample are close to linear dependences. For a relative length L/d of 18, the force reaches P=1100 kN, the maximum contact stress is 2000 MPa, which is close to the initial flow stress of tool steels in the hardened state at room temperature (2500...3000 MPa).

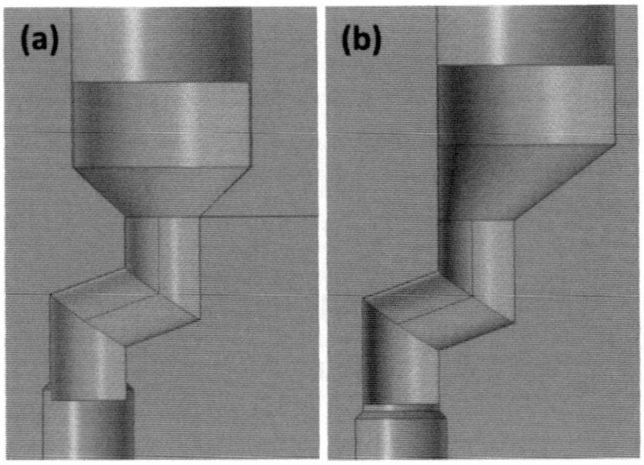

Figure 1. A scheme of the direct extrusion combined with ECAP-PC: die with axisymmetric deforming section (a); die with non-axisymmetric deforming section (b)

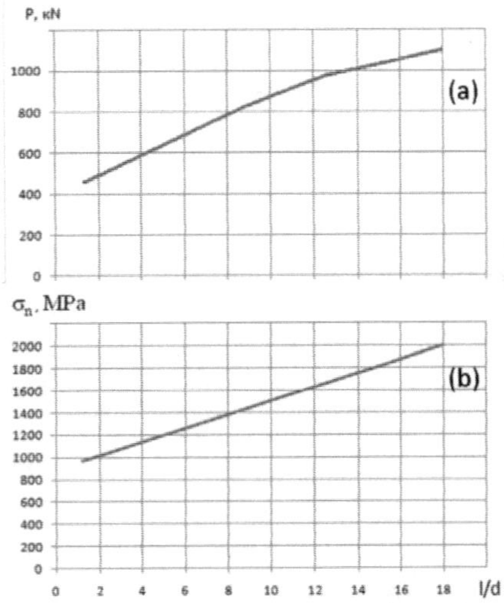

Figure 2. Dependences of the maximum values of the deformation force (a) and normal shear stress (b) on the relative length of the billet (die with non-axisymmetric deforming section)

3.2. Experimental results

Experimental pressing of the 6063 alloy billets with a diameter of 30 mm and a length of 40 mm was carried out on the basis of the results of computer simulation. As a result of deformation processing, defect-free billets were obtained that did not contain surface defects and cracks. The maximum force value obtained by experimental measurement during experimental pressing in the die with a deforming section for direct extrusion was 421.8 kN and differed by 3.4% from the force value obtained by numerical simulation.

The specimens were made from deformed ingots to determine the mechanical properties and electrical conductivity and to assess the hardening effect of the 6063 alloy achieved after processing according to the extrusion scheme combined with ECAP-PC (table 1). For comparison of the obtained level of properties of alloy ingots after processing according to the extrusion scheme combined with ECAP-PC, table 1 shows the properties of this material subjected to standard hardening treatment T6, as well as properties after processing by ECAP-PC. It has been found that the long-sized billets subjected to processing in the die-set, made in accordance with the schemes shown in figure 1, demonstrated higher strength than the billet after special hardening heat treatment T6. The strength of the ingots obtained in various dies for extrusion does not differ significantly, however, there are noticeable differences in the uniform (before necking) deformation of samples obtained from the billets pressed in the die with a non-axisymmetric deforming section for direct extrusion. An increased value of uniform deformation after ECAP indicates the formation of a more homogeneous UFG structure, which is also very important for enhancement of fatigue properties of materials [4].

Table 1. Mechanical properties and electrical conductivity of the 6063 alloy

Treatment	$\sigma_{0.2}$, MPa	σ_B, MPa	δ_P, %	δ, %,	μ, MS/m
Direct pressing in the die with axisymmetric deforming section + ECAP-PC at 100 °C	203±5	224±3	3,0±0,3	12,0±0,3	29,98±0,3
Direct pressing in the die with non-axisymmetric deforming section + ECAP-PC at 100 °C	210±6	230±3	6,0±0,3	11,0±0,3	30,01±0,2
ECAP - PC, 4 cycles at 100 °C [3]	256±2	264±2	-	12,5±0,3	30,04±0,2
T6	145±5	195±3	-	8,0±0,2	30,16±0,2

The values of specific electrical conductivity of the billets after one cycle of processing by the new combined method and four cycles of ECAP-PC are practically the same. This indicates that already in one processing cycle, along with the formation of a developed substructure, dynamic aging occurs in the material, accompanied by a decrease in the concentration of alloying elements in the aluminum solid solution, to a level comparable to that achieved after multi-cycle deformation processing.

The conducted electron microscopic studies of the microstructure of the billet showed that after direct pressing combined with ECAP-PC the initial coarse-grained structure transformed into a lamellar-type structure (figure 3), which had a pronounced orientation relative to the shear direction. The average transverse size of the formed (sub)grains was 430 nm, and the longitudinal size was up to 1500 nm, respectively. Thus, it is the significant refinement of the initial microstructure of the alloy achieved during processing that ensured the achievement of a strength level in the 6063 alloy that exceeds the strength of course grained material after standard hardening treatment and acceptable electrical conductivity.

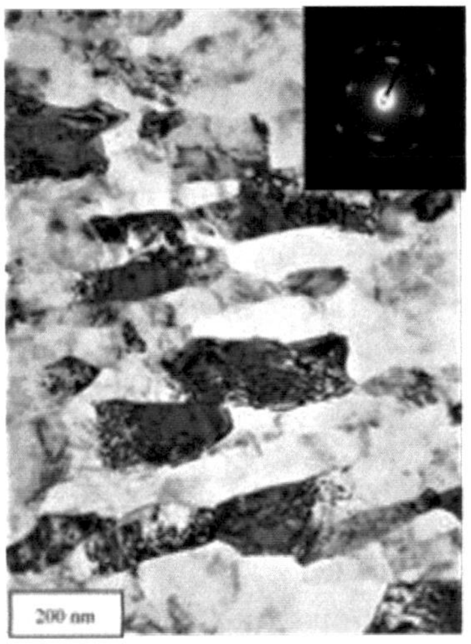

Figure 3. Microstructure of the 6063 alloy after the direct extrusion combined with ECAP-PC

Conclusion

1. Processing the aluminum alloy 6063 using a new technique by direct extrusion combined with ECAP-PC was studied by computer simulation and full-scale experimental characterization.
2. It was shown that a single cycle of direct extrusion combined with ECAP-PC provided a noticeable increase in strength and ductility of the 6063 alloy as compared with the standard hardening heat treatment (T6). This indicates that the proposed technique is promising for processing of aluminum alloys.

Acknowledgments

E.V. Bobruk and M.Yu. Murashkin would like to thank the Russian Science Foundation for financial support of this research on project No. 17-19-01311

References

[1] Valiev R Z, Langdon T G 2006 *Prog. Mater. Sci.* **51** 881
[2] Raab G I 2005 *Mater. Sci. Eng. A* **410–411** 230
[3] Bobruk E, Sabirov I, Kazykhanov V, Valiev R, Murashkin M 2014 *IOP Conf. Ser: Mater. Sci. Eng.* **63** 012116
[4] Estrin Y, Vinogradov A *Int. J. Fatigue* 2010 **32** 898

IOP Conf. Series: Materials Science and Engineering 1014 (2021) 012005 doi:10.1088/1757-899X/1014/1/012005

Portevin-Le Chatelier effect in austenitic steel Fe-18%Cr-8%Ni

Yu Borisova*, **D Yuzbekova and A Mogucheva**

Belgorod State University, Belgorod, 308015, Russia

*Corresponding author: borisovayuliya94@mail.ru

Abstract. The study of the mechanical behavior of the austenitic Fe-18% Cr-8% Ni steel was carried out in the temperature range of 530°C - 680°C at a strain rate of 1×10^{-3} s^{-1}, as well as at a temperature of 620°C and strain rates of 1×10^{-2} s^{-1} - 1×10^{-4} s^{-1}. At these conditions, the PLC effect was observed. Type A serrations were observed in the temperature range 530 - 680°C. A decrease in the strain rate at a temperature of 620°C leads to a change in the type of serrations from type A to type B.

1. Introduction

High-strength austenitic steels are the subject of research and development in many countries due to the growing demand for special steels for the production of high-quality equipment [1]. Austenitic stainless steels are used in the designs of boilers for new generation coal-fired power units operating at supercritical steam parameters (with a steam temperature above 620°C and a pressure over 25 MPa) [2]. These steels are widely used as structural materials due to their excellent oxidation resistance and high creep strength. In particular, metastable austenitic steels such as Fe-18% Cr-8% Ni type stainless steel exhibit excellent tensile properties in a wide temperature range [3].

However, at high operating temperatures, austenitic stainless steels are susceptible to dynamic strain aging (DSA) associated with interactions between dissolved atoms and mobile dislocations during plastic deformation [4]. This effect leads to the appearance of undesirable roughness on the surface of the products. This phenomenon depends on temperature and strain rate. DSA can have a large impact on mechanical properties such as strength and ductility [5]. The aim of this work is to study the manifestation of the Portevin-Le Chatelier (PLC) effect in the Fe-18% Cr-8% Ni austenitic steel.

2. Material and methods

The studied Fe-18% Cr-8% Ni austenitic stainless steel was produced by chill casting. Steel ingots were forged at a temperature of 1180°C. Then, the steel was subjected to solution treatment at 1150°C, for one hour followed by quenching into water.

The microstructure investigation was performed by scanning electron microscopy (SEM) with automated indexing of electron back-scattering diffraction (EBSD) patterns using a Quanta Nova NanoSEM scanning microscope and transmission electron microscopy (TEM) using a Jeol JEM-2100 microscope. Flat dog-bone-shaped tensile specimens with a 16 mm gauge length and a cross-section of 1.5×3 mm^2 were used. Tensile tests in the temperature range from 500°C to 680°C and a strain rate of 1×10^{-3} s^{-1} and at a temperature of 620°C and strain rates from 1×10^{-5} s^{-1} to 1×10^{-1} s^{-1} were performed by using an INSTRON 5882 universal testing machine.

3. Results and discussion

The studied steel after heat treatment is characterized by completely annealed austenitic grains with an average size of ~ 14 μm (figure 1). Numerous annealing twins can be seen within grain interiors. The fraction of high-angle boundaries and the average grain misorientation angle were determined by EBSD analysis. The fraction of high-angle boundaries is 97.3%, while the fraction of special or twin-type boundaries Σ3 is 56.3%. The average misorientation angle is 50.7°. The dislocation density inside the grains is about 2.7×10^{12} m^{-2}. TEM studies showed the presence of second phase particles which have been identified as primary Nb(C,N) particles. These particles are distributed homogeneously and have an average size of about 174 nm.

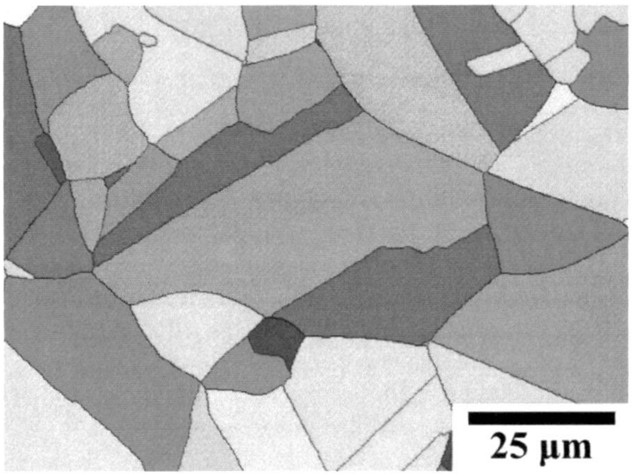

Figure 1. Microstructure of the Fe-18% Cr-8% Ni steel after solution treatment.

The uniaxial tensile tests of the studied steel performed at a strain rate of 1×10^{-3} s^{-1} showed that the deformation of the specimens at 530 – 680°C was accompanied by an unstable plastic flow, with serrations on the strain-stress curves (figure 2 (a)). This phenomenon in austenitic stainless steels is generally associated with the PLC effect due to DSA [6].

In the temperature interval of 590 - 680°C, only type A serrations were observed on the deformation curves, which are characterized by a sharp increase in stress followed by a drop to or below the general level of the flow stress. It is known that type A serration is a consequence of plane sliding, which is expressed by the propagation of deformation bands over long distances, and is associated with the interaction of mobile dislocations and interstitial atoms, such as nitrogen and carbon [7]. It should be noted that the frequency of the flow stress surges increases with increasing test temperature, while their amplitude decreases.

Since the operating temperature of these steels is ~ 620°C, the evaluation of the effect of the strain rate on the shape of deformation curves was evaluated at a given temperature. The manifestation of the PLC effect was observed in the range of strain rates from 1×10^{-2} s^{-1} to 1×10^{-4} s^{-1}. Type A serration was observed at a higher strain rate, while type B serration at low strain rates. Type B serration is associated with the slip of dislocations in deformation bands.

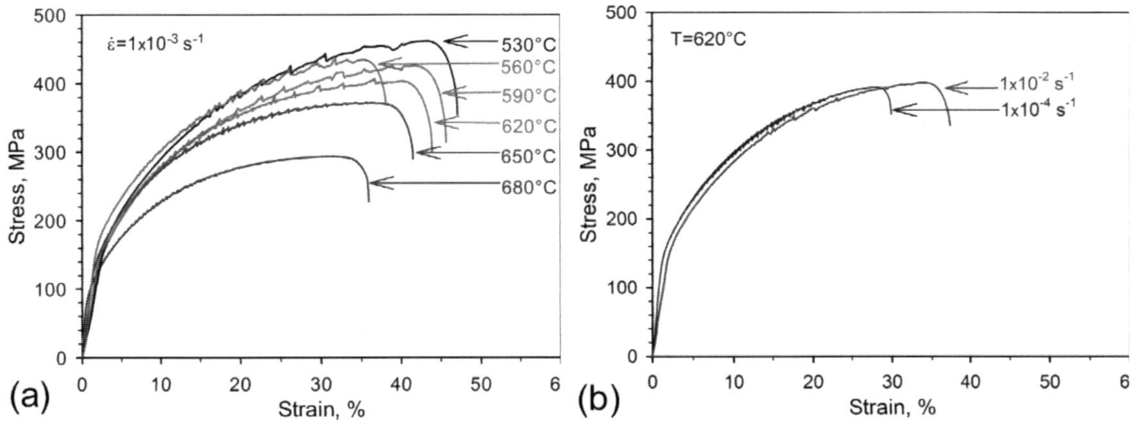

Figure 2. Deformation curves obtained at a strain rate of $\sim 1\times10^{-3}$ s^{-1} and temperatures of 530–680°C (a) and at 620°C in the strain rate interval from 1×10^{-2} s^{-1} to 1×10^{-4} s^{-1} (b).

The appearance of an unstable plastic flow takes place only after reaching a certain value of the critical strain [8]. The dependence of the estimated critical strain on temperature and strain rate is shown in figure 3. The critical strain values tend to decrease with increasing temperature, demonstrating "normal" behavior (figure 3 (a)). At a temperature of 620°C, an "inverse" critical strain dependence was observed (figure 3 (b)). It is suggested that such dependence is associated with the effect of strain rate/temperature on the dislocation movement and the cyclic segregation of atoms [8]. Both dependences in figures 3(a) and 3(b) are typical for the DSA process.

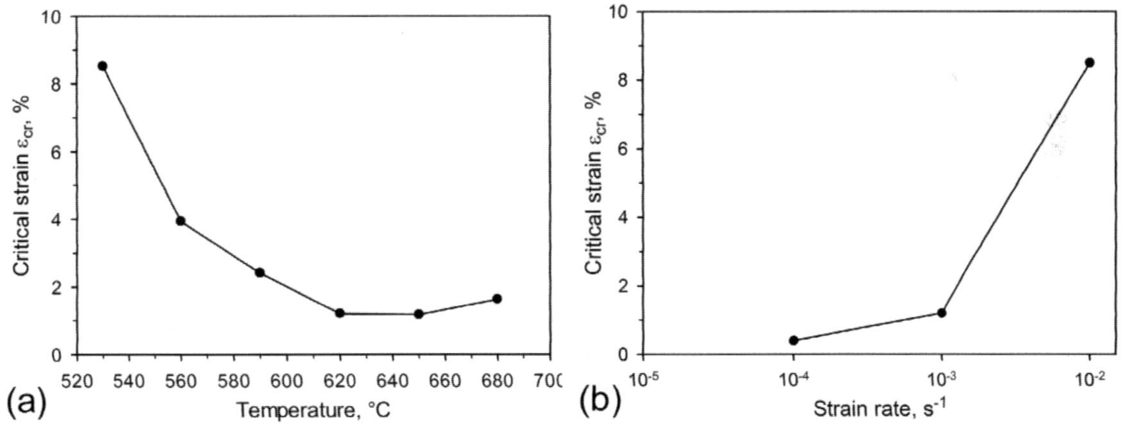

Figure 3. The critical strain as a function of the temperature (a) and strain rate (b).

4. Conclusions

The main conclusions based on the present study could be summarized as follows:

1. The temperature and strain rate intervals of the occurrence of the PLC effect in the Fe-18% Cr-8% Ni steel have been determined as 530 - 680°C at a strain rate of 1×10^{-3} s^{-1} and 1×10^{-2} s^{-1} to 1×10^{-4} s^{-1} at a temperature of 620°C.

2. Type A serrations were observed on the stress- strain curves at temperatures of 530-680°C and a strain rate of 1×10^{-3} s^{-1}. Type B serrations were observed at a temperature of 620°C and a strain rate of 1×10^{-4} s^{-1}.

3. The increase in temperature and decrease in the strain rate leads to an increase in the frequency of serrations and a decrease in their amplitude.

4. The critical strain for the appearance of the unstable plastic flow gradually decreases with decreasing strain rate or increasing temperature.

Acknowledgement

The reported study was funded by RFBR and NSFC according to the research project № 20-58-53053. The main results were obtained by using the equipment of the Joint Research Center, «Technology and Materials», Belgorod State University.

References

[1] Viswanathan R and Bakker W T 2000 *EPRI Technical report*. TR-114750
[2] Abe F 2016 *J. of Pressure Vessel Techn.* **138** 4
[3] Xi T, Babar Shahzad M, Xu D, Zhao J, Yang C, Qi M, Yang K 2016 *Mater. Sci. Eng.: A*. **675** 243
[4] Mogucheva A A, Nikulin I A, Kaibyshev R O, Skorobogatykh V N 2010 *Metal Sci. and Heat Treat.* **52** (3-4) 128
[5] Hong S G, Lee S B 2005 *J. Nucl. Mater.* **340** 307
[6] Rodriguez P 1984 *Bull. Mater. Sci.* **6** 653
[7] Müller A, Segel C, Linderov M, Vinogradov A, Weidner A, Biermann H 2016 *Metal. and Mater. Tran. A* **47** (1) 59
[8] Brechet Y and Estrin Y 1996 *Scr. Mater.* **35** 217

IOP Conf. Series: Materials Science and Engineering 1014 (2021) 012006 doi:10.1088/1757-899X/1014/1/012006

Influence of ECAP on the structure and properties of Ti18Zr15Nb and Ti10Mo8Nb6Zr alloys for medical application

A A Churakova[1,2*], D V Gunderov[1,2], G I Raab[2], S D Prokoshkin[3], V A Sheremetyev[3], Paulo N Lisboa Filho[4], João Pedro[4], Ana Paula Rosifini Alves Claro[4]

[1]Institute of Molecule and Crystal Physics, Ufa Federal Research Center RAS, Ufa, Russia
[2]Ufa State Aviation Technical university, Ufa, Russia
[3]NUST «MISiS», Moscow, Russia
[4]São Paulo State University, São Paulo, Brazil

*Corresponding author: churakovaa_a@mail.ru

Abstract. Studies of the influence of ECAP on the structure and properties of the metastable β-Ti alloys Ti18Zr15Nb and Ti10Mo8Nb6Zr have been carried out. The XRD data shows that the BCC β is the main phase in the initial alloys and alloys after ECAP. As a result of ECAP, the strength and yield stress of the alloys increases significantly due to the refinement of the structure. After ECAP, UTS and YS of the Ti10Mb8Nb8Zr alloy increase more noticeably than the Ti18Zr15Nb alloy, but after 7 ECAP cycles according to the regimes used (n = 6 T 250 °C + n = 1 350 °C), the Ti10Mb8Nb8Zr alloy becomes brittle. The increment in the yield point of the Ti18Zr15Nb alloy after ECAP processing together with the preservation of the β-state, allows one to expect an increase in the functional properties of the alloy.

1. Introduction

Titanium and its alloys are the most preferred materials for medical implants, due to their unique biocompatibility, good corrosion resistance, high specific strength [1, 2]. However, the large difference between the Young's modulus of the implant material and Young's modulus of the bone leads to "stress shielding" phenomenon, and bone resorption around the implant.

Metastable β-Ti alloys, including the Ti18Zr15Nb (at.% hereinafter) alloy, developed at NUST MISiS [3-6] and the Ti10Mo8Nb6Zr alloy, developed by the team of São Paulo State University, are the preferred materials for medical implants. These alloys do not contain toxic components, have low Young's modulus (high biomechanical compatibility with bone tissue) [3-5]. In the Ti18Zr15Nb alloy, recoverable thermoelastic $\beta \leftrightarrow \alpha''$ martensitic transformation provides the superelasticity and shape memory effects [3].

It was shown in [7-9] that due to SPD in pure titanium and its alloys, a significant increase in their properties for biomedical application is possible. The SPD method, which makes it possible to obtain relatively large samples for practical use, is equal channel angular pressing [9]. The results of studies of ultra-fine-grained (UFG) CP Ti and Ti-alloys processed via equal channel angular pressing (ECAP) are presented in particular in the review [9]. New modifications of ECAP processing is an ECAP-Conform (ECAP-C) technique that allows fabricating an UFG microstructure in long-length rods (up to few meters long) suitable for commercial applications [7,8]. A number of previous works also showed the possibility of increasing due to ECAP the strength and functional characteristics of the Ti-6Al-4V titanium alloy for medical applications [10] and low-modulus β-Ti alloy based on Ti-Mo [10].

However, studies of the effect of ECAP on the Ti18Zr15Nb and Ti10Mo8Nb6Zr alloys have not been conducted.

2. Experimental

ECAP was performed on a die-set with a diameter of the channel of 10 mm and an angle of the channels' intersection of 120°, route B_C. ECAP of Ti18Zr15Nb was performed at T = 250 °C, $n = 4$ and $n = 7$. A relatively low ECAP temperature (250 °C) was used to prevent the β-α transition. ECAP of Ti10Mo8Nb6Zr was carried out at a temperature of T = 250 °C, $n = 4$ and another sample - six ECAP cycles at a temperature of T = 250 °C, $n = 6$ and the seventh ECAP cycle at a temperature of T = 350 °C (on the last cycle they had to raise the temperature to 350 °C due to the hardening of the sample in the previous 6 cycles).

X-ray studies were carried out on a *Rigaku Ultima IV* diffractometer under Cu radiation. The tensile tests at room temperature were carried out on flat samples with a gauge section of 1x0.25x4 mm at a tension rate of 3.10-3 s^{-1} using an INSTRON tensile testing machine. Three or two samples were tested per one condition.

3. Results and discussion

Analysis of the XRD data shows that the BCC β is the main phase in the initial alloys and alloys after ECAP. Weak diffraction lines of the secondary α''-martensite phase or α are observed as well in Ti18Zr15Nb and Ti10Mo8Nb6Zr alloys, but the amount of α''-martensite phase or α-phase is insignificant both in the initial state of the alloy and after ECAP. The X-ray line width of the β-phase abruptly increases after ECAP (Table 1) pointing to an increase in dislocation density or/and grain refinement.

Table 1. Results of mechanical tests of alloys before and after ECAP and data of XRD at half-width of (FWHM) of the 110 β-phase line.

State	UTS	YS	ε, %	FWHM(deg) 110_β
Ti18Zr15Nb Initial	630	510	27	0.28
Ti18Zr15Nb ECAP n=4	880	855	6	0.63
Ti18Zr15Nb ECAP n=7	990	960	4	0.66
Ti10Mb8Nb8Zr Initial	710	430	30	0.29
Ti10Mb8Nb8Zr ECAP n=4	1280	1270	6	0.57
Ti10Mo8Nb6Zr ECAP n=7	1515	1420	5	0.51

The results of mechanical tests of the alloys before and after ECAP are presented in Figure 1 and Figure 2.

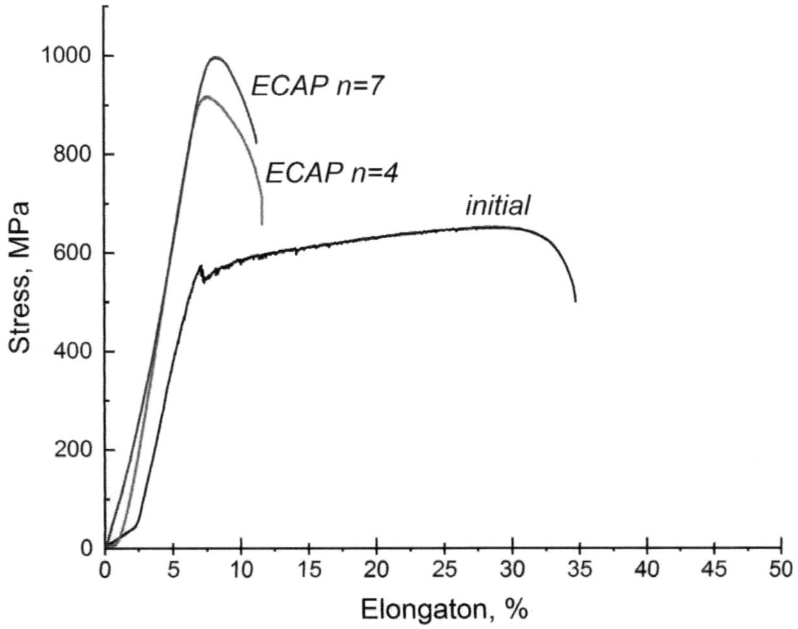

Figure 1. Results of mechanical tests of the Ti18Zr15Nb alloy before and after ECAP

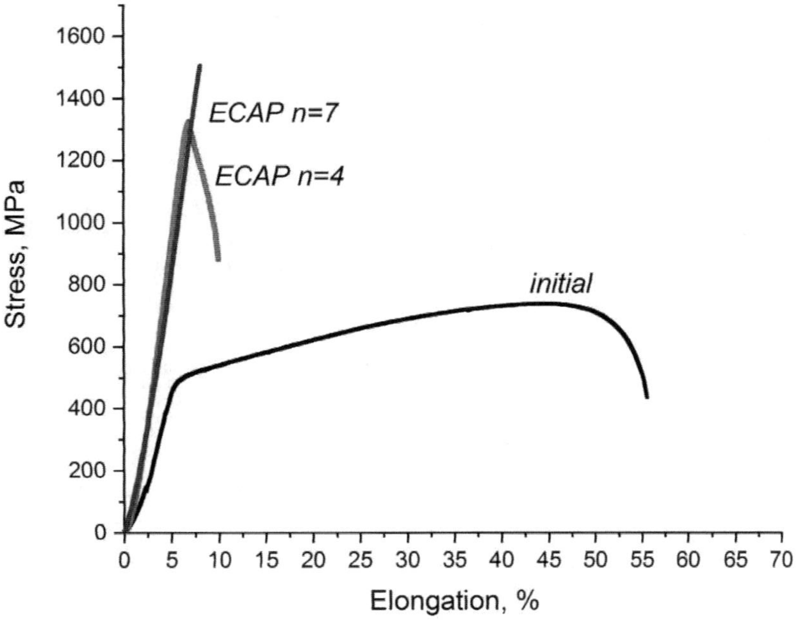

Figure 2. Results of mechanical tests of the Ti10Mo8Nb6Zr alloy before and after ECAP

Mechanical test data showed that as a result of ECAP, the strength and especially the yield stress of the alloys increases significantly due to the refinement of the structure. As the number of ECAP cycles n increases from 4 to 7, the strength properties of the alloys continue to increase. The yield stress of the Ti10Mo8Nb6Zr alloy increases especially significantly (almost threefold) as a result of ECAP $n = 7$. But the material becomes brittle. If in the initial state the strength properties of the initial Ti18Zr15Nb and initial Ti10Mb8Nb8Zr alloys are close, then after ECAP, UTS and YS of the Ti10Mb8Nb8Zr alloy increase significantly more noticeably than the Ti18Zr15Nb alloy. The nature of this result requires additional investigation, including fine TEM studies of the structure with phase analysis. The increment in the yield point of the Ti18Zr15Nb alloy after ECAP processing together with the preservation of the β-state, allows one to expect an increase in the functional properties of the alloy.

4. Conclusions

Analysis of the XRD data shows that the BCC β is the main phase in the initial alloys Ti10Mo8Nb6Zr, Ti18Zr15Nb and alloys after ECAP. As a result of ECAP, the strength and yield stress of the alloys increases significantly. As the number of ECAP cycles n increases from 4 to 7, the strength properties of the alloys continue to increase. After ECAP, UTS and YS of the Ti10Mb8Nb8Zr alloy increase much more noticeably than the Ti18Zr15Nb alloy, but the Ti10Mb8Nb8Zr alloy after ECAP n= 7 becomes brittle.

Acknowledgments The work is supported by the Russian Science Foundation under a joint funding scheme with NITU MISiS project No. 20-69-47029 (Ti18Zr15Nb alloy), and Russian Foundation for Basic Research, Project No. 19-58-80018 (BRICS_t) (Ti10Mb8Nb8Zr alloy).

References

[1] Brunette D M *Titanium in medicine: material science, surface science, engineering, biological responses, and medical applications. Engineering materials* 2001 Berlin New York: Springer. xiii, 1019.

[2] Geetha M, Singh A K, Asokamani R and Gogia A K 2009 *Progress in Materials Science* **54** (3) 397-425.

[3] Konopatsky A S , Dubinskiy S M , Zhukova Y S, Sheremetyev V A, Brailovski V, Prokoshkin S D, Filonov M.R. 2017 *Materials Science and Engineering A* **702** 301-311.

[4] Soldatenko A S, Karachevtseva M A, Sheremetyev V A, Kudryashova A A, Arkhipova, A Y, Andreev V A, Prokoshkin S D, Brailovski V, Moisenovich M M, Shaitan K V 2019 ***Moscow University Biological Sciences Bulletin*** **74(4)** 250-255.

[5] Kudryashova A, Sheremetyev V, Lukashevich K, Cheverikin V, Inaekyan K, Galkin S, Prokoshkin S, Brailovski V 2020 *Journal of Alloys and Compounds* **843** pp 1560-66.

[6] Sheremetyev V, Kudryashova A, Cheverikin V, Korotitskiy A, Galkin S, Prokoshkin S and Brailovski V 2019 *Journal of Alloys and Compounds* **800** 320-26.

[7] Raab G I, Valiev R Z, Gunderov D V, Lowe T C, Misra A and Zhu Y T 2009 *Mater. Sci. Forum* **584–586** 80–85.

[8] Gunderov D V, Polyakov A V, Semenova I P, Raab G I, Churakova A A, Gimaltdinova E I, Sabirov I, Segurado J, Sitdikov V D, Alexandrov I V, Enikeev N A and Valiev R Z 2013 *Materials Science and Engineering A* **562** 128-36.

[9] Valiev R Z, Zhilyaev A P, Langdon T G *Bulk Nanostructured Materials: Fundamentals and Applications* 2014 John Wiley & Sons, Inc., Hoboken, New Jersey, 456.

[10] Semenova I P, Polyakov A V, Polyakova V V, Valiev R Z, Huang Y, Langdon T G 2016 *Advanced Engineering Materials* **18(12)** 2057-62.

[11] Semenova I, Medvedev A 2011 *Materials Science Forum* **667-669** 731-36.

IOP Conf. Series: Materials Science and Engineering 1014 (2021) 012007 doi:10.1088/1757-899X/1014/1/012007

Effect of thermomechanical treatment on the microstructure and mechanical properties of a low-carbon low-alloy steel

A S Dolzhenko*, A N Belyakov and R O Kaibyshev

Belgorod State National Research University, Pobeda, 85, Belgorod, 308015, Russia

*Corresponding author: dolzhenko_a@bsu.edu.ru

Abstract. The effect of thermomechanical treatment on the microstructure, tensile strength and impact toughness of low-carbon low-alloy steel is considered. The developed microstructure is characterized by grain elongation along the rolling direction. Both the yield strength and the ultimate tensile strength increase with a decrease in test temperature. The ultimate strength at room temperature is 797 MPa, and at 77 K the ultimate strength increases to 1198 MPa. The steel samples subjected to tempforming exhibit impact toughness of KCV = 410 J/cm^2 at room temperature. A decrease in the test temperature to 233 K leads to a slight decrease in the KCV value to 340 J/cm^2. It should be noted that the impact specimens are not completely fractured even at a test temperature of 233 K, suggesting higher real values of impact toughness.

1. Introduction

High-strength low-alloy steels are one of the most widely used classes of materials. One of the significant disadvantages of such steels is their relatively high temperature of the ductile-brittle transition, below which the impact toughness of the steels drops sharply, i.e., the steel becomes brittle, which can lead to a sudden catastrophic failure of the construction.

Currently, special attention is paid to the development of structural steels and alloys adapted to the climatic conditions of the Far North, for use in facilities for the production, storage and transportation of gas and oil on the Arctic shelf, as well as for constructions for various engineering applications, operated under low temperature conditions. Comprehensive studies of fracture mechanisms and conditions of ductile-brittle transition had been being carried out from the middle of the last century [1–3]. This was caused by the need to create structures designed for operation at low temperatures such as arctic pipelines, tankers and ships, tanks and storage vessels, transportations of liquid gases, etc.

An interesting approach to increase the impact toughness and decrease the temperature of the ductile-brittle transition in carbon steels was proposed by Japanese scientists [4]. This approach consists in the formation of a lamella type structure with a transverse grain size of about 100 nm and a uniform distribution of dispersed nanoscale particles of secondary phases by means of warm rolling under conditions of tempering. Such thermo-mechanical processing was called tempforming, which allows us to get a promising combination of mechanical properties in low alloyed steels. Besides its high impact toughness, processed steel has high strength due to the reduction of grain size and precipitation hardening.

2. Experimental

IOP Conf. Series: Materials Science and Engineering 1014 (2021) 012007 doi:10.1088/1757-899X/1014/1/012007

A low-carbon low-alloy steel with a chemical composition of Fe - 0.03C - 0.23Si - 0.19Cr - 1.54Mn - 0.04Nb - 0.016Al - 0.12Mo - 0.21V (all in mass%), was subjected to initial heat treatment (annealing at 1423 K for 1 hour followed by oil quenching) and, then, rolling at 923 K to a total strain of 1.5 (tempforming). The structural observation were performed on the RD-ND sections (RD is the rolling direction and ND is the normal direction), using a Quanta Nova Nanosem 450 scanning electron microscope (SEM) equipped with an electron back scattering diffraction pattern (EBSP) analyzer incorporating an orientation imaging microscopy (OIM) system. The specimens for structural investigations were electro-polished using an electrolyte containing 10% perchloric acid and 90% acetic acid at a voltage of 20 V at room temperature. The OIM images were subjected to cleanup procedure setting minimal confidence index of 0.1. The mean grain size was evaluated on the OIM images as the average distance between high-angle boundaries with misorientations of $\theta \geq 15°$. The tensile tests were carried out by using an Instron 5882 testing machine on specimens with gauge dimensions of 12 mm in length, 3 mm in width, 1.5 mm in thickness cut out with the tensile direction parallel to the rolling direction. The specimens were tested at a crosshead rate of 2 mm/min. Standard Charpy V-notch specimens (length = 55 mm, width = 10 mm, thickness = 10 mm, notch depth = 2 mm) were tested using an Instron 450 J impact machine with an Instron Dynatup Impulse data acquisition system at temperatures ranging from 77 K to 293 K [5]. The specimens for impact tests were cut from the rolled plates so that the impact direction was parallel to the normal direction.

3. Results and discussion

3.1. Microstructures

The initial microstructure of the low-carbon low-alloy steel is shown in figure 1a. The mean grain sizes and the fraction of high-angle grain boundaries are about 11 μm and 0.44, respectively. The tempforming leads to the evolution of a fine grained microstructure consisting of grains elongated along the RD (figure 1b). The mean transverse grain size is 2.9 μm. The fraction of high-angle grain boundaries is 0.29. The tempformed steel is characterized by strong ⟨001⟩ ||ND and ⟨111⟩ ||ND fiber textures (corresponding to red and blue colors, respectively, in figure 1b).

Figure 1. Microstructures developed in a low-carbon low-alloy steel subjected to initial heat treatment (a) and tempforming at 923 K (b). The colors correspond to the crystallographic direction along the normal direction (ND). High-angle and low-angle boundaries are indicated by thick black and thin black lines, respectively.

3.2. Tensile properties

The engineering stress–elongation curves obtained for the low-carbon low-alloy steel samples by tension at different temperature are presented in figure 2. Following the onset of plastic flow, the tempformed specimens exhibit a negligible short stage of strain hardening and a well-defined peak stress followed by a plateau that is discontinuous yielding. The same type of curves was observed in previous studies on tempforming [6, 7].

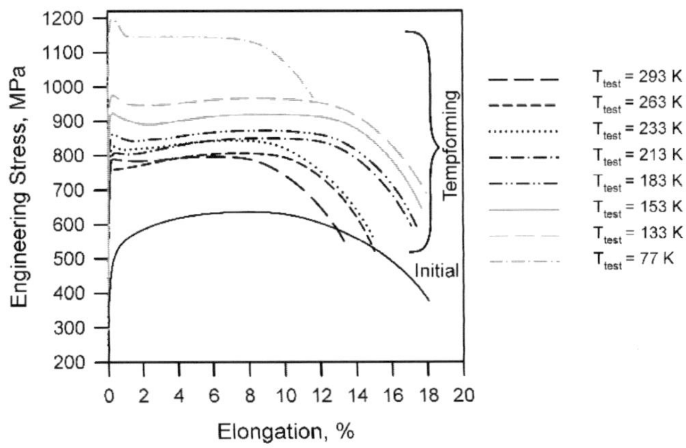

Figure 2. Tensile curves of a low-carbon low-alloy steel subjected to initial heat treatment and tempforming at 923 K.

Both the yield strength and the ultimate tensile strength increase with a decrease in test temperature. The ultimate strength at room temperature is 800 MPa, and at 77 K the ultimate strength increases to 1200 MPa. The total elongation comprises 13.5% at a temperature of 293 K. A decrease in temperature to 183 K leads to an increase in elongation to 17.9%. Then the total elongation decreases to 11.6% as the test temperature decreases to 77 K (table 1).

Table 1. Mechanical properties of a low-carbon low-alloy steel at different temperatures.

Treatment	Offset yield strength, MPa	Ultimate tensile strength, MPa	Elongation, %
Initial	465	635	18
Tempforming (T_{test} = 293 K)	780	800	13.5
Tempforming (T_{test} = 263 K)	755	805	15
Tempforming (T_{test} = 233 K)	820	840	15
Tempforming (T_{test} = 213 K)	790	850	17
Tempforming (T_{test} = 183 K)	855	870	17.5
Tempforming (T_{test} = 153 K)	920	925	17.7
Tempforming (T_{test} = 133 K)	970	975	17.9
Tempforming (T_{test} = 77 K)	1190	1200	11.6

3.3. Impact Toughness

Figure 3 shows the effect of thermomechanical treatment on the impact toughness of low-carbon low-alloy steel. The steel samples subjected to tempforming exhibit an impact toughness of KCV = 410 J/cm^2 at room temperature. A decrease in the test temperature to 233 K leads to a slight decrease in the KCV value to 340 J/cm^2. A further decrease in the test temperature leads to an increase in KCV to 425

J/cm^2. It should be noted that the impact specimens are not completely destroyed even at a test temperature of 183 K, suggesting higher real values of impact toughness. Upon further cooling to 77 K, the impact toughness decreases to 19 J/cm^2. The high fracture toughness is attributed to delamination, when fracture is accompanied by splitting along the rolling plane with high absorption energy.

The specimens after impact tests are shown in figure 3b. It is seen that the Charpy specimens in the tempformed condition exhibit delaminations, i.e., the cracks branch along the impact test specimens without their complete rupture that suggest a rather high absorbed impact energy.

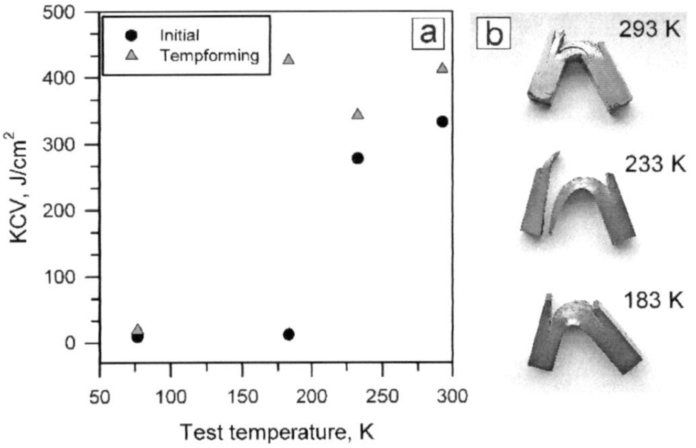

Figure 3. Effect of test temperature on impact toughness of low-carbon low-alloy steel subjected to initial heat treatment and tempforming at 923 K (a); and specimens of low-carbon low-alloy steel tempformed at 923 K after impact tests at the indicated temperatures (b).

4. Conclusion

The effect of thermomechanical treatment on the microstructure and mechanical properties of low-carbon low-alloy steel was studied. The main results can be summarized as follows:

1. Tempforming results in the formation of a fine grain microstructure with an average transverse grain size of 2.9 μm and strong ⟨111⟩ ||ND and ⟨001⟩ ||ND fiber textures.

2. Tempforming strengthens the steel remarkably. The ultimate tensile strength at room temperature is 797 MPa, and the total elongation comprises 13.5%

3. Tempforming leads to a significant increase in the values of impact toughness at room and negative temperatures. The impact toughness of the tempformed steel reaches its maximum value at a temperature of 183 K (KCV = 430 J/cm^2).

Acknowledgments

The study was financially supported by the Russian Science Foundation (Project No. 20-19-00497). Authors are grateful to the staff of the Joint Research Center, Technology and Materials, of the Belgorod State National Research University for their technical assistance.

References

[1] McEvily A J 2001 *Metal Failures* (New York: Willey-Interscience)
[2] Morris Jr J W, Lee C S and Guo Z 2003 *ISIJ Intern.* **43** 410
[3] McMeeking R M and Parks G M 1979 *Elastic-Plastic Fracture* (Philadelphia: American Society for Testing and Materials)
[4] Kimura Y, Inoue T, Yin F and Tsuzaki K 2008 *Science* **320** 1057
[5] Mishnev R, Dudova N, Dudko V and Kaibyshev R 2018 *Mater. Sci. Eng.* A **730** 1
[6] Dolzhenko A, Yanushkevich Z, Kopteva K, Belyakov A and Kaibyshev R 2019 *Conf. Proc on METAL 2019 - 28th Int. Conf. on Metallurgy and Materials* 632
[7] Lugovskaya A, Yanushkevich Z, Odnobokova M, Belyakov A and Kaibyshev R 2017 *AIP Conf. Proc.* 1909(2017)020120

Effect of multiple forging and annealing on microstructure and mechanical properties of a high-manganese steel

P D Dolzhenko[1*], R Z Valiev[2], A N Belyakov[1], E V Bobruk[2] and M S Tikhonova[1]

[1]Belgorod State National Research University, Pobeda, 85, Belgorod, 308015, Russia
[2]Ufa State Aviation Technical University, K. Marxa, 12, Ufa, 450008, Russia

*Corresponding author: dolzhenko_p@bsu.edu.ru

Abstract. The effect of multiple forging and annealing on the microstructure and mechanical properties of a high-manganese steel is considered. An austenitic high-manganese steel, Fe-0.03C-28Mn-1.5Al (all in wt.%), with an average grain size of about 37 μm was used as the initial material in this study. Multiple forging at room temperature was carried out up to a total true strain of 2. Multiple forging was accompanied by deformation twinning and resulted in significant strengthening. The steel samples subjected to multiple forging demonstrate an increase in the strength properties with an increase in the total strain, while elongation decreases. After multiple forging to a total strain of 2, the samples were annealed at temperatures from 673 to 1073 K for 30 minutes. An increase in the annealing temperature leads to a decrease in the strength and a significant increase in plasticity. Annealing at temperatures of $T \geq 873$ K leads to the formation of a recrystallized microstructure.

1. Introduction

Austenitic high manganese steels exhibiting the twinning-induced plasticity (TWIP) effect are of great interest in materials science and are considered advanced structural materials for use in the automotive industry[1,2]. TWIP steels exhibit an excellent combination of mechanical properties due to the fact that the twins nucleated during the plastic deformation act as obstacles, reducing the dislocation mean free path during deformation, leading to high strain hardening rates [3]. However, the widespread use of TWIP steels is limited. This is partly due to their relatively low yield stress compared to other advance high strength steels (AHSS) [4]. Many works are devoted to such methods of improving the yield strength as cold rolling and annealing [4–6] in a wide temperature range, as well as high-pressure torsion (HPT) [7] and other methods of austenitic steel processing [8]. However, the multiple forging of high-manganese steels at room temperature and subsequent annealing in a wide temperature range has received little attention. In this paper, we investigate the effect of multiple forging and subsequent annealing on the microstructure and mechanical properties of a high-manganese TWIP steel.

2. Experimental

High-manganese TWIP steel with a chemical composition of Fe-0.03C-28Mn-1.5Al (all in wt.%) was subjected to annealing at 1423 K for 1 hour followed by rolling at 1423 K with about a 60% reduction. Then, the prismatic steel samples with the initial dimensions of $10 \times 12 \times 15$ mm^3 were subjected to multiple forging at room temperature to a total strain of 2. Multiple forging was carried out with a change in the deformation axis by 90 degrees at a true strain of 0.4 per pass. After multiple forging, the samples were annealed at temperatures in the range from 673 to 1073 K for 30 minutes. Microstructure studies were carried out using a scanning electron microscope (SEM). The specimens for structural investigations were electro-polished using an electrolyte containing 10% perchloric acid and 90% acetic acid at a voltage of 20 V at room temperature. The average grain size was estimated by the intercept line method. The tensile tests of the flat-type specimens with a gauge length of 4 mm and a cross section of 1.0×0.5 mm were carried out at room temperature and at an initial strain rate of 10^{-3} s^{-1}.

3. Results and discussion

3.1. Deformation behavior

Several true stress-strain curves for five sequential forgings to a total strain of 2 at room temperature are shown in figure 1. Following the yielding at about 245 MPa, the first compression pass is characterized by a gradual increase in the flow stress to about 820 MPa. Further deformation leads to a gradual increase in the flow stresses, which tends to saturate at a level of about 1000 MPa after total strains exceeding 1.6.

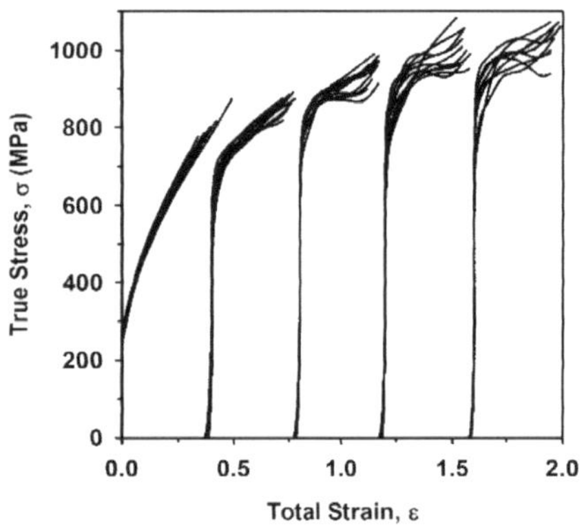

Figure 1. True stress–strain curves obtained during multiple forging of high-Mn steel at room temperature.

3.2. Microstructures

The microstructure of high-manganese TWIP steel in the initial state (figure 2a) is characterized by equiaxed austenite grains with an average size of about 37 μm. The fraction of high-angle boundaries (HAB) in this state is 0.67, and the fraction of Σ3 CSL boundaries is 0.30. The CSL boundaries have a misorientation angle of 60° and are associated with annealing twins [8]. In the process of multiple forging to a true strain of 0.4 (figure 2b), a network of low-angle boundaries (LAB) with a misorientation angle of less than 15° is formed, the fraction of high-angle boundaries (HAB) decreases by a factor of 2. The deformation twins form bundles with individual twin thickness of 470 nm.

An increase in the total strain to 2 (figure 2c) is also accompanied by mechanical twinning and, as a result, the refinement of structural elements to sizes less than 50 nm. Heat treatment at 673K and 773K for 0.5 h does not change remarkably the microstructure evolved during multiply forging. An increase in the annealing temperature to 823K results in a partially recrystallized microstructure as shown in figure 2d. New fine austenitic grains with a size of 1 μm with quite a low dislocation density, as well as work hardened regions with high density of deformation twins were found in the microstructure after 0.5 h of annealing at 823 K. The fraction of Σ3 CSL boundaries is 0.14. An increase in the annealing temperature to 873K leads to the formation of a recrystallized structure with an average grain size of 900 nm (figure 2e), as a result of static recrystallization, while the fraction of special Σ3 boundaries increases by 2 times, and the HAB fraction is 0.98.

Annealing at a temperature of 1073 K for 0.5 h leads to the formation of austenite grains with an average size of 5.7 μm (figure 2). This microstructure consists of equiaxed grains with numerous annealing twins. The fraction of Σ3 CSL boundaries comprises 0.42.

Figure 2. Microstructure of high-Mn steel in the initial state (a), deformation microstructures after multiple forging at room temperature to total strains of 0.4 (b), 2 (c), and microstructures after multiple forging to a total train of 2 and 30 min annealing at 823 K (d), 873 K (e) and 1073 K (f). The colors correspond to the crystallographic direction along the forging axis (FA) or the rolling axis (RA). High-angle boundaries, low-angle sub-boundaries, and twin boundaries are shown with bold black lines, thin white lines, and red lines, respectively.

3.3. *Tensile properties*

This steel in the initial state demonstrates a yield strength equal to 255 MPa, and a tensile strength equal to 465 MPa with an elongation of 44% (figure 3a). Forging to a true strain of ε = 0.4 leads to an increase in yield stress to 810 MPa, the sample exhibits little uniform elongation and necking soon after reaching the yield stress, and the total elongation is 9%.

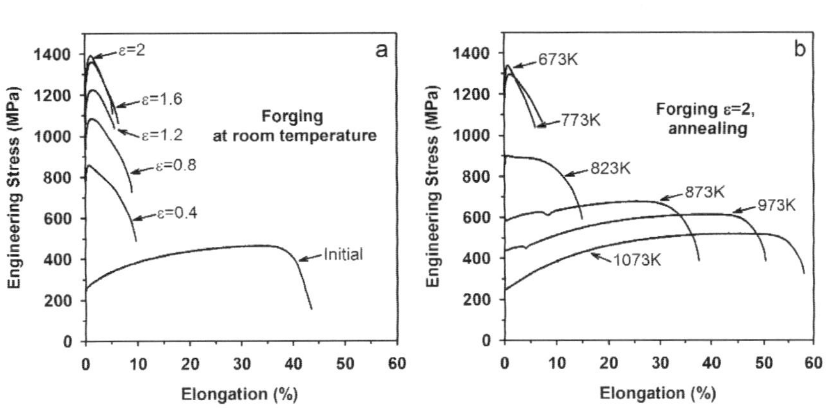

Figure 3. Tensile stress-elongation curves at room temperature for austenitic high-Mn steel subjected to multiple forging at room temperature to the indicated total true strains from ε - 0.4 to ε -2.0 (a), and then annealed for 30 min at the indicated temperatures following multiple forging to ε = 2 (b).

The samples subjected to multiple forging to larger total strains show similar behavior in the tensile stress-elongation curves, necking occurs almost immediately upon reaching the yield point. As the total forging strain increases, the strength characteristics increase, and the elongation decreases. After $\varepsilon = 2$, the yield strength is 1260 MPa, the ultimate strength is 1390 MPa, and the elongation is 5%.

Annealing at temperatures of 673 K and 773 K does not lead to significant changes in the tensile properties of TWIP steel subjected to multiple forging to a total strain of 2. An increase in the annealing temperature to 823 K leads to an increase in the elongation up to 15%. A further increase in the annealing temperature leads to an expansion of the region of uniform elongation, while the strength characteristics decrease. On the tensile curves of high-manganese steel annealed at temperatures of 873 K and 973 K, there is a tooth characterized by a drop in flow stresses when reaching 635 MPa and 468 MPa, respectively. The nature of this phenomenon should be studied in further research. Annealing at a temperature of 1073 K leads to a decrease in the yield point and ultimate strength to 245 MPa and 520 MPa, respectively, with an elongation of 58%.

4. Conclusion

The present study elaborates further the urgent task of obtaining structural materials with a combination of high strength and good plasticity by means of ultrafine grain refinement by severe plastic deformation [9]. The main results can be summarized as follows:

1. Multiple forging of high-manganese steel is accompanied by a gradual increase in the flow stress as the total true strain increases. The flow stress reaches saturation of about 1000 MPa at total strains above 1.6.

2. Multiple forging is accompanied by the formation of deformation twins and a network of low-angle subboundaries. Ultrafine structural elements with a size below 50 nm evolve as the total strain increases to 2. These microstructures do not remarkably up on subsequent annealing for 30 min at temperatures of $T \leq 773$ K, whereas an increase in the annealing temperature to 823 K leads to the development of recrystallization process. A completely recrystallized microstructure develops during 30 min of annealing at $T \geq 873$ K.

3. The steel samples subjected to multiple forging demonstrate an increase in the strength properties with an increase in the total forging strain while elongation decreases. The yield stress reaches 1260 MPa at a total strain of 2. Annealing for 30 min at $T \leq 773$ K does not lead to significant changes in the mechanical properties, whereas annealing at higher temperatures decreases the yield stress and increases elongation.

Acknowledgments
The reported study was funded by RFBR according to the research project № 19-38-50113.

References

[1] Chen L, Zhao Y and Qin X 2013 *Acta Metall. Sin. (English Lett.)* **26** 1
[2] Bouaziz O, Allain S, Scott C P, Cugy P and Barbier D 2011 *Curr. Opin. Solid State Mater. Sci.* **15** 141
[3] Vidilli A L, Otani L B, Wolf W, Kiminami C S, Botta W J, Coury F G and Bolfarini C 2020 *J. Alloys Compd.* 154806
[4] Rahman K M, Vorontsov V A and Dye D 2015 *Acta Mater.* **89** 247
[5] Dini G and Zamani D 2017 *Metall. Mater. Trans. A* **48** 3164
[6] Tewary N K, Ghosh S K, Mandal A, Chakrabarti D and Chatterjee S 2020 *Metall. Mater. Trans. A* **51** 4483
[7] Kim J G, Enikeev N A, Abramova M M, Park B H, Valiev R Z and Kim H S 2017 *Mater. Sci. Eng. A* **682** 164
[8] Belyakov A, Tikhonova M, Dolzhenko P, Sakai T and Kaibyshev R 2019 *Adv. Eng. Mater.* **21** 1800104
[9] Ovid'ko I A, Valiev R Z and Zhu Y T 2018 *Prog. Mater Sci.* **94** 462.

Preparation of Si_3N_4-YAG mixtures for the spark plasma sintering method

P V Andreev[1*], P D Drozhilkin[2], L S Alekseeva[2], E Ye Rostokina[1], K O Karazanov[2], M S Boldin[2] and S S Balabanov[1]

[1]G.G. Devyatykh Institute of Chemistry of High-Purity Substances of the Russian Academy of Sciences, Nizhny Novgorod, 603951, Russia
[2]Physical-Technical Research Institute, Lobachevsky State University of Nizhny Novgorod, Nizhny Novgorod, 603950, Russia

[*]Corresponding author: andreev@phys.unn.ru

Abstract. In order to obtain ceramics based on Si_3N_4 with improved physical and mechanical properties, methods of creating composites for their subsequent consolidation by the spark plasma sintering method were investigated. Four different methods of producing mixtures based on the Si_3N_4 and $Al_5Y_3O_{12}$ (YAG) precursor were considered in order to obtain a homogeneous distribution of the YAG sintering additive on the surface of the Si_3N_4 particles. It is shown that the YAG and Y-Si-Al-O-N phases are formed by co-precipitation, deposition in a gelatin matrix and Pecini methods. The spray drying method does not lead to the formation of the YAG phase. This may indicate a high homogeneity of the YAG coating on the surface of the Si_3N_4 particles, preventing the growth of crystalline particles.

1. Introduction

Sintering of pure Si_3N_4 is complicated by its covalent nature. For this reason, the main modern methods of sintering Si_3N_4 ceramics use activation additives, which at high temperatures form a liquid phase at the grain boundaries of the main component and promote consolidation of ceramics. Binary and triple eutectic compounds based on the Y_2O_3–Al_2O_3 are widely used as sintering additives [1]–[6]. The additives increase the density of ceramic samples, but adversely affect the properties of the completed ceramic samples due to the fact that the sintering additive forms an amorphous phase [7], [8].

Different ways of precipitation from solutions are used for the preparation of Si_3N_4 mixtures with uniformly distributed sintering additives, to achieve a more homogeneous and finer microstructure of ceramics than mechanical mixing. [9]–[15]. Our preliminary studies have also shown the high efficiency of precipitation methods for the implementation of the technological process. [16], [17].

The purpose of presented research is optimization of the technology of Si_3N_4 high-strength ceramics fabrication by the method of spark plasma sintering (SPS). Various methods of the synthesis of composites based on Si_3N_4 powder and yttrium aluminum garnet (YAG, $Al_5Y_3O_{12}$) precursor are presented. The phase composition and microstructure of the obtained samples were studied.

2. Materials and methods

Mixtures based on silicon nitride with YAG precursors, synthesized by various methods were used as the object of study. Silicon nitride powder (Si_3N_4) produced by AlfaAesar GmbH & Co KG (Germany) with a particle size of up to 44 microns was used as a raw material. The composition of the mixtures was chosen based on the expected ratio of Si_3N_4-YAG equal to 91.5–8.5% wt.

The method of co-precipitation (1) is the joint deposition of yttrium and aluminum hydroxide from solutions of the corresponding nitrates $Y(NO_3)_3$ and $Al(NO_3)_3$. Silicon nitride powder is added to an aqueous solution of nitrates at room temperature with continuous stirring. Aqueous solution of ammonia was added to the obtained suspension by drops until pH = 8 with continuous stirring. The resulting mixture was dried at 80 °C.

The Pechini method (2) included the production of citrate gels from solutions of nitrates $Y(NO_3)_3$ and $Al(NO_3)_3$ with the addition of citric acid (CA). To the aqueous solution of yttrium and aluminum nitrate the CA solution was added in such an amount that the molar ratio of CA to the total content of nitrate ions was 1 to 1. Silicon nitride powder was added to the gel obtained at the previous stage with continuous stirring. The resulting mixture was dried with continuous stirring to the condition of a thick gel. Then the gel was annealed at 350 and then 500 °C for 2 hours.

The precipitation in the gelatin matrix (3) was the creation of a thick gel based on gelatin. Gelatin was added to a hot (80 °C) solution of yttrium and aluminum nitrate (at the rate of ~ 2 g per 1 g of the output product) and kept at this temperature for 2 hours with continuous stirring. Powder of silicon nitride was added to the obtained solution. The produced mixture was cooled to 4 °C until the formation of a gel, then cut into 3-5 mm pieces. The cut gel was covered with an aqueous ammonia solution and kept for 24 hours at 4 °C. After that, the gel was washed first with distilled water, then with ethanol. After that, the gel was dried at 100 °C for 8 hours.

The deposition of the sintering additive by spray drying (4) is based on a mixture of $Al_5(NO_3)_3(OH)_{12}$ and $Y(OH)_2(CH_3COO)$ colloidal solutions with Si_3N_4 powder. The aqueous suspension contained 52.5% wt. Si_3N_4-YAG in the conversion to the solid phase, for stabilization purposes an additional 1.5% wt. of ammonium polyacrylate was introduced. After mixing in a planetary mill, the suspension was spray dried at 150 °C.

As a result, four mixtures were obtained with the YAG precursor distributed over the surface of Si_3N_4 particles. To form YAG, the powders were annealed in stages at 300, 500, 800 and 1000 °C for 2-6 hours with dispersion in an agate mortar between the stages.

The obtained mixtures were studied by scanning electron microscopy using a JEOL JSM-6490 instrument and X-ray diffraction analysis on a Shimadzu XRD-7000 diffractometer.

3. Results and discussion

X-ray diffraction studies of mixtures right after synthesis showed the presence of crystalline phases α-Si_3N_4 and β-Si_3N_4 , regardless of the method of preparation. Diffraction patterns of the mixture samples annealed at 1000 °C are shown in Figure 1. The crystal phase composition of the mixtures obtained by methods 1-3 is identical and contains the crystal phases of YAG and Y-Si-Al-O-N in addition to the initial phases α-Si_3N_4 and β-Si_3N_4. The mixture obtained by method 4 (spray drying) differs because it does not contain the YAG crystal phase. This may indicate a more homogeneous deposition of the sintering additive precursor on the Si_3N_4 particles.

Researches carried out on a scanning electron microscope also did not show significant differences in the structure of mixtures prepared by methods 1-3, both before and after annealing. However, the structure of the mixture prepared by the spray drying method (4) contains spherical particles 5-100 microns in size. SEM images of mixtures obtained by the deposition method in the gelatin matrix (3) and the spray drying method (4) are shown in Figure 2.

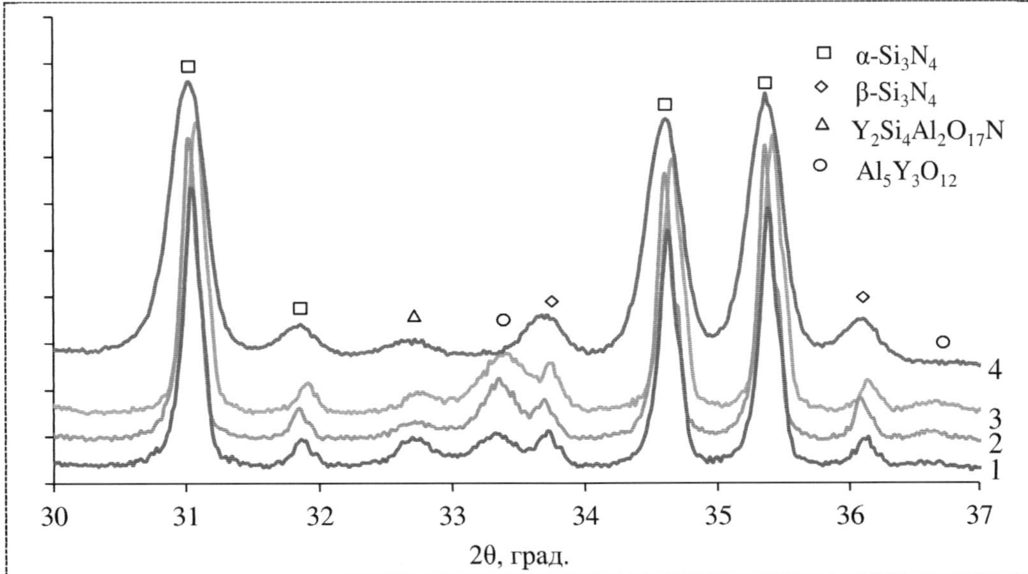

Figure 1. X-ray diffraction patterns of mixtures annealed at 1000 °C for the YAG ($Al_5Y_3O_{12}$) phase formation. The number corresponds the mixture preparation methods.

Figure 2. SEM-images of mixtures prepared by different methods: deposition in the gelatin matrix (a) and spray drying (b).

4. Conclusion

Four methods for fabrication of -Si_3N_4-YAG mixtures by applying a sintering additive precursor to the surface of Si_3N_4 particles were considered. The phase composition of the mixtures after the oxide phase formation during annealing was studied. The appearance of the Y-Si-Al-O-N phase in the samples of the mixtures obtained by all considered methods has been shown. The YAG phase is characteristic of the mixtures synthesized by co-precipitation, Pechini and precipitation in gelatin matrix methods. The YAG phase ($Al_5Y_3O_{12}$) was not formed in the mixture obtained by the spray drying method. This may indicate a high homogeneity of the mixture. Particularly, the absence of large agglomerates of the precursor phase and its predominant formation as a coating on the surface of the silicon nitride particles, therefore the formation of YAG crystalline grains do not occur during annealing.

In summary, the spray drying method is a promising method of fabrication Si_3N_4-based mixtures for sintering ceramics with increased physical and mechanical characteristics.

Acknowledgments
The reported study was funded by RFBR, project number 19-33-60084.

References

[1] Loehman R E and Rowcliffe D J 1980 *J. Am. Ceram. Soc.* **63** 144
[2] Yoon S-Y, Akatsu T, and Yasuda E 1996 *J. Mater. Res* **11** 120
[3] Sun E Y, Becher P F, Plucknett K P, Hsueh C-H, Alexander K B, Waters S B, Hirao K, Brito M E 1998 *J. Am. Ceram. Soc.* **81** 2831
[4] Getman O I, Holoptsev V V, Panichkina V V, Plotnikov I V, and Soolshenko V K 2002 *Sci. Sinter.* **34** 223
[5] X. Xu, Nishimura T, Hirosaki N, Xie R-J, Zhu Y, Yamamoto Y, Tanaka Hidehiko 2005 *J. Am. Ceram. Soc.* **88** 934
[6] Xu W, Yin Z, Yuan J, Wang Z, and Fang Y 2017 *Mater. Sci. Eng. A* **684** 127–134
[7] Deeley G G, Herbert J M, and Moore N C 1961 *Powder Metall.* **4** 145–151
[8] Podobeda L G 1979 *Sov. Powder Metall. Met. Ceram.* **18** 59
[9] Vrolijk J W G A, Willems J W M M, and Metselaar R 1990 *J. Eur. Ceram. Soc.* **6** 47
[10] Sordelet D J, Akinc M, Panchula M L, Han Y, and Han M H 1994 *J. Eur. Ceram. Soc.* **14** 123
[11] Manalert R and Rahaman M N 1996 *J. Mater. Sci.* **31** 3453
[12] Li J-G, Ikegami T, Lee J-H, Mori T, and Yajima Y 2000 *J. Eur. Ceram. Soc.* **20** 2395
[13] Lu Q, Dong W, Wang H, and Wang X 2002 *J. Am. Ceram. Soc.* **85** 490
[14] Wang H M, Simmonds M C, Huang Y Z and Rodenburg J M 2003 *Chem. Mater.* **15** 3474
[15] Wang J, Zheng S, Zeng R, Dou S, and Sun X 2009 *J. Am. Ceram. Soc.* **92** 1217
[16] Golovkina L S, Orlova A I, Nokhrin A V, Boldin M S, Lantsev, Chuvil'deev V N, Sakharov N V, Shotin S V, Zelenov A Yu 2018 *J. Nucl. Mater.* **511** 109
[17] Golovkina L S, Orlova A I, Nokhrin A V, Boldin M S, Lantsev E A and Chuvil'deev V N 2019 Spark Plasma Sintering of Fine-Grain Ceramic–Metal Composites Based on Garnet-Structure Oxide $Y_{2.5}Nd_{0.5}Al_5O_{12}$ with Mo, W, and Ni *Spark Plasma Sintering of Materials* (Springer International Publishing) pp 459–470

Evolution of strengthening factors during long-term aging at 650 °C in advanced 10% Cr heat-resistant steel

N Dudova*, R Mishnev, R Kaibyshev

Belgorod State National Research University, Pobeda, 85, Belgorod, 308015, Russia

*Corresponding author: dudova@bsu.edu.ru

Abstract. The effect of long-term aging for 1000...~40,000 h at 650 °C on the tensile strength at ambient temperature was studied in a low-nitrogen and high-boron 10%Cr martensitic steel. In order to establish the reason for increasing the yield stress and ultimate tensile strength after 10,000 h of aging, the evolution of strengthening factors was analyzed. A decrease in the substructure and solid solution strengthening during long-term aging is compensated by an increase in the dispersion strengthening due to the precipitation of V-rich MX carbonitrides.

1. Introduction

New generation fossil power plants with higher thermal efficiency and reduced emission of carbon dioxide require the heat-resistant materials that operate at higher temperatures. The most heated components of the power units will be made of austenitic steels and nickel-based superalloys. Nevertheless, the 9-12% Cr martensitic steels will be the base materials due to their excellent combination of high creep strength, good fatigue and oxidation resistance, and low cost [1]. One of the advanced approaches to improve the 100,000 h creep strength up to a minimum value of 100 MPa at 650 °C is a steel alloying modification by increasing the B content and decreasing the N content [2,3].

The creep resistance of these steels is associated with the stability of non-equilibrium structure, which is called the tempered martensite lath structure (TMLS). The complex TMLS consists of prior austenite grains, packets, blocks and laths with a high dislocation density in the lath interiors [2,3]. The main dispersion strengthening phases are nanoscale boundary $M_{23}C_6$-type carbides and MX carbonitrides homogeneously distributed in the laths. The role of precipitates of secondary phases ($M_{23}C_6$, MX, Laves phase) in the high stability of TMLS under creep condition is the subject of research interest. It was recently revealed that long-term aging led to a slight increase in the yield strength and ultimate tensile strength [4]. The aim of this work is to examine the effect of long-term aging at 650 °C on the evolution of strengthening factors in advanced 10% Cr steel with low N and high B contents at ambient temperature.

2. Experimental

A 10% Cr steel with the following chemical composition (in wt.%) 0.1C, 0.06Si, 0.1Mn, 10.0Cr, 0.17Ni, 0.7Mo, 0.05Nb, 0.2V, 0.003N, 0.008B, 2.0W, 3.0Co, 0.002Ti, 0.006Cu, 0.01Al and Fe-balance was examined. The vacuum induction-melted steel was subsequently hot-forged by Lasmet, Chelyabinsk. The steel was subjected to standard heat treatment: normalization at 1060 °C/30 min and tempering at 770 °C/3 h. Small specimens for tensile tests were cut from the grip portions of creep tested specimens at 650 °C, subjected only to long-term thermal aging for 1000; 10,000; 28,286 and 39,437 hours [4]. Tensile tests were carried out using flat specimens with a gauge length of 4 mm and cross-sectional dimensions of 1 mm x 1 mm using an Instron 5882 testing machine at ambient temperature and a strain rate of ~10^{-3} s^{-1}. The structural characterization was performed using a Jeol JEM-2100 transmission electron microscope (TEM) with an INCA energy dispersive X-ray spectrometer. The transverse lath widths were measured on TEM micrographs by the linear intercept method. The dislocation densities were estimated by counting individual dislocations in the (sub)grain/lath interiors per unit area on at least six arbitrarily selected typical TEM images for each data point.

3. Results and discussion

3.1. Tensile strength and microstructure after long-term aging

Aging of the 10% Cr steel at 650 °C for 1000…~30,000 h leads to a slight increase in the yield strength (YS) and ultimate tensile strength (UTS) (Figure 1a). Maximum rise of YS and UTS (+3…6%) to 520 and 695 MPa, respectively, occurs for 10,000 h aged steel. A ~40,000 h aging results in the UTS of 650 MPa that is 4…6% lower as compared with the tempered condition, whereas the YS returns to the initial level of 500 MPa. Therefore, the 10%Cr steel obviously demonstrates higher tensile strength characteristics after aging at 650 °C for 10,000 h.

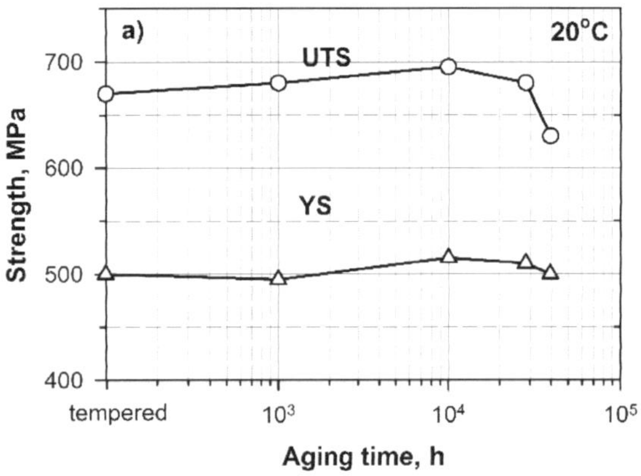

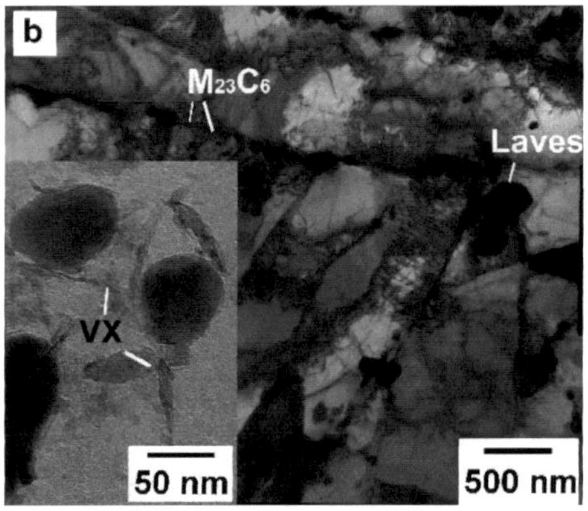

Figure 1. Effect of long-term aging at 650 °C on the YS and UTS at 20 °C (a). Microstructure of the 10% Cr steel after long-term aging for 10,000 h at 650 °C (b).

The TMLS of the 10% Cr steel is stable under long-term aging condition at 650 °C (Figure 1b) and does not transform into the subgrain structure up to ~40,000 h. Lath width increases less than 2 times (Table 1). Dislocation density in the lath interiors continuously decreases by 4 times.

Secondary phases in the tempered steel are presented by nanoscale $M_{23}C_6$ carbides (70 nm) located on the boundaries and fine Nb-rich MX carbonitrides (30 nm) homogeneously distributed in the laths. Both $M_{23}C_6$ and MX precipitates are resistant to coarsening. An insignificant coarsening of $M_{23}C_6$ carbides from 70 nm to 96 nm after ~ 40,000 h occurs. Nb-rich MX are highly stable.

Aging leads to Laves phase ($Fe_2(W,Mo)$) precipitation at the boundaries (Figure 1b). Consequently, depletion of W and Mo from the ferritic matrix occurs (Table 1). The onset of the Laves phase coarsening appears after 1000 h and accelerates after 10,000 h.

Between 1000 and 10,000 h of aging, fine V-rich MX carbonitrides precipitate in the lath interiors (Figure 1a). The mean size of these particles estimated by TEM observation of carbon replicas is 26.5 nm. The volume fraction of V-rich MX phase is 0.00862% as calculated by Thermo-Calc. However, structure observations reveals approximately 0.1% of V-rich MX particles.

Table 1. Change in the structural parameters of the 10% Cr steel during long-term aging at 650 °C.

Structural parameters	Duration of aging (h)				
	0	1000	10,000	28,286	39,437
Dislocation density, x10^{14} (m^{-2})	1.70	1.23	1.06	0.61	0.43
Lath width (μm)	0.38	0.409	0.507	0.566	0.614
Concentration of element in the matrix					
(at. fraction): Cr	0.109	0.105	0.104	0.103	0.1004
W	0.0062	0.0034	0.003	0.0026	0.0024
Mo	0.004	0.0034	0.0029	0.0024	0.0018
Co	0.0304	0.0304	0.0304	0.0304	0.0304
Mean size of particles (nm) [a]:					
$M_{23}C_6$	70	72	81.9	83.6	96
Laves phase	-	145.6	197.6	298.8	319
NbX	30	35	31.3	31.6	35
VX	-	-	26.5	40	58

[a] The volume fraction of $M_{23}C_6$ is 2.048 /2.107%, Laves – 0/1.6%, NbX – 0.05/0.06%, VX – 0/0.1% (before / after aging, respectively.)

3.2. *Strengthening factors*

To reveal the reason for the increase in the steel's strength after long-term aging for 10,000 h, the strengthening factors were analyzed. Estimation was carried out in accordance with a model describing the YS of high-chromium steels [5,6]. The YS is presented as:

$$\sigma_Y = M\sqrt{(\tau_A^{\ 2} + \tau_B^{\ 2})} \quad (1)$$

where M is the Taylor factor (2.9), τ_A is the strengthening from dislocations, τ_B is the strengthening from obstacles.

Dislocation strengthening is determined as: $\tau_A = \alpha_1 \mu b \sqrt{\rho}$, where α_1 is a constant (0.2); μ is the temperature depending shear modulus (at 20 °C it is 83 GPa); b is the Burgers vector ($2.48 \cdot 10^{-10}$ m); ρ is the dislocation density.

Strengthening from the obstacles can be defined as a sum of the strengthening from Peierls-Nabarro barriers (τ_{PN}), low angle boundaries (τ_{BD}), solute atoms (τ_{SS}) and dispersion particles (τ_{prec}) estimated according to Ref. [5,6]: $\tau_B = \tau_{PN} + \tau_{BD} + \tau_{SS} + \tau_{prec}$. Dispersion strengthening is determined by the Orowan-Ashbi model as [5,7]: $\tau_{prec} = 0.045(\mu b / \lambda)\ln(r/b)$, where λ is the mean distance between particles, r is the mean radius of particles.

The estimated strengthening factors are shown in Figure 2 (a-c). The evolution of strengthening during long-term aging can be described as follows:

- the dislocation strengthening continuously decreases with aging time, however, in the range of 1000…10,000 h the intensity of decreasing is the smallest;
- the solid solution strengthening by substitutional atoms of Cr, W, Mo, and Co is slightly reduced due to precipitation of W- and Mo-rich particles of Laves phase;
- the substructural strengthening by low angle boundaries of martensitic laths is continuously decreased starting from 1000 h of aging due to the lath widening;
- the dispersion strengthening during aging for 1000…10,000 h is increased on 16% as compared with tempered condition. Although the strengthening from $M_{23}C_6$ and Laves phases is reduced, precipitation of V-rich MX particles compensates for this decrease.

The calculated values of YS (Figure 2d) remain stable between 1000 and 10,000 h of aging and are close to the experimental YS. This fact is resulted from the increasing precipitation strengthening.

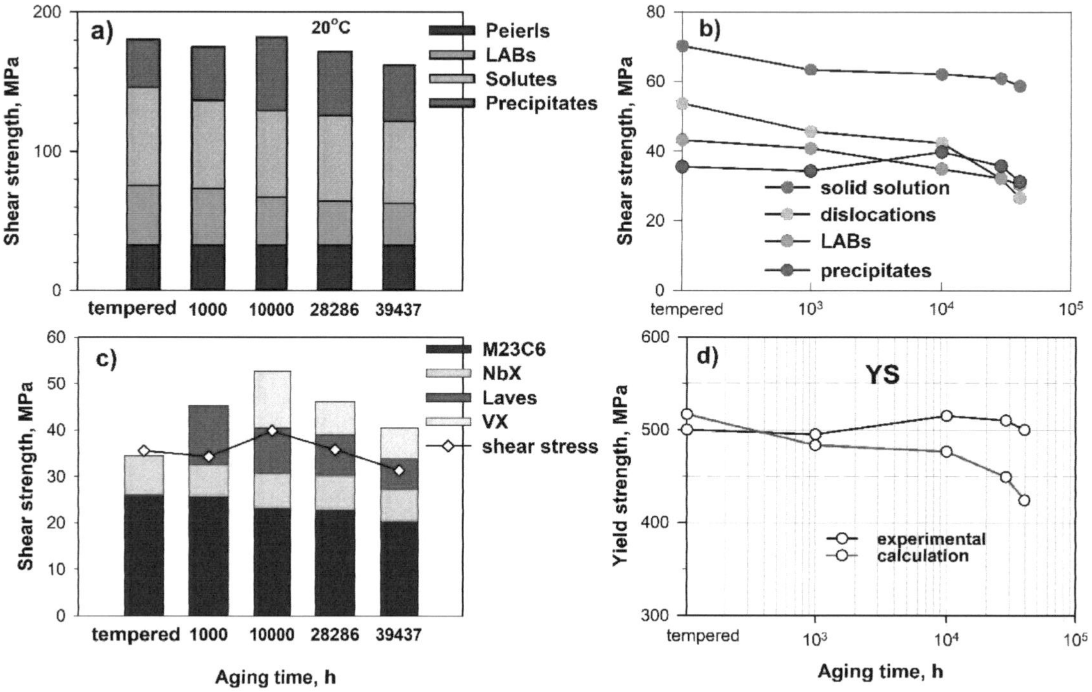

Figure 2. Change in the strengthening factors (a-c) during long-term aging at 650 °C of the 10% Cr steel. Change in the calculated YS at 20 °C after aging in comparison with experimental data (d).

4. Conclusion

The stabilization of the calculated yield strength of the low-nitrogen and high-boron 10% Cr steel is revealed after 1000...10,000 h aging at 650 °C that is attributed to an increase in dispersion strengthening. Precipitation of fine V-rich MX carbonitrides during aging for 1000...10,000 h can contribute to the strengthening of this steel.

Acknowledgments

The authors are grateful to the staff of the Joint Research Center, "Technology and Materials", Belgorod State University, for providing the equipment for instrumental analysis.

References

[1] Abe F, Kern T-U, Viswanathan R 2008 *Creep-Resistant Steels* (Cambridge: Woodhead Publishing)
[2] Abe F 2011 *Procedia Engineering* **10** 94
[3] Dudova N, Mishnev R, Kaibyshev R 2019 *Mater. Sci. Eng. A* **766** 138353
[4] Dudova N, Mishnev R, Tikhonova M, Kaibyshev R 2019 *IOP Conf. Ser.: Mater. Sci. Eng.* **672** 012062
[5] Barrett R A, O'Donoghue P E, Leen S B 2018 *Mater. Sci. Eng. A* **730** 410
[6] Li Q 2003 *Mater. Sci. Eng. A* **361** 385
[7] Ashby M F 1968 in: Gordon and Breach (Eds.). Proceedings of the Second Bolton Landing Conference on Oxide Dispersion Strengthening (New York: Science Publishers)

Evolution of the tempered lath structure of the 12%Cr steels with low N and high B contents during creep

A E Fedoseeva[1*], I S Nikitin[1], A E Fedoseev[1], R O Kaibyshev[1]

[1]Belgorod National Research University, Pobeda 85 308015 Belgorod, Russia

[*]Corresponding author: fedoseeva@bsu.edu.ru

Abstract. 9-12%Cr martensitic steels are promising materials for elements of boilers, tubes, pipes, heaters and steam blades for fossil power plants, which are able to operate at ultra-supercritical parameters of steam. Creep tests were carried out for two 12%Cr and 9%Cr steels at 650°C. After tempering at 750-770°C, a tempered lath structure was revealed in all the steels studied. The 12%Cr-Ta steel had the smallest lath width among the steels studied that provided the longest creep rupture time after creep test at 650°C/120 MPa. During creep, the applied stress causes the lath growth and the formation of subgrains instead of martensitic laths. The experimental sizes of subgrains and equilibrium subgrains, calculated on balance of driving and retarding forces and using the dependence of the size of subgrains on the applied stress during creep, were compared to estimate the contributions of secondary particles and applied stress to the growth of subgrains during tempering and creep in the different steels.

1. Introduction

12%Cr martensitic steels are considered to be materials for the elements of steam blades in fossil power plants operating at ultra-supercritical parameters of steam [1-2]. The main disadvantage of 12%Cr martensitic steels is the low structural stability during creep at elevated temperatures, i.e. the tempered martensite lath structure (TMLS) evolves fast into a subgrain structure [2]. This reduces the creep resistance of 12%Cr steels and limits their application in fossil power plants [1,2]. 9%Cr steels usually demonstrate high structural stability for long-term thermal exposure that provides high creep resistance [3-5]. A new approach to alloying of 9-12%Cr steels, which consists in adding Co, decreasing the N content, and increasing the B content [6-10], can eliminate the poor structural stability of the 12%Cr steel during creep. The addition of Co increases the volume fraction of the secondary phase particles and contribute to the solid solution strengthening [3,4,6]. Decreasing N content together with increasing B content eliminates the formation of coarse BN nitrides and slows down the particle coarsening [7-9]. Additionally, new 12%Cr steels contain 0.8%Cu and 0.07%Ta as austenite-stabilizing elements [9]. Moreover, Cu acts as nucleation sites for Laves phase [7-9]. The aim of this study is to report the evolution of the tempered martensite lath structure of two Co-containing 12%Cr steels with low N and high B contents during creep at 650°C in comparison with a 9%Cr steel containing Co with standard N and B contents [4].

2. Experimental procedure

Two 12% Cr steels, denoted here as 12%Cr and 12%Cr-Ta, were prepared by vacuum induction melting, and a 9% Cr steel, denoted here as 9%Cr, was prepared by air induction melting. The chemical compositions of the Co-containing 12%Cr steels with low N and high B contents and the 9%Cr steel with standard N and B contents are listed in Table 1.

Table 1. Chemical compositions of 12%Cr and 9%Cr steels.

	C	**Cr**	**Co**	**Mo**	**W**	**V**	**Cu**	**Nb**	**B**	**N**	**Ta**
12%Cr	0.09	11.3	3.9	0.63	2.4	0.24	0.78	0.07	0.009	0.003	-
12%Cr-Ta	0.11	11.4	3.0	0.62	2.5	0.23	0.76	0.04	0.01	0.003	0.07
9%Cr	0.10	9.5	3.0	0.5	1.8	0.20	-	0.05	0.005	0.05	-

All steels were solution treated at 1050-1070°C for 1 h, cooled in air, and subsequently tempered at 750-770°C for 3 h. Cylindrical specimens with a gauge length of 60 mm and a diameter of 6 mm were crept until rupture at 650°C under applied stresses of 180-100 MPa with a step of 20 MPa. A JEOL-2100 transmission electron microscope (TEM) was used for structural characterization of tempered and crept structures [3-5]. The volume fractions of the secondary phase particles were calculated using the Thermo-Calc software [3-5].

3. Results and Discussion

3.1 Tempered state

The microstructures of the Co-containing 12%Cr and 9%Cr steels after heat treatment, obtained by TEM are represented in Figure 1. The microstructural parameters after heat treatment for both steels are summarized in Table 2.

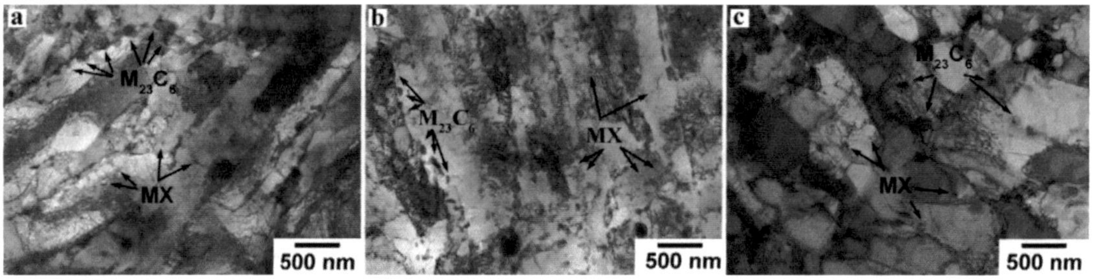

Figure 1. TMLS of the 12%Cr (a), 12%Cr-Ta (b) and 9%Cr (c) steels after heat treatment.

Table 2. Structural parameters of the 12%Cr and 9%Cr steels after heat treatment.

	PAG size, μm	Fraction of δ-Fe,%	Lath width, nm	Dislocation density, m^{-2}	Particle size, nm		Volume fraction, %	
					$M_{23}C_6$	MX	$M_{23}C_6$	MX
12%Cr	51±2	6±1	305±30	1.3×10^{14}	61±5	29±3	1.59	0.06
12%Cr-Ta	48±2	9±2	285±30	2.0×10^{14}	55±5	20±3	1.81	0.07
9%Cr	10±1	0	380±30	2.0×10^{14}	90±5	30±3	1.80	0.25

After tempering at 750-770°C, a tempered martensite lath structure (TMLS) was observed in both 12%Cr martensitic steels (Figure 1a,b), whereas a mixture of TMLS and subgrain structure was revealed in the 9%Cr steel (Figure 1c). The width of the martensitic laths in all steels was similar and comprised 300-400 nm, whereas the dislocation density in the 12%Cr steel was lower than that in the 12%Cr-Ta and 9%Cr steels (Table 2). In all steels, Cr-rich $M_{23}C_6$ carbides are located at grain boundaries and V-, Nb- or Ta-rich MX carbonitrides randomly distributed in the matrix (Figure 1). The mean size of $M_{23}C_6$ particles in the 9%Cr steel was significantly larger than that in the 12%Cr steels (Figure 1, Table 2), whereas the mean size of MX carbonitrides was similar in all steels.

3.2 Creep properties

Figure 2 shows the creep rupture data of the steels at a temperature of 650°C. Both 12%Cr steels did not demonstrate any breakdown in the curve "Applied stress vs. Time to rupture", whereas a creep strength breakdown appeared in the 9%Cr steel after 2000 h of creep tests (Figure 2a). The addition of Ta, as well as a decrease in the N content to 0.003% together with an increase in the B content to 0.012 wt.% affects positively the creep rupture strength (Figure 2a). A decrease in the minimum creep rate and an increase in the time to rupture took place (Figure 2b).

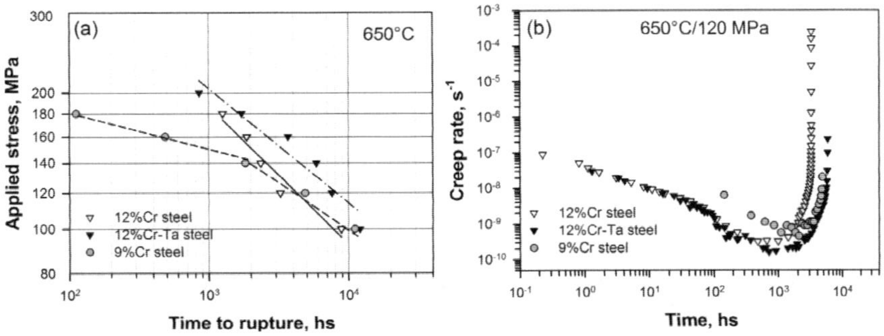

Figure 2. Time to rupture vs. stress curves (a) and minimal creep rate vs. time curves (b).

3.3 Evolution of TMLS during creep

After the creep test at 650°C/120 MPa, the microstructures of all steels strongly evolved. The TMLS partially transformed into a subgrain structure with a mean size of subgrains of 0.7 μm in the 12%Cr steels (Figure 3a,b); a 100% subgrain structure with a mean subgrain size of 1.4 μm instead of the TMLS was revealed in the 9%Cr steel (Figure 3c). The dislocation density in all steels decreased to 10^{13} m^{-2}. Grain boundary $M_{23}C_6$ and Laves phase particles and MX carbonitrides coarsened, especially, in the 9%Cr steel (Table 3). No evidence of Z-phase particles was observed in all of the steels studied.

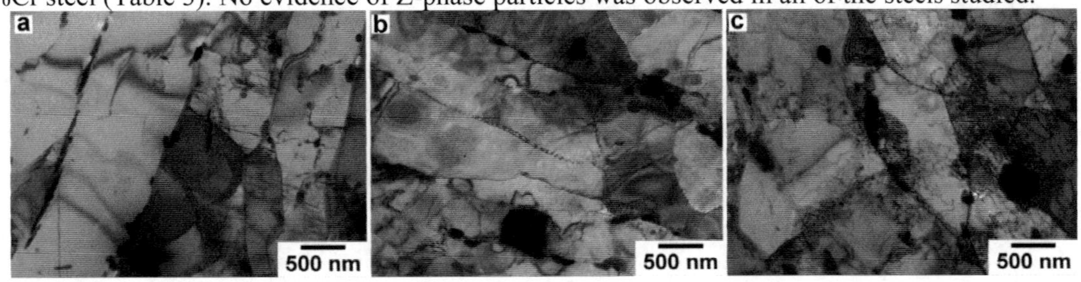

Figure 3. Evolution of TMLS in the 12%Cr (a), 12%Cr-Ta (b) and 9%Cr (c) steels after creep test at 650°C/120 MPa.

Table 3. Structural parameters of the 12%Cr and 9%Cr steels after creep test at 650°C/120 MPa.

	Lath width, nm	Subgrain size, nm	Dislocation density, m^{-2}	Particle size, nm			Volume fraction, %		
				$M_{23}C_6$	Laves	MX	$M_{23}C_6$	Laves	MX
12%Cr	665±30	760±30	0.8×10^{14}	109±5	170±5	45±3	1.68	2.10	0.06
12%Cr-Ta	650±30	645±30	0.7×10^{14}	96±5	197±5	37±3	2.07	2.10	0.07
9%Cr	-	1400±30	0.4×10^{14}	200±5	280±5	70±3	1.96	1.32	0.25

The equilibrium subgrain size *D*, which may be achieved during creep, can be estimated (1) on the assumption of the driving and retarding forces are in a balance [3] and (2) using the dependence of the subgrain size on the applied stress during creep [2]. The calculated and experimental values of the subgrain size in the steels studied are represented in Table 4.

Table 4. Calculated and experimental values of the subgrain size.

	Tempered state		Crept state		
	D [3], μm	D_{exp}, μm	*D* [3], μm	D_{∞} [2], μm	D_{exp}, μm
12%Cr	0.33	0.31	0.57	1.24	0.76
12%Cr-Ta	0.27	0.29	0.49	1.24	0.65
9%Cr	0.41	0.38	0.92	1.24	1.40

The experimental subgrain sizes after tempering in all steels are in accordance with the equilibrium values estimated from the balance between the driving and retarding forces [3] (Table 4). The largest subgrain size in the 9%Cr steel is caused by the low Zener pressure from the particles due to their large size (Table 2). The smallest subgrain size of 0.27 μm occurs in the 12%Cr-Ta steel that is attributed to the finest particle sizes and the highest volume fraction of $M_{23}C_6$ carbides (Table 2).

After creep tests, the equilibrium subgrain sizes, estimated from the balance between driving and retarding forces, are insignificantly smaller than the experimental subgrain ones with the exception of the 9%Cr steel (Table 4). Boundary particles in the 12%Cr steels effectively prevent the lath growth, resulting in smaller experimental subgrain sizes than those calculated using the dependence of the subgrain size on the applied stress [2], though the applied stress leads to a strong evolution of the TMLS accelerating the migration of lath boundaries [2,7-9]. In the 9%Cr steel, the secondary particles do not control the lath growth during long-term creep that probably causes a breakdown of the creep strength [4]. An insignificant lath growth in the 12%Cr steels provides no creep strength breakdown appearance (Figure 2a). The 12%Cr-Ta steel demonstrates the longest creep rupture time in a creep condition of 650°C/120 MPa (Figure 2) that can be attributed to the remaining TMLS and the finest subgrain sizes.

4. Conclusions

The formation of a TMLS in all 12%Cr and 9%Cr steels during tempering at 750-770°C was revealed. The smallest lath width was observed in the 12%Cr-Ta steel, since fine boundary particles ensured, effective retarding of the lath boundary migration. During creep, the TMLS strongly evolves in all steels; the applied stress provides the formation of subgrains. However, the boundary particles effectively prevent subgrain growth in the 12%Cr steels compared with the 9%Cr steel.

Acknowledgments

The study was financially supported by Ministry of Science and Higher Education of the Russian Federation, the President's Grant for PhD – young scientists (grant number 075-15-2019-1165).

References

[1] Kern T U, Staubli M, Scarlin B 2002 *ISIJ International* **42** 1515
[2] Hald J, Korcakova L 2003 *ISIJ International* **43** 420
[3] Fedoseeva A, Kozlov P, Dudko V, Skorobogatykh V, Shchenkova I, Kaibyshev R 2015 *Phys. Met. Metall.* **116** 1047
[4] Fedoseeva A, Dudova N, Kaibyshev R 2016 *Mater. Sci. Eng. A* **654** 1
[5] Dudko V, Fedoseeva A, Belyakov A, Kaibyshev R 2015 *Phys. Met. Metall.* **116** 1165
[6] Helis L, Toda Y, Hara T, Miyazaki H, Abe F 2009 *Mater. Sci. Eng. A* **510–511** 88
[7] Abe F 2011 *Procedia Engineering* **10** 94
[8] Hald 2008 *J Int J Pres Ves Pip* **85** 30
[9] Klueh R L 2005 *International Materials Reviews* **50** 287

Degradation of the creep resistance of a Re-containing 10%Cr steel upon creep testing at low applied stress

A E Fedoseeva[1*], I S Nikitin[1], A E Fedoseev[1], R O Kaibyshev[1]

[1]Belgorod National Research University, Pobeda 85 308015 Belgorod, Russia

[*]Corresponding author: fedoseeva@bsu.edu.ru

Abstract. New Re-containing steel is a promising material for steam blades of fossil power plants worked at ultra-supercritical parameters of the steam. This steel was subjected to creep testing at 650°C under the applied stresses ranging from 200 to 100 MPa. The steel exhibits excellent creep resistance at high applied stresses; the time to rupture comprises 10,987 hours after the creep test at 650°C/140 MPa. However, a dramatic drop in the creep strength is observed at low applied stresses. A possible reason for this decrease in the creep resistance may be a significant coarsening of the Laves phase particles.

1. Introduction

Creep resistant 9-12%Cr martensitic steels are widely used as materials for fossil power plants [1]. The main reason for the high creep resistance of these 9-12%Cr steels is the slowing down the transformation of the tempered martensite lath structure (TMLS) into the subgrain structure under short-term and long-term creep conditions [2,3]. The growth of martensitic laths is accompanied by the coarsening of the Laves phase particles that is one of the reasons for the decrease in creep resistance [2-5]. Fine Laves phase particles diminish the creep rate at the primary creep stage, whereas the coarsening of the Laves phase particles provides an increase in the creep rate at the tertiary creep stage [4,5]. Suppression of the Laves phase coarsening shifts the onset of the acceleration creep stage to longer time that increases the creep rupture time. However, this task is very complicated. The addition of Co reduces the solubility of W atoms in ferrite [6,7] that accelerates the growth kinetics of Laves phase [6]. The addition of Re strongly affects the precipitation of Laves phase [8-11]. Thus, alloying with Re, together with an increase in the W content to 3% and a decrease in the Mo content to 0.13% in the 10% Cr-3%Co martensitic steel with low N and high B contents [3,10-12], positively affects the creep resistance at high applied stresses, at which the creep rupture time increases from about 1800 h [12] to 10,987 h [11] for the creep test at 650°C/140 MPa. This increment is related to a low coarsening rate of $M_{23}C_6$ carbides and the Laves phase particles and the retaining chains of these particles on the lath boundaries up to rupture [11]. However, at lower applied stresses, the creep strength breakdown is observed. The aim of the present work is to consider microstructural aspects of degradation of creep resistance of the martensitic 10% Cr-3% Co-3% W-0.2% Re steel under creep condition of 650°C/120 MPa.

2. Experimental procedure

A Re-containing 10% Cr-3% Co-3% W steel with a chemical composition of (in wt.%) Fe (bal.)–0.11C–9.85Cr–3.20Co–2.86W-0.13Mo–0.22Cu-0.03Si–0.14Mn–0.03Ni-0.23V–0.07Nb–0.002N–0.008B-0.17Re was prepared by vacuum induction melting. This steel was normalized at 1050°C for 1 h and tempered at 770°C for 3 h. Creep tests were carried out at 650°C under the applied stresses ranging from 100 to 200 MPa up to rupture. Under an applied stress of 120 MPa, the creep test was interrupted after achieving strains of 1% (1001 h), 2% (5035 h) and 2.1% (10052 h) corresponding to the transient, steady-state and tertiary creep stage. The structural characterization of the crept samples was carried out using a JEOL–2100 transmission electron microscope (TEM) and a Quanta 600 FEG scanning electron microscope (SEM).

3. Results and Discussion

3.1 Tempered state

After heat treatment, the formation of a TMLS with a lath width of 290 nm was revealed in the Re-containing 10% Cr-3% Co-3% W steel [10,11]. A high dislocation density of 2×10^{14} m^{-2} was observed inside martensitic laths [10,11]. Cr-rich $M_{23}C_6$ carbides with a mean size of 67 nm and NbX carbonitrides with a mean size of 35 nm precipitated during tempering.

3.2 Creep properties

Figure 1 shows the creep rupture data of the Re-containing 10% Cr steel at a temperature of 650°C.

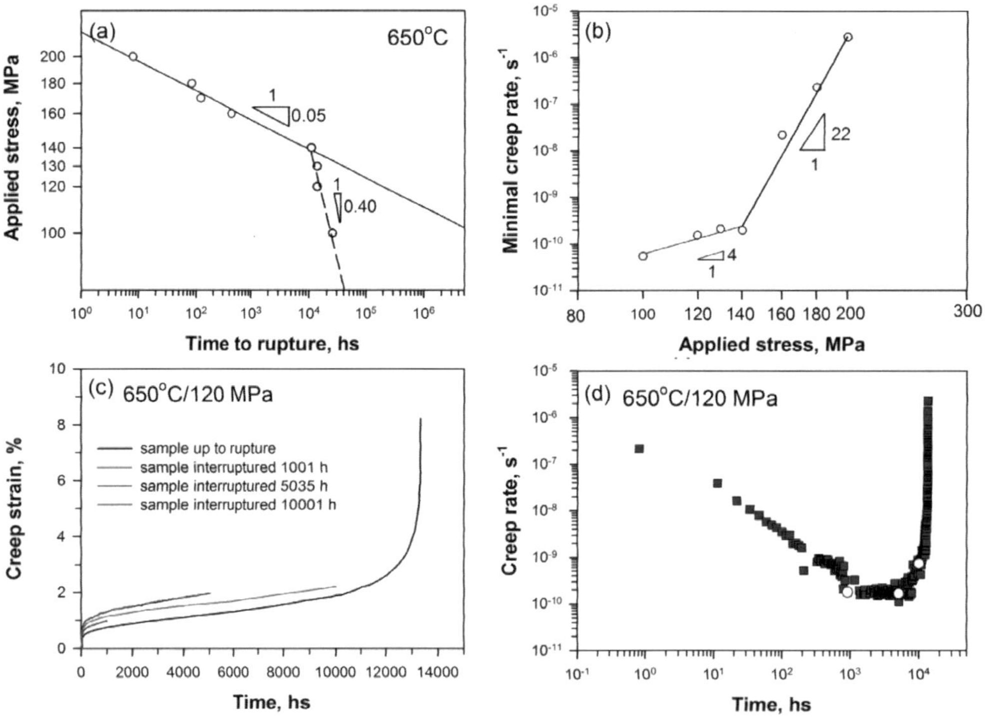

Figure 1. Time to rupture vs. stress (a) and minimal creep rate vs. applied stress (b) curves for creep tests at 650°C and applied stresses ranging from 200 to 100 MPa; strain vs. time (c) and creep rate vs. time (d) curves for creep test of 650°C/120 MPa.

Creep strength breakdown (Figure 1a) and power-law breakdown (Figure 1b) appeared at applied stresses below 140 MPa indicating a change in the creep mechanism from dislocation climb in the region of high applied stresses (> 140 MPa) and diffusion creep in the region of low applied stresses (< 140 MPa). The appearance of these breakdowns can be related to the growth of boundary carbides and particles of the Laves phase [13].

3.3 Evolution of TMLS during creep

After the creep test at 650°C/120 MPa, the microstructure of the studied steel strongly evolved (Figure 2). The lath width increased to 2.5 μm in the neck section of the ruptured specimen; the TMLS transformed into a subgrain structure with a subgrain size of 1.8 μm. The dislocation density decreased to about 10^{13} m^{-2} after 13,600 h of creep testing. W-rich Laves phase particles were observed already after 1001 h of creep testing at 650°C/120 MPa (Figure 2a). This corresponded to the equilibrium volume fraction of 1.6% of the Laves phase in the studied steels [10]. The average chemical composition of this phase was 8%Cr-30%Fe-62%W (in wt%). When the creep time increased, the Laves phase particles grew strongly, whereas the $M_{23}C_6$ carbides and NbX carbonitrides were dimension stable. The mean size of Laves phase in the ruptured specimen was about 250 nm, while very large particles with a size more than 2 μm, randomly distributed in the matrix, were also observed. Their fraction was about 5% of all Laves phase particles.

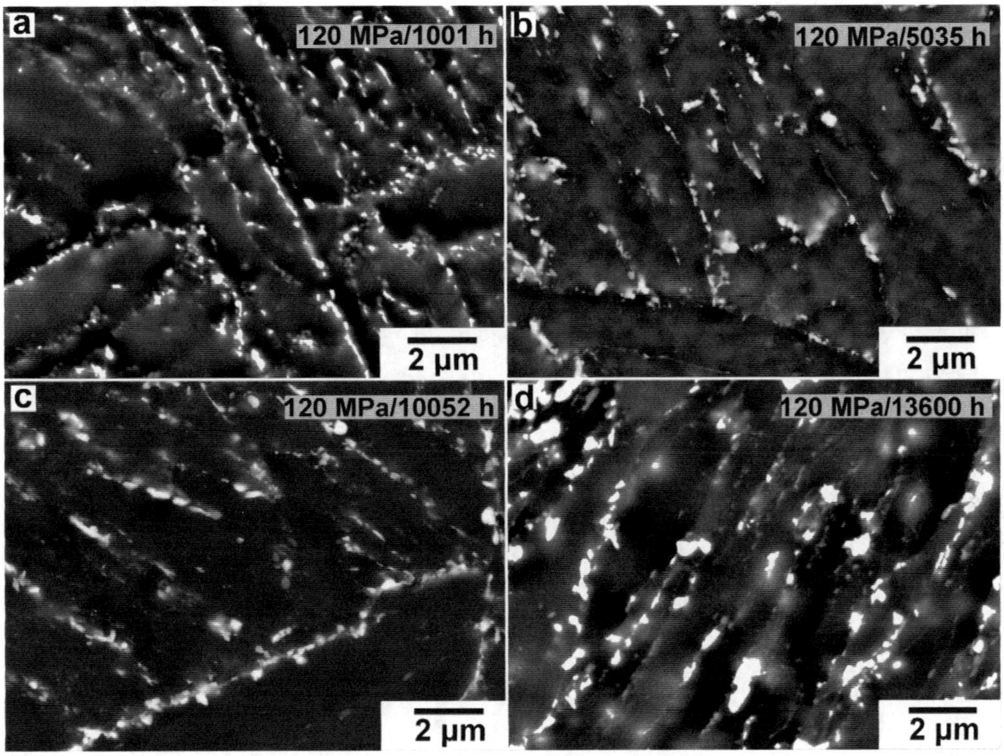

Figure 2. SEM images of the interrupted samples tested at 650°C/120 MPa with different durations.

3.4 Fracture mechanism

The fracture surface of specimens tested at an applied stress of 120 MPa is shown in Figure 3. A dimple rupture occurring on the transgranular fracture path was observed. Few very large voids in the center and rims of the ruptured specimen were revealed (Figure 3a). In the center of the ruptured specimen (Figure 3b), there were few nucleation sites, in which the microvoids grew to a large size before coalescing, and this produced a fracture surface that exhibited various dimple sizes. In the rim of the ruptured specimens (Figure 3d), small dimples were formed that was indicative for the activation of numerous nucleating sites; adjacent microvoids coalesced before they started to grow. Large dimples were nucleated by particles located at the bottom of these dimples (Figure 3b,d). These particles were considered to be Laves phase, which served as nucleation sites for coarse cracks.

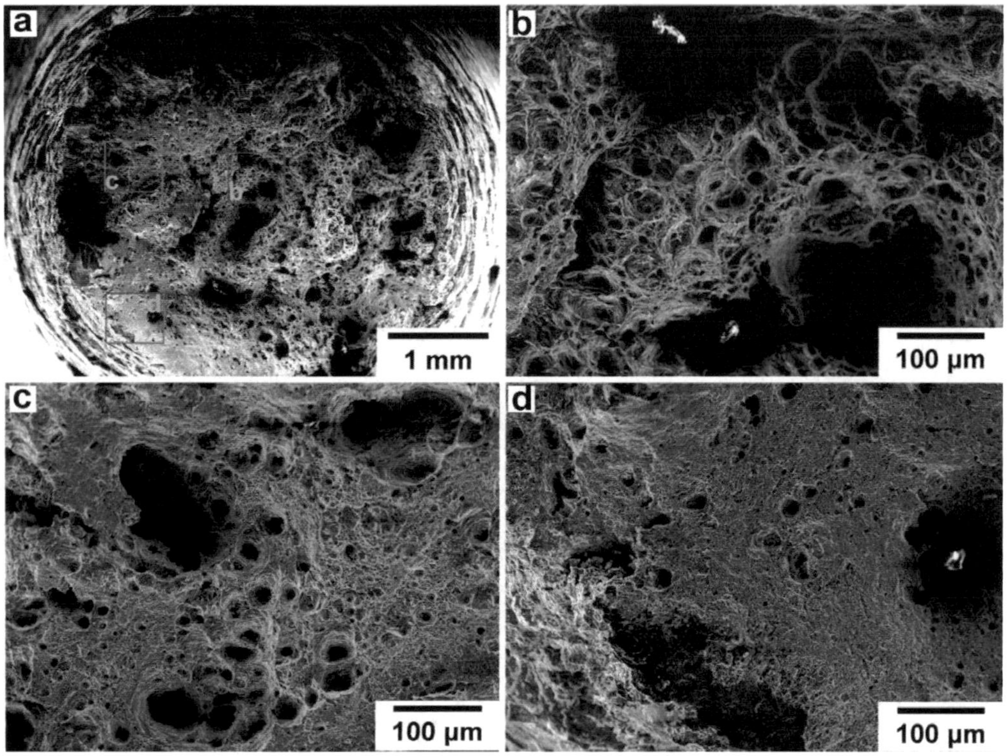

Figure 3. SEM micrographs of the rupture surface of specimens tested at 650°C/120 MPa.

4. Conclusions

A strong degradation of the creep resistance of the Re-containing 10% Cr-3% Co-3% W steel was observed upon creep testing under low applied stresses. Creep strength breakdown and power-law breakdown appeared at applied stresses below 140 MPa. Significant growth of Laves phase is considered to be the reason of this degradation because coarse particles of the Laves phase with sizes of about 2 µm can serve as nucleation sites for large voids and cracks.

Acknowledgments

The study was financially supported by Russian Science Foundation, grant number 19-73-10089.

References

[1] Bhadeshia HKDH 2001 *ISIJ International* **41** 626
[2] Kostka A, Tak K-G, Hellmig RJ, Estrin Y, Eggeler G 2007 *Acta Mater.* **55** 539
[3] Kaibyshev R, Mishnev R, Fedoseeva A, Dudova N 2017 *Materials Science Forum* **879** 36
[4] Abe F 2004 *Mater. Sci. Eng. A* **387-389** 565
[5] Abe F 2005 *Metall. Trans. A* **36A** 321
[6] Prat O, Garcia J, Rojas D, Sauthoff G, Inden G 2013 *Intermetallics* **32** 362
[7] Klueh RL 2005 *Int. Mater. Reviews* **50** 287
[8] Viswanathan R, Bakker W 2001 *J. Mater. Eng. Perform.* **10** 96
[9] Maruyama K, Sawada K, Koike J 2001 *ISIJ International* **41** 641
[10] Fedoseeva A, Nikitin I, Dudova N, Kaibyshev R 2019 *Mater. Letters* **269** 81
[11] Fedoseeva A, Nikitin I, Dudova N, Kaibyshev R 2020 *Mater. Letters* **262** 127183
[12] Mishnev R, Dudova N, Fedoseeva A, Kaibyshev R 2016 *Mater. Sci. Eng. A* **678** 178
[13] Armaki G, Chen R, Kano S, Maruyama K, Hasegawa Y, Igarashi M 2012 *Mater. Sci. Eng. A* **532** 373

IOP Conf. Series: Materials Science and Engineering 1014 (2021) 012013 doi:10.1088/1757-899X/1014/1/012013

Quantitative analysis of $\{100\}_{Al}$ plate/lath- and $<100>_{Al}$ rod-shaped precipitates in an aged Al-Cu-Mg-Si alloy using TEM

M R Gazizov[1,*], R Holmestad[2], C D Marioara[3] and R O Kaibyshev[1]

[1]Belgorod State University, Pobeda, 85, Belgorod, 308015, Russia

[2]Department of Physics, Norwegian University Science and Technology (NTNU), Trondheim, 7491, Norway

[3]Materials and Nanotechnology, SINTEF Industry, Trondheim, 7465, Norway

*Corresponding author: gazizov@bsu.edu.ru

Abstract. Precipitate parameters, such as length, diameter and thickness of rationally oriented $\{100\}_{Al}$ plate/lath-shaped and $<100>_{Al}$ rod-shaped precipitates in an age-hardenable Al-Cu-Mg-Si alloy have been examined quantitatively. An appropriate procedure of the plate- and rod-shaped precipitate separation was suggested to determine the precipitate parameters as well as their distribution histograms using bright- (BF) and dark-field (DF) transmission electron microscopy (TEM) images taken in the $<100>_{Al}$ and $<110>_{Al}$ zone axes (ZA).

1. Introduction

The size and morphology of precipitates are important parameters of the microstructure of age-hardenable aluminium alloys, because they determine their service properties of these materials. In the case of Al-Cu-Mg-Si alloys, the presence of rationally oriented plate- or/and lath-shaped precipitates with $\{100\}_{Al}$ habit planes and rod-shaped precipitates along the $<100>_{Al}$ direction consisting of diverse phase fragments takes place after aging [1,2]. Quantitative characterization of several types of precipitates has not been performed, despite this may help to better understand the effect of various thermomechanical parameters (temperature, time, strain, etc.) on the microstructure of precipitates in these alloys, as well as to estimate the precipitate strengthening contribution to the strength of the material.

Different analytical techniques can be used to characterize the precipitation behaviour [3,4]. However, conventional TEM [5,6] still remains an attractive technique because this method is relatively fast and provides direct observations of the precipitates to measure their sizes and analyze their distribution although it may give poorer statistics. The major drawback of different TEM techniques is that the output is 2D projection images, which may be difficult to interpret in 3D. In the Al-Cu-Mg-Si alloys, TEM images taken along different ZA do not allow to identify and separate rationally oriented $\{100\}_{Al}$ plates and $<100>_{Al}$ rods because these projected precipitates look similarly in representative TEM images. The aim of the present work has been to develop an appropriate procedure for quantitative characterization of precipitate parameters such as size, number density and volume fraction in an Al-Cu-Mg-Si alloy using conventional TEM.

2. Materials and methods

An aluminum alloy with the chemical composition of Al-4.9Cu-0.74Mg-0.51Si-0.48Mn-0.1Cr-0.08Ti-0.02Fe (wt. %) was cast and subjected to thermomechanical treatment (TMT) as described in [1,2]. It should be just noted final TMT steps included solution heat treatment at 500 °C for 1 h followed by water-quenching to room temperature and then aging at 170 °C in the time range up to 96 hours [1].

Details of the sample preparation procedure for TEM analysis are given in [1,2]. TEM analysis was performed on JEOL JEM-2100 and JEM-2100F microscopes operated at 200 kV. Observations were carried out with an electron beam parallel to an <100>$_{Al}$ and <110>$_{Al}$ ZA with DF images using $\vec{g}$ = 110$_{Al}$ and 100$_{Al}$, respectively. Foil thickness (φ) was measured by convergent beam electron diffraction (CBED) [6]. All the precipitate parameters were estimated taking the tilting of the TEM foils into account. The Crystal Studio V10 software was used to simulate different diffraction patterns (DPs) including spots of the first-, second- and third-order Laue zones. This simulation condition did not take into account double diffraction. The procedure of ZA calculation for the lattices of the precipitates, having orientation relationships (ORs), in the respective {100}$_{Al}$ and {110}$_{Al}$ projections was described in [9]. The simulated DPs were validated for the β''- and θ'-phases in the Al matrix by [9] and [10], respectively.

3. Results and Discussions

TEM images, showing a typical precipitate microstructure in the Al-Cu-Mg-Si alloy aged for 16 and 96 h, are given in figure 1a-b. A close examination reveals the presence of two precipitate groups categorized by morphology in the images taken along <100>$_{Al}$ and <110>$_{Al}$ ZA. The first group is associated with <100>$_{Al}$ rod-shaped precipitates normal to the image and clearly seen in cross-section in the {100}$_{Al}$ projections as dark and bright dots in the BF and DF images, respectively (marked as a_1 in figure 1a). These precipitates are of a hybrid type consisting of small fragments of the β'', β'-Cu, C and Q' phases, sometimes forming somewhat jagged bands along dislocation lines [1]. Some of these rods having in-plane orientation appear as elongated 'rectangles' marked by a_2 in figure 1a. However, a certain fraction of these elongated 'rectangles' are caused by projections of plate- or lath-shaped particles categorized as the second group of precipitates. These precipitates can be identified as discrete C laths and θ' plates with {100}$_{Al}$ habit planes. They are clearly seen in DF-TEM images taken along <110>$_{Al}$ ZA using an aperture in the forbidden $\overline{g}$=001$_{Al}$ position (figure 1b).

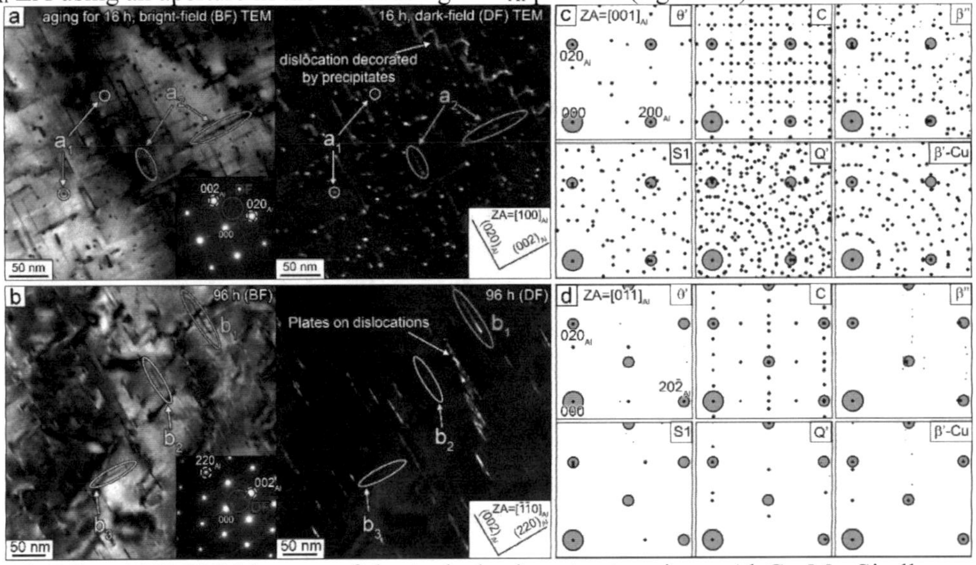

Figure 1. BF and DF-TEM images of the typical microstructure in an Al-Cu-Mg-Si alloy aged for 16 h (a) and 96 h (b); Superposition of simulated DPs for the Al matrix in <100>$_{Al}$ ZA (c) and <110>$_{Al}$ ZA (d) and for the precipitates with ORs from [1,2,4]. ZAs for the precipiatates were determined using a method in [9].

DP simulations, performed for the β''-, β'-Cu, C-, Q'-, S1 and θ'-phases previously found in the same Al-Cu-Mg-Si alloy and having respective ORs, are presented in figure 1d and e. The DPs for these phases were obtained for the directions corresponding to the $<100>_{Al}$ and $<110>_{Al}$ ZA. It is seen in figure 1b and c that DPs for the AlMgSi(Cu) phases (β''-, β'-Cu, C- and Q'-phase) show a dense distribution of the diffraction spots in ZA=$<100>_{Al}$.

Thus, the presence of aggregates with different precipitate morphologies and orientations does not allow separating them according to their shape using $\{100\}_{Al}$ projections and conventional TEM. The use of the $\{110\}_{Al}$ projections and a DF-TEM mode with $\vec{g}$=100_{Al} helps to highlight the plate-shaped particles. It is seen in figure 1d that the diffraction spots for the β''- and β'-Cu phases are weak. The C, Q', θ' and S1 phases have strong diffraction spots in $<110>_{Al}$ ZA (figure 1d), and the DF intensity is sufficient to distinguish the plate-shaped particles (marked by b_1 in figure 1b) due to their edge-on orientations. Previous high-resolution TEM analysis of the same alloy [1,2] showed that the C-phase appears as small fragments being predominantly inter-grown with other phases. It makes us conclude that bright 'rectangles' in the DF images are mostly θ' plates. The "disappearance" of rod-shaped particles in DF images using $\vec{g}$=100_{Al} are seen in figure 1b. This is due to the rods which are 45°-tilted to the $\{110\}_{Al}$ projection plane (marked as b_2) and the rods lying in the projection plane (b_3), which become dark in the respective DF images.

A joint analysis of $\{100\}_{Al}$ and $\{110\}_{Al}$ projections allows us to get all information about the precipitate parameters. In the case of the $\{110\}_{Al}$ projection, one third of the $\{100\}_{Al}$ plates of the laths/plates are seen to measure the plate diameter (D) and thickness (t). The number density of the plates can be estimated as

$$N_{V\_p} = N_p\left(A_S(\varphi + D)\right)^{-1} + 2N_pA_S^{-0.5} \qquad (1)$$

where N_p is the number of the counted plates in each TEM image (the total number of plates counted (n) was 241 in all TEM images), φ is the foil thickness and A_S is the respective TEM image area. This equation was developed using the methodology described in [5] and takes the precipitate orientation features in the respective projections into account. It is assumed that the number of the plates/laths crossed by the projection plane and seeing in the DF images taken along the $<110>_{Al}$ ZA is higher than the plates/laths 45°-tilted to the projection plane. In the case of $\{100\}_{Al}$ projections, one third and two third of the $<100>_{Al}$ rods are seen as 'dots' and 'rectangles', respectively. An equivalent circular rod diameter can be calculated as $D_{r_{eq}} = 2(S_r/\pi)^{0.5}$, where S_r is the rod cross-section area. The total number of cross-sections counted (n) was 1023 in all TEM images. The average length (L) of the elongated 'rectangles' can also be measured representing a common parameter for the rods and plates due to similar visibility for both precipitate types in the $\{100\}_{Al}$ projection. The L value is required to put into equation (1) with N_r (the total number of the counted rod cross-sections) instead of D and N_p, respectively, to estimate the number density of the rods (N_{V_r}). In this case, the equation (1) takes into account that the number of the rods crossed by projection plane and seeing in the BF TEM images taken along $<100>_{Al}$ ZA is higher than the rods parallel to the projection plane (orthogonal to the $<100>_{Al}$ ZA). An average L value was calculated using 1286 measurements in all TEM images. An average rod true length can be estimated using statistical joint analysis of length/diameter distributions for the rods and plates (L) obtained from the $\{100\}_{Al}$ projection and the single distribution of the plates (D) obtained from the $\{110\}_{Al}$ projection as

$$L_r = \left(N_{V\_p} + N_{V\_r}\right)^{-1} \sum_{i=1}^{n} L_i\left(N_i - N_{p\_i}\right) \qquad (2)$$

where N_{V_p} and N_{V_r} is the number density of the plates and rods, respectively, L_i is the bin width of the 'rectangle' length (L) or plate diameter (D) in the distribution histograms, N_i and N_{p_i} are the number densities of the rods and plates, respectively, in each bin width interval. The total volume fractions, F_V, of the rods and plates within the grain/subgrain can be calculated as

$$F_V = N_V \times f_i \qquad (3)$$

where N_V is the number density of the rods (N_{V_r}) or plates (N_{V_p}), f_i is the average volume of the rods and plates approximated as an ideal cylinder ($f_i = \pi\left(D_{r\_eq}/2\right)^2 L_r$ and $f_i = \pi(D/2)^2 t$ for the rods and plates, respectively).

All the measured and estimated precipitate parameters in the alloy aged for 96 h are shown in figure 2. During this quantification procedure, the parameters of, at least, one thousand rod cross-sections and 'rectangles' in the $\{100\}_{Al}$ projections and two hundred plate/laths in the $\{110\}_{Al}$ projections have been measured. It is seen that the geometrical size distributions of precipitates are quite similar for both rod- and plate-shaped precipitates, but the number density of the former is one order higher than for the latter. In general, our quantitative analysis shows that similar volume fraction values were reached for both precipitate groups (rods and plates) in the alloy aged at 170 °C for 96 h. A validation for all parameters is needed by comparing with the numbers obtained using other techniques, and this is the aim of future work.

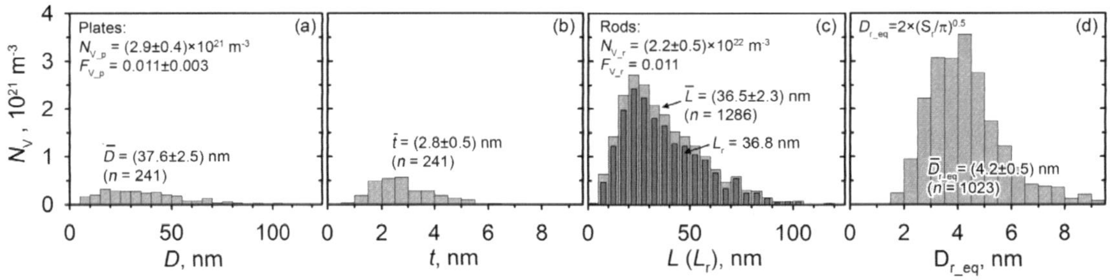

Figure 2. The plate diameter (a) and thickness (b) as well as true rod length (c) and equivalent circular diameter (d) distributions in the aged at 170°C for 96 h. N_{V_p}, N_{V_r}, $\bar{D}$, $\bar{t}$, $\bar{L}$ ('rectangle' length in $\{100\}_{Al}$ projection), $\bar{D}_{r\_eq}$ and F_{V_p} (for the plates) were averaged using at least three TEM images in respective projections.

Conclusions

By combining tilting and DF imaging, it has been possible, using conventional TEM, to determine precipitate parameters and separate $<100>_{Al}$ rods and $\{100\}_{Al}$ plates/laths in the Al-Cu-Mg-Si alloy aged at 170 °C for 96 h. It was found that the most obvious feature of precipitation behavior in this aged alloy is that the number density of rod-shaped precipitates is ten times higher than for the plates, despite the fact that they have similar volume fractions.

Acknowledgements

This work was financially supported by the Ministry of Science and Higher Education and Council on Grant No. 075-15-2019-994 and the Faculty of Natural Sciences at the Norwegian University of Science and Technology (NTNU), Project No. 81617879 using the NORTEM infrastructure (grant number NFR197405) at the TEM Gemini Centre.

References

[1] Gazizov M, Marioara CD, Friis J, Wenner S, Holmestad R and Kaibyshev R 2019 *Mater. Sci. Eng.* A **767** 138369
[2] Gazizov M, Marioara CD, Friis J, Wenner S, Holmestad R and Kaibyshev R 2020 *J. Alloy. Compd.* **826** 153977
[3] Dorin T, Deschamps A, De Geuser F and Sigli C 2014 *Acta Mater.* **75** 134
[4] Sunde JK, Marioara CD, van Helvoort ATJ and Holmestad R 2018 *Mater. Char.* **142** 458
[5] Nie J-F and Muddle BC 2008 *Acta Mater.* **56** 3490
[6] Gazizov M and Kaibyshev R 2017 *Mater. Sci. Eng.* **A702** 29
[7] Andersen SJ 1995 *Metall. Mater. Trans.* **26A** 1931
[8] Yang W, Wang M, Yanlin J and Zhang R 2011 *Metall. Mater. Trans.* **42A** 2917
[9] Wang SC and Starink MJ 2005 *Inter. Mater. Rev.* **50(4)** 193

Correlation of the tribological properties of LIPSS on TiN surface with 3D parameters of roughness

M Yu Gazizova[1*], N A Smirnov[2], S I Kudryashov[2], J Chen[2], Yu H Ahmadeev[3], I Yu Goncharov[1], V V Shugurov[3] and N A Prokopenko[3]

[1]Belgorod State National Research University, Pobedy Str., 85, Belgorod, 308015, Russia
[2]Lebedev Physical Institute, RAS, Leninskiy Pr., 53, Moscow, 119991, Russia
[3]Institute of High Current Electronics, SB RAS, Akademichesky Pr., 2/3, Tomsk, 634055, Russia

*Corresponding author: smolyakovamarina@mail.ru

AbstractThe relationships between friction coefficients and 3D spatial roughness parameters, such as Sci (surface core fluid retention index), Svi (surface valley fluid retention index) and Sbi (surface bearing index), have been investigated for the laser-induced periodic surface structures (LIPSS) on the TiN coatings. Two types of the surface structures (cross-like and parallel ones) with different roughness levels were obtained using a fs-laser. It has been shown that the proposed roughness parameters correlate with the friction coefficients measured for the LIPSSs.

1. Introduction

The reduction of friction coefficients for new composite materials and coatings, development of new approaches in the field of surface engineering is the main task of tribology nowadays. About 23% of total energy consumption in the world is referred to the frictional contact. And most of it, about 20%, is because of friction itself [1]. The problem of developing new wear-resistant materials has been in the second place. Today, technologies enabling to obtain low and ultra-low friction coefficients are needed in tribological contacts. For instance, based on a few studies, the surface structuring by creating a certain multilevel relief (roughness) including grooves, protuberances and pits on micro- and nano-level has proved to be an effective approach to control friction properties of the surfaces. The positive effect of surface structuring has been shown in several works. Successful applications of the surface structuring by laser irradiation methods are known to reduce friction and wear for cutting tools [2], mechanical seals [3], pistons and cylinders [4], bearings [5] implants [6]. However, the problem appears due to the method enabling to modify the surfaces of wear-resistant and hard materials. Thus, the widely used TiN coating, which allows increasing the wear resistance of the components refers to difficult-to-machine materials and, as a rule, its application is the final stage of processing. Reduction of the friction coefficient due to surface structuring for such materials and coatings becomes possible with the appearance of commercial fs-lasers. When working on the surface structuring, the problem appears because of a lack of knowledge to describe the relationships between the surface parameters and the tribological properties. Some of the surface roughness parameters cannot be directly related to friction. To estimate the roughness, a two-dimensional model and such parameters as Ra, Rz, Rsk were used. Unfortunately, these parameters cannot provide reliable surface characterization, because the same Ra and Rz values can be obtained for so completely different surfaces structures [7]. More reliable information about the real surface topography can be provided by spatial 3D roughness parameters. Taking this into account, the investigation of the spatial roughness parameters, which allows predicting the surface properties, including tribological parameters, is of great interest [8, 9].

The aim of this study is to find regularities between functional spatial parameters of roughness and tribological characteristics of TiN coatings with laser-induced periodic surface structures (LIPSS) in the oil friction conditions.

2. Experimental details

The VT1-0 titanium alloy with TiN coating has been chosen for the purpose of this study. TiN coatings were deposited by a plasma assisted vacuum-arc spraying method. The samples were treated using a PC-controlled three-dimensional motorized translation stage and scanned in air by 1030-nm 320-fs Yb-fiber pulses with a maximal energy of 10μJ in a TEM_{00} mode. Parallel and cross scan modes with a scanning speed of 100 mm/s and pulse energy, *E* of 1.5μJ, 3.2μJ, 6.2μJ were applied [10]. LIPPSs were analysed using a FEI Nova NanoSEM 450 scanning electron microscope (SEM) and an Olympus GX71 optical microscope. The surface topography was characterized with a Ntegra Aura atomic force microscope (AFM). Tribological tests were carried out in air using two lubricants: engine and cutting oils. The lubricants with a volume of 3 ml were added to the friction contact area before starting the test. The "pin-disk" testing scheme was used with a 6 mm pin that is made of steel 304. The counterbody load was 1 N. The rotation speed of the samples was 5 cm/s. The friction path was 450 m. The friction coefficient was measured with an accuracy of 0.001.

The regularities between the functional 3D spatial parameters and friction coefficients (CoF) in oil friction conditions have been analyzed in the frame work of this study. The following parameters were used: Sci (surface core fluid retention index), Svi (surface valley fluid retention index) and Sbi (surface bearing index). Sbi characterize the upper zone of the surface involved in wear phenomena. Sci characterizes the main void volume acting as a lubricant reserve. It is presumed that these voids locate at a distance of $0.05h$-$0.8h$, where h is the profile height. Svi characterizes the void volume of the deepest valleys.

3. Results and discussion

The formation of LIPSSs oriented perpendicularly to the polarization of laser radiation was analyzed using AFM [10]. The cross-like structures found in the samples can be classified to three different regions: (1) the untreated TiN region (figure 1a, red area), (2) the area corresponding to a single laser scanning pass (figure 1a, green area) and (3) the area with the intersection of the laser scanning lines (figure 1a, blue area). The surface roughness level is higher for the cross-like structures in comparison with that for the parallel ones (figure 1). Thus, Sa (arithmetical mean deviation of the assessed profile) and Sz (maximum peak to valley height of the profile) depending on the irradiation energy for parallel structures was measured to be 61-132 nm and 459-523 nm, respectively. The Sa and Sz ranges for the cross-like structures varies from 166 nm to 321 nm and from 417 nm to 1039 nm, respectively.

Tribological tests of the samples before and after surface modification showed that the wear processes have not been found regardless of the irradiation parameters, the surface structures formed after processing and, especially, the test conditions applied in this study. At the same time, a decrease in the friction coefficient after fs-laser irradiation has been identified. For instance, the friction coefficient was 0.14 for the TiN coating in test using both engine oil and cutting tool oil as a lubricant. Its values varied in the range from 0.04 to 0.13 after the surface structuring, except the sample with parallel structure processed with an energy of 1.5 mJ. The friction coefficient has been measured to be 0.55 for these samples in the test using engine oil as a lubricant. Thus, it can be concluded that there is a positive effect in the TiN coating surface structuring referring to the improvement in wear resistance and the reduction in friction coefficients.

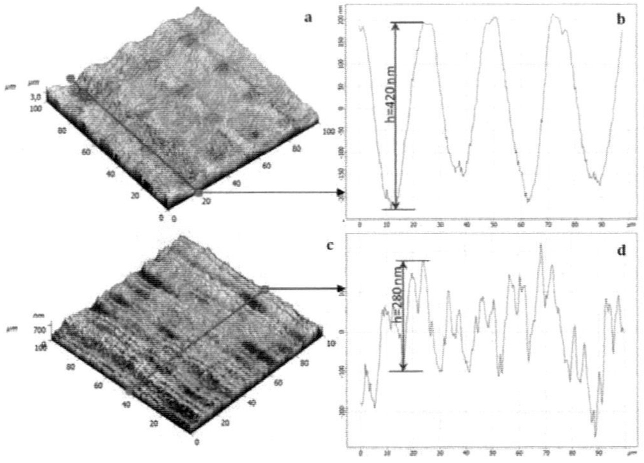

Values of surface roughness parameters after fs-laser processing with different laser energies

	Energy, mJ	S_m, mkm	S_{ci}	S_{vi}	S_{bi}
Parallel structures	1.5	6.0	1.61	0.10	0.29
	3.2	9.3	1.45	0.11	0.71
	6.2	4.3	1.58	0.10	0.17
Cross-like structures	1.5	25.0	1.54	0.07	1.67
	3.2	41.7	1.30	0.12	1.92
	6.2	54.5	1.19	0.13	1.22

Figure 1.Surface structure of TiN after fs-laser processing (a, c), cross-sectional profile of surface (b, d) and 3-D spatial parameters of LIPSS.

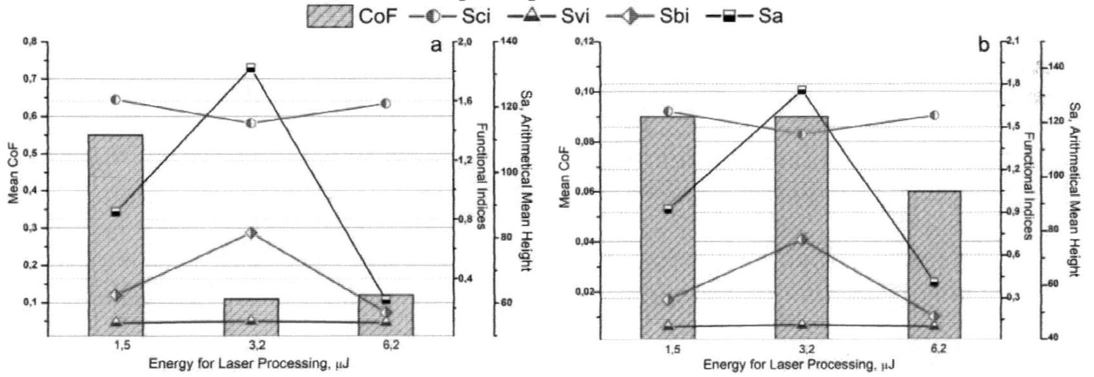

Figure 2.Regularities between the friction coefficients and 3D spatial roughness parameters for the surfaces with the parallel structures at friction in the engine oil (a) and cutting tool oil (b).

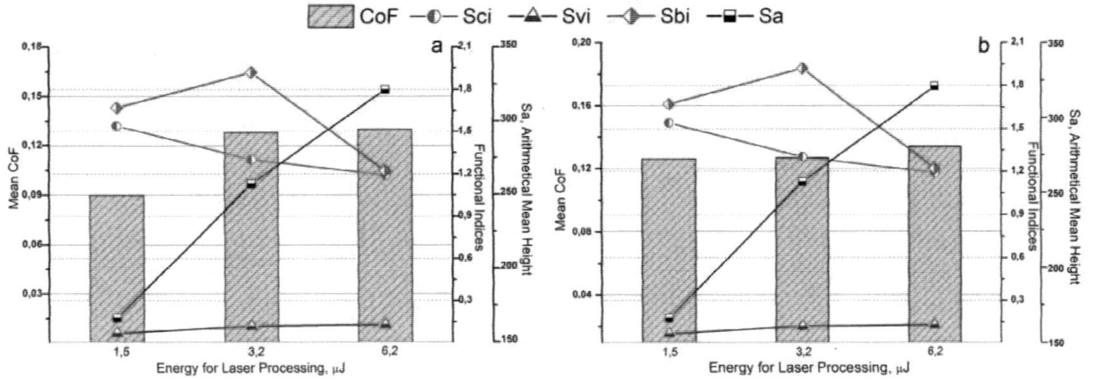

Figure 3.Regularities between the friction coefficients and 3D spatial roughness parameters for the surfaces with the cross-like structures at friction in the engine oil (a) and cutting tool oil (b).

Based on the AFM data analysis, the 3D spatial parameters of the LIPSS (figure 1) were calculated. These parameters were compared with the tribological tests (figures 2 and 3). For the samples with the parallel structures tested in the engine oil, the minimum friction coefficient was measured when the Sbi values increased, i.e. when the contact areas increased (figure 2a). On the contrary, with friction in the cutting tool oil, a reduced in the contact area allows a decrease in the coefficient of friction, respectively, the Sbi value decreases in comparison with other 3D spatial parameters (figure 2b). In this case, there was a clear correlation between Sbi and the friction coefficient, namely, higher Sbi values correlated with lower CoF. The dependence of Sci change is inversely proportional to Sbi change, namely, the smaller the contact area, the smaller the volume of retained fluid. It should be noted that the Sa values measured for the parallel structures correlated with the Sbi values. The Svi index for all the investigated surfaces with the parallel structures showed no changes for different LIPSS, meaning that it did not significantly contribute to the friction coefficient changes.

An increase in the Sci values decreased the friction coefficients for the samples with the cross-like structures using the engine oil and the cutting tool oil (figure 3). There is also a dependence between the friction coefficient with the Svi value: that the lower Svi the lower friction coefficient. Thus, it can be presumed that liquid retention parameters are decisive for the friction test condition of the cross-like structures.

The change in the aforementioned behavior of the samples with cross-like structures occurred when the test speed and the friction path increased by 3 times (from 5 cm/s to 15 cm/s) and by approximately 4 times (from 450 m to 2000 m), respectively. It was found that the lower Sbi correlated with the lower friction coefficient. The effect of the Sci decreased. According to the tests results at low test speeds using the cutting tool oil, smaller contact areas (i.e., the lower Sbi values) corresponded to lower friction coefficients. However, at the test speed of 15 cm/s, the indicated dependence for parallel structures was not completely retained, namely, the maximum friction coefficient corresponded to the highest Sbi, but the smallest Sbi value corresponded to the structure obtained using the laser pulse energy, E of 6.2 μJ, where CoF = 0.05. The sample irradiated at E = 1.5 μJ has the lowest friction coefficient of 0.04. This happened due to the close Sbi values and friction coefficients, in other words, the detected difference was insignificant. Similar dependences were found for the cross-like structures.

4. Conclusion

The chosen 3D spatial roughness parameters can be used to predict the tribological properties (friction coefficient) for wear-resistant coatings with LIPSS in the friction conditions with lubricants. The functional parameter of roughness determining the surface valley fluid retention, Svi played the major role, when the engine oil was used. Differences in the dependence of the influence of the functional parameters of roughness on the friction coefficients under the friction conditions in the engine oil and the cutting tool oil can be apparently explained by the different viscosity of lubricants and the type of the additives used in them. Structures with smaller contact areas were better in the case of using the cutting tool oil, i.e. when the surface bearing index Sbi demonstrated the lowest values.

Acknowledgments

The study was carried out with a grant from the Russian Science Foundation (project No.19-79-00295).

References

[1] Holmberg K, Erdemir A 2017 Friction 5(3) 263–284
[2] Kawasegi N, Sugimori H, et. al 2009 Precision Engineering 33(3) 248–254
[3] Wan Y, Xiong D S 2008 Journal Matererials Processing Technology 197 (1–3) 96–100
[4] Etsion I, Sher E 2009 Tribology International 42(4) 542–547
[5] Qiu M, Minson B R, Raeymaekers B 2013 Tribology International 67 278–288
[6] Langhorn J, Borjali A, Hippensteel E et al 2018 Tribology International 124 178–183
[7] Sedlaček M, Podgornik B, Vižintin J 2012 Tribology international 48 102–112
[8] Sedlaček M, Podgornik B, Vižintin J 2016 Tribology Transactions 60 260-266
[9] Gualtieri E, Pugno N, et. Al 2011 Journal of Nanoscience and Nanotechnology 11 9244–9250
[10] Gazizova M Yu, Smirnov N A, Kudrayshov S I et al 2020IOPConf. Ser.: IOP Conference Series: Materials Science and Engineering 862

IOP Conf. Series: Materials Science and Engineering 1014 (2021) 012015 doi:10.1088/1757-899X/1014/1/012015

Grain structure evolution during friction-stir welding of 6061-T6 aluminum alloy

A Kalinenko[1*], **I Vysotskiy**[1], **S Malopheyev**[1], **S Mironov**[1] **and R Kaibyshev**[1]

[1] Belgorod State National Research University, Pobeda 85, Belgorod, 308015, Russia

[*]Corresponding author: kalinenko@bsu.edu.ru

Abstract. In this work, electron backscattered diffraction (EBSD) was used to examine the evolution of the grain structure during friction stir welding of aluminum alloy 6061-T6. In the entire FSW range studied, the microstructure evolution was found to be governed by continuous recrystallization. An increase in the welding temperature resulted only in microstructural coarsening.

Keywords – Friction-stir welding, Heat-treatable aluminum alloys, Electron backscatter diffraction

1. Introduction

From a thermo-mechanical standpoint, friction-stir welding (FSW) represents a unique combination of very large strains, high temperature and relatively high strain rate [1]. Such extreme deformation conditions lead to significant changes in the grain structure. Many researchers suppose that these changes mainly take place in the thermo-mechanical affected zone (TMAZ) and the stir zone (SZ).

In the SZ, deformation and temperature reach their highest values. Accordingly, the most significant microstructural changes occur in this zone and, accordingly, this microstructural region attracts the greatest interest of researchers. An extensive research in this area has demonstrated a significant grain refinement [2, 3]. The microstructures formed in the SZ are usually characterized by equiaxial grain morphology, as well as a relatively high proportion of high-angle boundaries (HABs). Remarkably, the misorientation distribution in this area is often found to be nearly random [4]. The dislocation density is usually low [5-9]. All these microstructural features indicate an occurrence of recrystallization during FSW. The grain size in the SZ is known to be very sensitive to the FSW conditions [10]. Typically, the average grain size increases with the tool rotation speed but decreases with a tool translation speed. This effect is most likely associated with the concomitant variations of welding temperature.

On the other hand, the microstructure formed in the TMAZ has received relatively low attention in scientific literature. It is only known that the grains in this area are significantly elongated, orienting themselves approximately parallel to the SZ [1]. It is accepted that the material in this microstructural region experiences large shear deformation, and the change in grain morphology is explained by the geometric effect of deformation. It is also assumed that a developed subgrain structure is formed within the elongated grains [1].

Considering a transition nature of TMAZ between the original material and the final stir zone, it should be emphasized that the microstructural changes in this region are a key issue in ascertaining of the recrystallization mechanism operating during FSW. In this work, electron backscatter diffraction

(EBSD) was applied to examine the TMAZ microstructure evolved during FSW of 6061 aluminum alloy.

2. Material and experimental procedure

The material used in this study was a commercial aluminum alloy AA6060-T6 with nominal chemical composition (wt.%) of 0.88 Mg, 0.66 Si, 0.72 Fe, 0.26 Cu, 0.12 Mn, 0.12 Cr, 0.09 Zn, and balance Al. The material was produced by semi-continuous casting, homogenized at 380 °C for 1 h and then extruded at the same temperature to 75% of area reduction. To obtain the peak-aged condition, the extruded material was T6-tempered, i.e., solutionized at 550 °C for 1 h, water quenched at then artificially aged at 160 °C for 8 h. Sheets with a thickness of 3 mm was friction-stir butt welded using an AccurStir 1004 FSW machine. The welding tool consisted of a shoulder having a diameter of 12.5 mm and an M5 cylindrical probe of 2.7 mm in length. Two FSW trials were performed. To investigate the microstructure evolution at relatively low welding temperature, FSW was conducted at the spindle rate of 500 rpm and a feed rate of 380 mm/min. To explore the microstructural changes at relatively high welding temperature, FSW was performed at the spindle rate of 1100 rpm and a feed rate of 125 mm/min. The produced welds were denoted throughout as 500-380 weld and 1100-125 weld, respectively. In all cases, the tool tilting angle of 2.5° was employed and stainless steel was used as a backing plate.

Microstructural observations were conducted by EBSD technique by using a FEI Quanta 450 Nova field-emission-gun scanning electron microscope equipped with TSL OIM^{TM} software. A suitable surface finish for EBSD was obtained by electro-polishing in a solution of 25% nitric acid in methanol. A 15° criterion was employed to differentiate low-angle boundaries (LABs) from high-angle boundaries (HABs).

3. Results and discussion

To understand the mechanisms of evolution of the grain structure during FSW, microstructural changes in the TMAZ of both welds were analyzed. To this end, the appropriate EBSD maps were acquired and shown in Figures 1 and 2. From the maps, it was seen that the microstructure evolution in both cases was associated with progressive development of LABs.

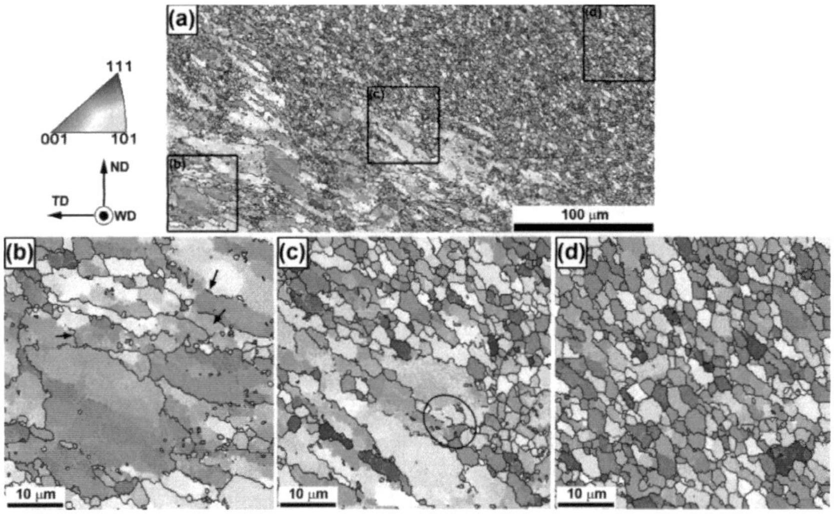

Figure 1. EBSD orientation map taken from TMAZ of the 500-380 weld. LABs and HABs are depicted as white and black lines, respectively.

In the cold periphery of the deformation zones of both welds, under the condition of relatively low temperatures and low strains, an extensive formation of deformation-induced LABs was found (Figures

1b and 2b). The deformation-induced boundaries formed a developed subgrain structure in the grain interior (Figures 1b and 2b). With approaching to the SZ (and the concomitant increase in temperature and the accumulated strain, the LABs rapidly accumulated misorientation over 15° thus transforming into HABs (several examples are arrowed in Figures 1b and 2c). Accordingly, the subgrains tended to transform into grains (several examples are circled in Figures 1c and 2c). The observed process fits a definition of *continuous recrystallization*.

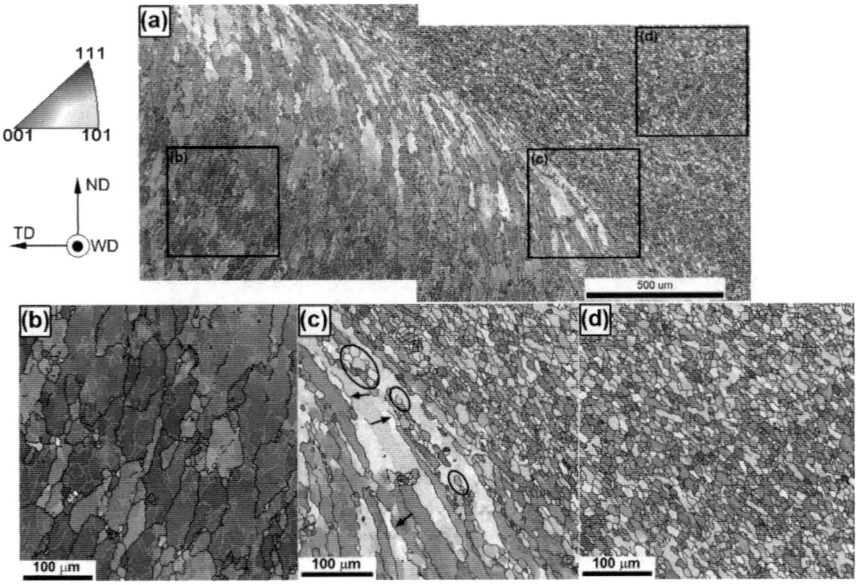

Figure 2. EBSD orientation map taken from TMAZ of the 1100-125 weld. LABs and HABs are depicted as white and black lines, respectively.

Moreover, the original grains were significantly sheared in a direction of material flow, thereby essentially thinning the grain cross-section; this effect was most pronounced in the high-temperature weld (Figure 2c). This phenomenon is usually attributed to a so-called *geometrical effect of strain*.

A development of both processes eventually resulted in a formation of fine-grained structure in the stir zone (Figures 1d and 2d). Therefore, the grain structure development in both cases was concluded to be driven by a superposition of continuous recrystallization and the geometrical effect of strain. The sole difference between two welds was a somewhat coarser stir zone microstructure evolved at higher welding temperature (with the mean grain intercept of 8.6 μm vs 2.4 μm).

4. Summary and conclusions

In this work, grain structure evolution in friction-stir welded 6061-T6 aluminum alloy was studied by EBSD. In the entire studied FSW range, the microstructure evolution was shown to be governed by the *continuous recrystallization*. An increase in the welding temperature promoted only a subtle microstructural coarsening.

Acknowledgments

This study was financially supported by the Russian Science Foundation, grant No. 19-49-02001. The authors are also grateful to the staff of the Joint Research Center "Technology and Materials" at Belgorod State National Research University for assistance in experimental works.

References

[1] Mishra R S, Ma Z Y 2005 *Mater. Sci. Eng. R* **50** 1-78
[2] Esparza J A, Davis W C, Murr L E 2003 *J. Mater. Sci.* **38** 941– 952
[3] Kwon Y J, Shigematsu I, Saito N 2002 *J. Jap. Inst. Met.* **66** 1325-1332
[4] Salem H G 2003 *Scripta Mater.* **49** 1103–1110
[5] Sato Y S, Urata M, Kokawa H, Ikeda K, Enomoto M 2001 *Scripta Mater.* **45** 109-114
[6] Ma Z Y, Mishra R S 2003 *Acta Mater.* **51** 3551–3569
[7] Charit I, Mishra R S 2003 Mater. Eng. A. **359** 290-296
[8] Sato Y S, Kurihara Y, Park H S C, Kokawa H, Tsuji N 2004 *Scripta Mater.* **50** 57–60
[9] Salem H G, Reynolds A P, Lyons J S 2004 *J. Mater. Eng. Perform.* **13** 24-31
[10] Chang C I, Lee C J, Huang J C 2004 *Scripta Mater.* **51** 509–514

Effect of thermal treatment on the structure and hardness of the laser powder bed-fused Ti-Zr-Nb shape memory alloy

V Kalinicheva[1], V Sheremetyev[1*], M Letenneur[2], A Konopatsky[1], S Prokoshkin[1], V Brailovski[2]
[1]National University of Science and Technology "MISiS", Moscow, Russia
[2]Ecole de technologie superieure, Montreal, QC, Canada

*Corresponding author: sheremetyev@misis.ru

Abstract. A new biomedical Ti-18Zr-14Nb (at.%) alloy was processed using the laser powder bed fusion additive manufacturing technology and subjected to post-fusion annealing in the 400 to 900 °C temperature range. The meltpool and grain structure features, the phase composition and hardness of samples fabricated using different laser energy densities and build rates, while having identical printed material densities, were analyzed. It was shown that the microstructure of the alloy differs depending on the printing parameters used. The higher the energy density, the coarser in size and the more elongated towards the build plate grains are formed during the process. Subsequent annealing results in the formation of different structures depending on the printing conditions, and the smaller the as-built grain size, the lower the onset temperature of recrystallization, which starts up to 100 °C lower for the finer grain structure than for the coarser grain structure of this study.

1. Introduction

The development of novel materials and complex parts with a controlled structure using the 3D and 4D additive manufacturing technologies represents a global trend in modern materials science [1-2]. 4D-technologies are considered a breakthrough, since they combine the potential of digital additive technologies with that of functional materials, such as shape memory alloys (SMA). These technologies allow printing intricate parts capable of modifying their geometry and properties in response to some external stimuli, such as temperature or stress [1].

Special interest represents the use of layer-by-layer material consolidation technologies, in particular laser powder bed fusion (LPBF), to manufacture personalized implants, since this technology allows an easy scaling up from a medical image to a prototype, and then to a product of mass customization [3]. Owing to their unique functional properties, Ti-Zr-Nb SMAs belong to the group of especially promising materials for bone implants [3-5]. These alloys exhibit a unique combination of biomechanical compatibility (low elastic modulus and superelastic behavior) with excellent corrosion resistance and non-toxicity of all the constitutive elements. Furthermore, the functional properties of these alloys can be additionally improved by optimizing their microstructure and phase state using thermal and/or thermomechanical treatments [6-7].

To guarantee adequate mechanical properties of printed parts, modern research in the field of additive technologies is primarily aimed at the fabrication of these parts with densities approaching those of their wrought equivalents [3, 8-9]. However, it is known that different microstructures and textures can intentionally be formed in as-printed parts of the same porosity [9], and these structure features can further be modified using post-fusion thermal treatments. Such an approach appears to be especially promising for the improvement of functional properties of such structure-sensitive materials as SMAs [10]. Given the importance to study the effect of post-fusion thermal treatment conditions on the structure and phase state and therefore, on the mechanical properties of laser powder bed-fused Ti-18Zr-14Nb alloys, the present study endeavors to provide a preliminary understanding of this relationship.

2. Experimental Procedure

As a precursor material, a 50 mm-diameter, 600 mm-long ingot of the Ti-18Zr-14Nb alloy supplied by *Flowserve Corp. Comp* (USA) was used. The ingot was gas-atomized by *TLS Technik GmbH* (Germany). Next, a *TruPrint1000* (TRUMPF, Germany) LPBF system equipped with a 200 W ytterbium fiber laser (spot diameter 30 μm) was employed to build a series of samples. Among them, two 10x10x10 (mm^3) cube samples, #1 and #2, with identical densities (99.7%), but printed using different laser exposure conditions, were selected (Table 1). For the density measurements, the Archimedes' method (ASTM B962) was used. The chemical composition of major elements and their distributions were measured using a *TESCAN VEGA LMH* scanning electron microscope equipped with an energy-dispersive X-ray microanalyzer. The oxygen and nitrogen contents were determined using *LECO TC600* infrared cells. Chemical analysis showed that the LPBF process alters the chemical composition of the feedstock powder by decreasing its content in titanium and increasing in oxygen, and the higher the laser energy density, the greater the oxygen intake during the fusion process (Table 1).

Table 1. Printing conditions of sample #1 and #2 and chemical composition.

Sample	Power, W	Speed, m/s	Hatch distance, mm	Energy density, J/mm³	Building rate, cm³/h	Measured density, %	Chemical composition				
							in atomic, %			in weight, %	
							Ti	Zr	Nb	O	N
1	175	1420	0.06	68.5	9.2	99.7	65.6	18.8	15.5	0.218	0.019
2	130	985	0.11	40.0	11.7	99.7	65.6	19.0	15.4	0.200	0.010
Powder							68.6	17.8	13.6	0.121	0.012

Samples for structural characterization and hardness measurements were cut from the printed cubes and subjected to post-fusion annealing (PFA) for 0.5 h at 400-900 °C in argon atmosphere, and then quenched in water. A *Union Versamet-2* optical microscope (OM) equipped with a *Nikon D90* camera and a *TESCAN VEGA LMH* scanning electron microscope equipped with an electron backscatter diffraction (EBSD) unit were used for microstructural analysis. X-ray diffraction phase analysis of the selected samples was performed using an *Ultima IV Rigaku* diffractometer (monochromatic CuKα-radiation). The Vickers hardness was measured using a *Metkon* microhardness tester with an applied load of 1 kg and a holding time of 10 sec.

3. Results and Discussion

Figure 1 shows the grain structure images from faces parallel to the build direction (vertical planes) just after LPBF (as-built) and after post-fusion annealing at different temperatures. These images are obtained using optical microscopy and EBSD techniques. It can be seen from Figure 1a-d, that the as-built samples, depending on the LPBF parameters, manifest differences in the melt pool morphologies, grain sizes and elongations: the higher the laser energy density, and the lower the build rate, the larger the melt pools, the coarser the grains and the more pronounced the elongations: grains of sample #1 are bigger and more elongated than those of sample #2 (see Table 1 for the values of energy density and building rate). The LPBF process results therefore in the formation of a columnar microstructure with average grain sizes in the vertical direction of ~98 μm for sample #1 and ~66 μm, for sample #2. The grains grow epitaxially with the continuation of their crystallographic orientations across the meltpool boundaries [10].
Regarding sample #2, PFA at 700 °C of this sample leads to an increase in the average grain size from 66 to 89 μm, and to a weakening of the meltpool boundaries (Figure 1e-f). It is assumed that PFA of sample #2 at this temperature corresponds to the initial stage of the local recrystallization process, while PFA of this sample at 800 °C leads to an active phase of recrystallization and the formation of a microstructure containing a large number of 50-150 μm-sized equiaxed grains (Figure 1j). Regarding sample #1, the

active phase of recrystallization of this sample does not start even after PFA at 800 °C, and the grain structure preserves its columnar morphology. These differences can be explained by the fact that the building rate of sample #2 is higher than that of the sample #1, and therefore, the solidification cooling rate is also higher, which results in finer and more stressed as-built microstructure. It has a cumulative effect that starts the processes of recrystallization of sample #2 at a lower temperature than that of sample #1.

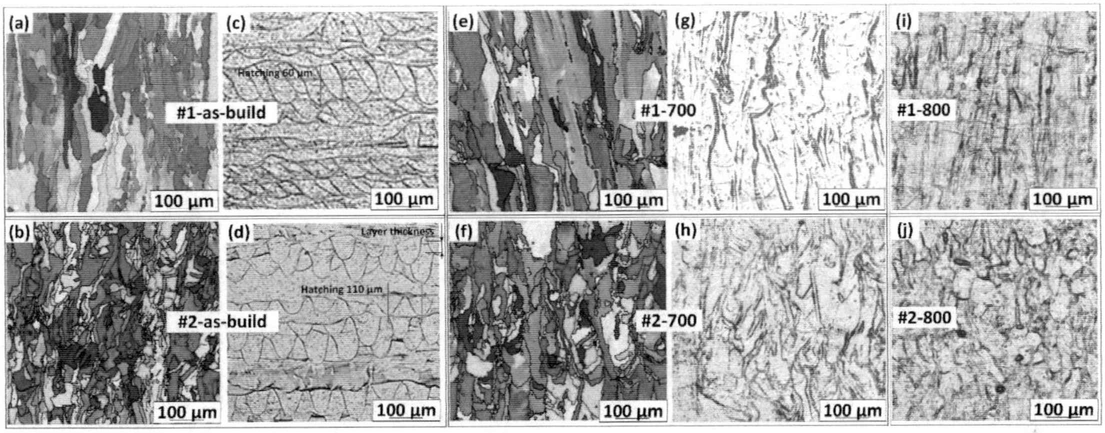

Figure 1. Images (vertical planes) of the Ti-Zr-Nb alloy microstructure after LPBF (a-d) and after LPBF +PFA at 700 °C (e-h), 800 °C (i, j): sample #1 (a, c, e, g, i) and sample #2 (b, d, f, h, j).

The X-ray diffractograms obtained for the samples in the as-built state and annealed at different temperatures are shown in Figure 2. The main phase constituent in all the cases is the BCC β-phase. After PFA at 400 °C, some amounts of cooling- or annealing-induced α''- and α-phases are also observed. It is worth noting the presence of TiO_2 (rutile) peaks in all diffractograms. It can be assumed that TiO_2 (rutile) is formed as particles or in the form of oxide on the surface of nano- or micropores. This assumption requires additional study using electron microscopy methods.

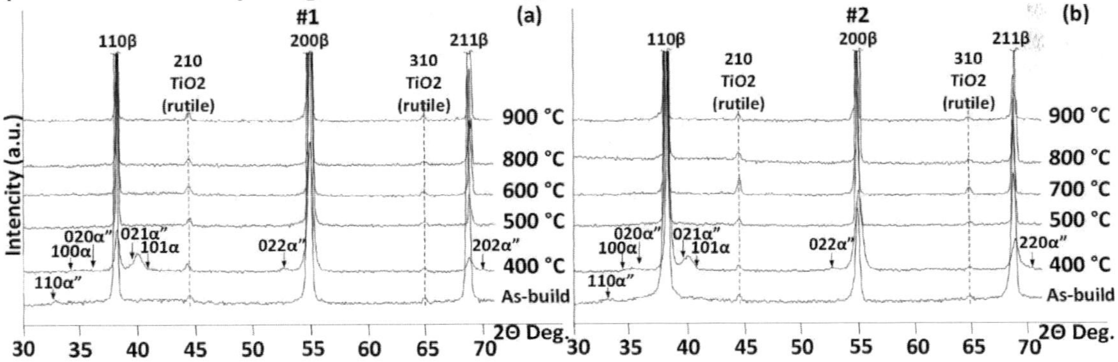

Figure 2. Results of X–ray diffraction analysis of Ti-Zr-Nb alloy after LPBF and LPBF+PFA: sample #1 (a) and sample #2 (b).

In the initial, as-built, state, the alloy hardness of both samples is ~280 HV (Figure 3). Annealing at 400 °C leads to a sharp increase in hardness up to 290-300 HV, which is associated with the formation of α- and α''- phases in the alloy [6]. An increase in the PFA temperature to 500-600 °C gradually reduces the hardness to ~250 HV. The lowest hardness, associated with the $\alpha'' \leftrightarrow \beta$ transformation-related phenomena [3, 6], is achieved after PFA at 550-600 °C for sample #2 and after PFA at 600-700 °C, for sample #1. Then followed a slight increase in hardness, which is explained by a decrease in the softening contribution of β-phase with increased PFA temperature. Finally, hardness in both samples decreases again, which is caused by the onset of the recrystallization process. Note that ~100 °C difference between the temperatures corresponding to the lowest hardness (550 °C for sample #2 and 650 °C for sample #1) and between the onset temperatures corresponding to the last decrease in hardness (700 °C for sample #2 and 800 °C for sample #1), indicate that all the processes occurring during PFA of the fine-grained sample #2 occur at temperatures ~100 °C lower than for the coarse-grained sample #1, which correlates with the results of the microstructure analysis (Figure 1).

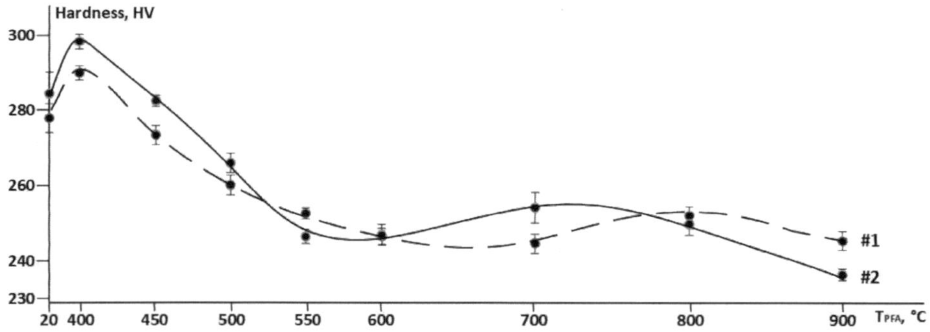

Figure 3. Results of hardness measurements of Ti-Zr-Nb alloy after LPBF and LPBF+PFA.

Summary

The effect of the laser powder bed-fusion (LPBF) process with two different sets of printing parameters and subsequent annealing at 400-900 °C on the structure and hardness of a new biomedical Ti-18Zr-14Nb shape memory alloy is studied. The morphology of the grain structure depends on the energy density and the building rate. Sample #1 printed with higher energy density and lower building rate contains a columnar microstructure with an average grain size in the vertical directions of ~98 μm. On the contrary, sample #2 printed with lower energy density and higher building rate contains less elongated grains with an average grain size of ~66 μm. The onset of the recrystallization process in sample #2 with a finer microstructure occurs at ~700 °C, and its active phase, at 800 °C, which is ~100 °C lower than for sample #1. The results of the hardness measurements correlate well with the results of the microstructural analysis.

Acknowledgments

The present work is supported by the Russian Science Foundation (project № 20-79-00299).

References

[1] Ma J, Franco B, Tapia G, Karayagiz K, Johnson L, Liu J 2017 *Scientific Reports* **7** pp 1–8.
[2] Zafar M, Zhao H 2019 *Metals and Materials International* **34** pp 1–22.
[3] Kreitcberg A, Brailovski V, Prokoshkin S 2018 *Journal of Materials Processing Technology* **252** pp 821-829.
[4] Kim H, Fu J, Tobe H, Miyazaki S 2015 *Shape Mem. Superelasticity* **1** pp 107–116.
[5] Sheremetyev V, Petrzhik M, Zhukova Y, Kazakbiev A, Arkhipova A, Moisenovich M, Prokoshkin S, Brailovski V 2020 *Journal Biomed Mater Res. B* **18(3)** pp 647–62.
[6] Kudryashova A, Sheremetyev V, Lukashevich K, Cheverikin V, Inaekyan K, Galkin S, Prokoshkin S, Brailovski V 2020 *Journal of Alloys and Compounds* **843** pp 1560-66.
[7] Kreitcberg A, Sheremetyev V, Tsaturyants M, Brailovski V, Prokoshkin S 2019 *S Shape Mem. Superelasticity* **5** p 33.
[8] Letenneur M, Brailovski V, Kreitcberg A 2019 *Journal Manuf. Mater. Process.* **3** p 21.
[9] Letenneur M, Kreitcberg A and Brailovski V 2020 *Journal Manuf. Mater. Process.* **4** p 11.
[10] Kreitcberg A, Brailovski V, Sheremetyev V, Prokoshkin S 2017 *Shape Mem. Superelasticity* **3** pp 361-72.

IOP Conf. Series: Materials Science and Engineering 1014 (2021) 012017 doi:10.1088/1757-899X/1014/1/012017

Prediction of the strength of high entropy alloys of the Al-Cr-Nb-Ti-V-Zr system by the solid-solution hardening model

D N Klimenko*, **N D Stepanov**, **S V Zherebtsov**

Laboratory of Bulk Nanostructured Materials, Belgorod State University, Belgorod, 308015, Russia

*Corresponding author: klimenko@bsu.edu.ru

Abstract. An approach based on the phase prediction of both the phase composition and the solid solution hardening was investigated. This approach showed good agreement between the calculated and experimental values of the yield strength at 20 and 600 °C.

1. Introduction

High entropy alloys (HEAs) usually consist of at least 5 elements in approximately equimolar proportions [1]. It was believed that the high entropy of such alloys resulted in the formation of face-centered cubic (FCC) or body-centered (BCC) solid solutions [2]. Although this idea has not been fully worked out, some of HEAs have a single-phase microstructure with a very good combination of properties [3]. For example, HEAs based on refractory elements demonstrate remarkable high temperature strength, outperforming currently used nickel-based superalloys [4-5], suggesting the potential application of these HEAs as material for the production of high temperature parts.

The main strengthening mechanisms of metal and alloys comprise grain boundary strengthening, solid solution strengthening, precipitation hardening and dislocation strengthening. Considering refractory high entropy alloys (RHEA) with a single-phase BCC metallic solid solution (or with only a very small amount of second phases) it can be suggested that solid solution hardening (SSH) contributes mainly to the yield strength in RHEAs. The range of models based on classical approaches for binary systems and multicomponent alloys [8,9] for calculation of SSH in HEAs were proposed in [4-7]. The combination of solid-solution hardening models with empirical rules and CALPHAD modeling can serve as a basis for predicting the phase composition of alloys and development an approach to creating high-strength HEAs. Therefore, the goal of this work was evaluation of this approach for searching Al-Cr-Nb-Ti-V-Zr system high-strength alloys.

2. Materials and experimental details

Model alloys for the testing model were chosen in three stage. Grid search in the Al - Cr - Nb - Ti - V – Zr system was done in the range of concentrations of 0-15% with a step of 1% for Al and Cr and in the range of concentrations of 0-45% with a step of 5% for other elements. At the first stage, a set of alloys was chosen using the criteria of the single-phase alloy formation: atomic size mismatch $\delta<5.4\%$, valence electron concentration VEC< 6.87, enthalpy of mixing ΔH_mix = -16.25 – 4 kJ/mole [10]. Then, the solid-solution hardening was calculated at 20 and 600 °C for the selected alloys. The calculations solid solution hardening calculations were carried out using an approach described in [7]. At the last stage, thermodynamic modeling of equilibrium phase diagrams was conducted using the Thermo-Calc 2019 software with the HEA3 (high entropy alloys) database.

The model alloys were produced by vacuum arc melting, using proper mixtures of pure metals with purities higher than 99.9 wt.%, in Ti-gettered argon atmosphere. The measured compositions of the model alloys are listed in Table 1. The alloys were remelted five times to improve their homogeneity. The obtained ingots were sealed in vacuumed (10^{-2} torr) quartz tubes, and then homogenized at 1200 °C for 10 h. After homogenization annealing, the ingots were cut off using an electrical discharge machine into samples for compression test and microstructure investigation.

Table 1. Chemical compositions (at.%) of the model alloys.

Alloys	Al	Cr	Nb	Ti	V	Zr
A1	13.6	13.1	23.4	22.4	25	2.5
A2	11.8	1.9	18.8	44.5	15.8	7.2
A3	14.6	11.5	6.2	37	21	9.7
A4	11.2	6.3	8	49.2	13.9	11.4
A5	11.7	0	9.7	44.8	29.1	4.7
A6	11.8	0.9	22.1	30.5	20.4	14.3
A7	13.6	3	20.6	34	4	24.8
A8	9	0	20.3	36.2	13.8	20.7

Rectangular specimens in size 8×5×5 mm^3 were compressed using an Instron 300LX testing machine equipped with a radial heating furnace. The tests were carried out at temperatures 22 or 600 °C with an initial strain rate of 10^{-4} s^{-1} till 50% height reduction (or till fracture).

The microstructure of the homogenized alloys was studied using X-ray diffraction (XRD) and scanning electron microscopy (SEM) equipped with energy-dispersive X-ray spectrometry (EDS). Specimens for structural investigations were finished with OP-S suspension (the abrasive particle size of 0.04 μm). The chemical composition of the alloys was measured using an SEM-EDS with a scanning area of 2×2 mm^2.

Microhardness was measured using a Wolpert group 402mvd microhardness tester. The load and dwell time were 300 g and 10 s, respectively. The microhardness value was averaged on measurements in five distinct regions.

3. Results of prediction and experimental studies

The solid solution hardening of the model alloys was calculated using the equations given in Ref. [7]:

$$\sigma = F_\tau Z M \mu$$

$$M = (x_1 \,....\, x_n)\begin{pmatrix} 0 & (\xi\sqrt{\{\eta_1'^2\}^2+\{\alpha\delta_1^2\}^2})^{4/3} & (\xi\sqrt{\{\eta_1'^n\}^2+\{\alpha\delta_1^n\}^2})^{4/3} \\ (\xi\sqrt{\{\eta_2'^1\}^2+\{\alpha\delta_2^1\}^2})^{4/3} & 0 & (\xi\sqrt{\{\eta_2'^n\}^2+\{\alpha\delta_2^n\}^2})^{4/3} \\ (\xi\sqrt{\{\eta_n'^1\}^2+\{\alpha\delta_n^1\}^2})^{4/3} & (\xi\sqrt{\{\eta_n'^2\}^2+\{\alpha\delta_n^2\}^2})^{4/3} & 0 \end{pmatrix}\begin{pmatrix} x_1 \\ \cdots \\ x_n \end{pmatrix}$$

$$\delta_i^j = \frac{1}{a}\left[\frac{da_i^j}{dx_i^j}\right]$$

$$\eta_2'^1 = \frac{1}{\mu}\left[\frac{d\mu_i^j}{dx_i^j}\right] \Big/ \left(1 + 0.5\frac{1}{\mu}\left[\frac{d\mu_i^j}{dx_i^j}\right]\right)$$

$$\frac{d\mu_1^2}{dx_1^2} = \left((x_1, x_2 \ldots, x_n)\begin{pmatrix} \mu_1 \\ \mu_2 \\ \mu_n \end{pmatrix}(x_1 - \delta x, x_2 + \delta x \ldots, x_n)\right)\frac{1}{\delta x}$$

where x_i denotes the atomic concentration of the i-component, μ_i denotes the shear modulus of the i-component, α =12.2, ξ = 2, F_τ=3 is the Taylor factor and Z= 0.0026. The values of the atomic radii, shear modulus and bulk modulus of the elements were used in the calculations are given in Table 2. The values of SSH were specified after production of the model alloys using their measured chemical compositions (Table 1). The result of calculating SSH for the model alloys at 20 °C and 600 °C are given in Table 3.

Meanwhile, XRD analysis showed that only alloys A2, A4 and A7 comprised a single phase structure with a BCC lattice. Other alloys contained intermetallic phases. The actual phase composition is in good agreement with the results of modeling by Thermo-Calc.

Table 2. Values of atomic radii, shear modulus and bulk modulus used in calculations.

Elements	Atom radius at 20/600 °C (pm)	Shear modulus at 20/600 °C (GPa)	Bulk modulus at 20/600 °C (GPa)
Al	143.17/145.55	95/95	86/7/67.7
Cr	124.91/125.49	103.7/93.7	136.4/114.9
Nb	142.9/143.24	37.3/38.8	170.5/165.2
Ti	146.15/148.34	74/43	97/81.52
V	131.6/131.74	48.1/39.1	157.1/139.3
Zr	160.25/160.89	60.2/37.8	91.7/79.2

The correspondence between the calculated values of the yield strength of the model alloys and its actual value is shown in Figure 2. The results of prediction agreed well with the experimental values. For most of the model alloys, the difference between the calculated and experimental values of the yield strength did not exceed 15%.

Table 3. Mechanical properties of model alloys at 20 and 600 °C.

Alloys	Compressive yield strength at 20 °C (MPa)	Microhardness (GPa)	Yield stress calculated from microhardness at 20 °C (MPa)	Predicted SSH at 20 °C (MPa)	Compressive yield strength at 600 °C (MPa)	Predicted SSH at 600 °C (MPa)
A1	1070	5.42	1316	1417	1093	1284
A2	948	5.80	1410	902	1069	769
A3	1608	6.38		1804	1385	1549
A4	1337	5.41	1314	1205	1096	967
A5	1295	5.30		1196	1113	1026
A6	1290	4.79		1353	1048	1107
A7	998	4.86		1221	895	929
A8	1049	5.45	1323	1335	1122	1001

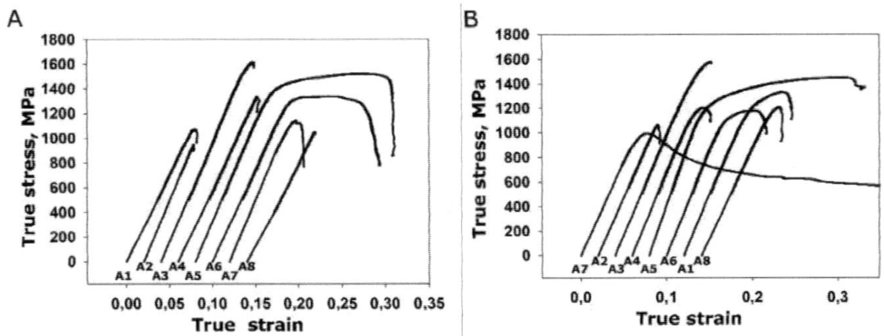

Figure 1. Compressive stress-strain curves for the model alloys at 20 °C (A) and 600 °C(B).

Figure 1 shows the compressive true strain-true stress curves for the model alloys at 20 and 600 °C. The detailed data on the mechanical properties of the model alloys are listed in Table 3. The yield strengths of the model alloys at 20 °C are in a range from 998 MPa for A7 alloy to 1608 MPa for A3 alloy. But most of the model alloys show the yield strengths at room temperature around 1300 MPa. Ductility over 1% was showed only by A5, A6 and A7 alloys (16.6, 13.8 and 3.6% respectively). A1, A2, A4 and A7 alloys fractured in the elastic region. For those specimens, which fractured brittle at room temperature, the yield strengths were recalculated from the microhardness values. The ratio between microhardness and yield strengths was evaluated using the corresponding values for the more ductile alloys (i.e. A5, A6 and A7). This ratio was found to be 0.2428 and the estimated yield strengths values for A1, A2, A4 and A8 are 1316, 1410, 1314 and 1323 MPa, respectively. At 600 °C as well as at 20 °C, the highest and the lowest values of the yield strengths were showed by A7 and A3 alloys

(1385 and 895 MPa, respectively). The yield strengths of the other model alloys were around 1100 MPa. Only A1, A5 and A7 alloys showed ductility over 5% (17.2%, 5.5% and more 50%, respectively), but none of the alloys broke out in the elastic region. The best strength/ductility ratio at both 20 °C and 600 °C was demonstrated by A1 and A6 alloys.

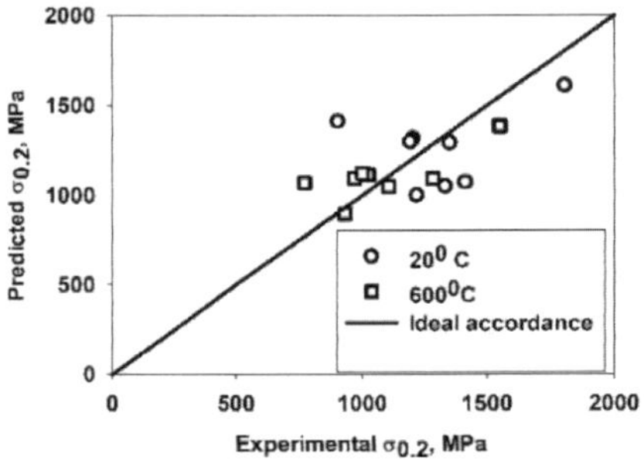

Figure 2. Experimental and predicted yield strengths at 20 and 600 °C.

4. Conclusion

The prediction of the solid solution hardening based on the atom radii and elastic modules misfit showed good agreement with the experimental values of the yield strength at 20 and 600 °C. The combination of approaches for predicting the phase composition of HEA(CALPHAD and thermodynamically rules) with the calculation of solid-solution hardening is promising for quick searching for high entropy alloys with a given phase composition and strength properties. For most of the model alloys, the difference between the predicted and experimental values of the yield strength at both 20 °C and 600 °C was found to be less than 15%.

Acknowledgments

This research was carried out with the financial support of the Russian Science Foundation, Project No. 19-79-30066

References

[1] Yeh J-W, Chen S-K, Lin S-J, Gan J-Y, Chin T-S, Shun T-T, Tsau C-H and Chang S-Y 2004 *Adv. Eng.Mater.* **8** 299-303
[2] Zhang Y, Zhou Y J,Lin J P, Chen G L and Liaw P K 2008 *Adv. Eng.Mater.* **10** 534-8
[3] Zhang Y, Zuo T T, Tang Z, Gao M C, Dahmen K A, Liaw P K and Lu Z P 2014 *Prog.Mater.Sci.***61** 1-93
[4] Wu Y D, Cai Y H, Wang T, Si J J, Zhu J, Wang Y D and Hui X D 2014 *Mater.Lett* **130** 277-80
[5] Stepanov N D, Yurchenko N Y, Skibin D V, Tikhonovsky M A and Salishchev G A 2015 *J. Alloys Compd*. 266-80
[6] Toda-Caraballo I, del-Castillo P E J R-D 2015 *Acta Mater.* **85** 14-23
[7] Toda-Caraballo I 2017 *Scripta Mater.* **127** 113-7
[8] Wang Z, Fang Q, Li J, Liu B and Liu Y 2018 *J.Mater. Sci. Technol.* **32** 349-54
[9] Coury F G, Kaufman M and Clarke A J 2019 *Acta Mater* **175** 66-81
[10] Fleischer R L 1963 *Acta Metall.* **11** 203-9
[11] Labush R 1972 *Acta Metall.***20** 917-27
[12] Guo S 2015 *J. Matter. Sci. Technol.* **31** 1223-30

B2 precipitates formation in Al-containing CoCrFeMnNi-type high entropy alloy

M Klimova*, D Shaysultanov, A Semenyuk, S Zherebtsov, G Salishchev, N Stepanov

Belgorod State University, Belgorod 308015, Russia

*Corresponding author: klimova_mv@bsu.edu.ru

Abstract. The non-equiatomic $Al_{7.5}Co_{21.75}Cr_{5.5}Fe_{21.75}Mn_{21.75}Ni_{21.75}$ high entropy alloy was studied. After cold rolling and annealing at 700-900 °C, a dual-phase ultrafine-grained structure composed of the partially recrystallized fcc matrix and B2 particles was produced; complete recrystallization of the fcc phase was observed after annealing at 1000 °C. The B2 precipitates formation provided a significant strengthening of the alloy in comparison with the single-phase state. The alloy demonstrated an attractive combination of mechanical properties, presenting yield and ultimate tensile strengths of 900 MPa and 1200 MPa, respectively, along with a ductility of 22%.

1. Introduction

A new class of metallic materials, the so-called high-entropy alloys (HEAs), has become a topic of great interest in materials science due to the possibility for attaining unique structures and properties [1]. High-entropy alloys based on 3d transition elements with a face-centered cubic (fcc) structure have demonstrated attractive properties, for example, extremely high ductility and toughness at room and cryogenic temperatures [2]. However, CoCrFeMnNi-based alloys generally have a rather low yield strength at room temperature [3]. The properties of the alloy can be improved by the proper selection of alloying elements along with microstructure modification using thermomechanical treatment, resulting in the formation of an ultrafine-grained recrystallized microstructure [4].

The effect of Al addition to alloys based on Co-Cr-Fe-Ni system has been extensively studied in [5,6]. For example, the as-cast microstructure of FeCoNiCrMn alloys with increasing Al content changed from the initial single fcc structure (until 8 at.% of Al) to a duplex fcc + body-centered cubic (bcc) structure (until 16 at.% of Al) and then to a single bcc structure [7]. However, along with an increase in the strength, a significant drop in the ductility of the Al-containing alloys is observed.

A good strength-ductility combination can be achieved by producing alloys with heterogeneous microstructures, consisting of a soft matrix and hard intermetallic secondary phases [8]. The possibility of B2 strengthening in fcc HEAs based on 3d transition metals was shown in [9,10]. HEAs with dual-phase precipitation-hardened structures can be expected to possess an enhanced balance of mechanical properties. However, the optimal chemical compositions and processing conditions for obtaining proper fcc + B2 microstructures in HEAs is poorly studied. Therefore, the CoCrFeNiMn-type high-entropy alloy, doped by ~7.5 at% of Al was thermomechanically processed to study the effect of Al addition on the microstructure formation and mechanical properties.

2. Materials and Methods

The non-equiatomic $Al_{7.5}Co_{21.75}Cr_{5.5}Fe_{21.75}Mn_{21.75}Ni_{21.75}$ alloy (subscripts indicate the concentration of the respective elements in at.%) was produced using mixtures of pure (≥99.9 wt.%) elements by vacuum induction melting. The information on the similar alloy without Al was presented in [11]. The as-cast samples were cold rolled at room temperature to a thickness reduction of 80% and annealed at temperatures of 700–1000 °C for 1 h followed by air cooling. The structure of the alloy was studied by transmission electron microscopy (TEM) using a JEOL JEM-2100 microscope equipped with an energy dispersive spectrometry (EDS) detector and a Nova NanoSem scanning-electron microscope (SEM) with back-scattered electron (BSE) detector and an electron-backscatter-diffraction (EBSD) camera. Samples for SEM and EBSD analysis were prepared by careful mechanical polishing. Samples for TEM analysis were prepared by conventional twin-jet electro-polishing of mechanically pre-thinned 100 μm foils in a mixture of 90% CH_3COOH and 10% $HClO_4$ at a potential of 30 V at room temperature. Vickers microhardness tests were conducted at room temperature using a load of 300 g. Tensile tests of dog-bone flat specimens (gauge measured $6\times3\times1.5$ mm^3) were performed at room temperature using an Instron 5882 universal testing machine with an initial strain rate of 10^{-3} s^{-1}.

3. Results and Discussion

The alloy in the as-cast condition demonstrated a coarse-grained single-phase microstructure without any secondary phases. The grains had curved irregular boundaries; the average grain size was ~150-200 μm. Cold rolling led to the development of a typical for fcc 3d HEAs lamellar-type microstructure formed by twin and (sub)grain boundaries aligned along the rolling direction. Deformation to large strain resulted in a significant strengthening of the alloy; the hardness increased from 165 Hv in the as-cast condition to 385 Hv after 80% thickness reduction.

The effect of cold-rolling and annealing at 700–1000 °C for 1 h on the microstructure of the Al-containing CoCrFeNiMn alloy is shown in Figure 1. Thermomechanical processing resulted in the recrystallization of the matrix and the precipitation of second-phase particles. After annealing at 700 °C, a partially recrystallized structure consisted of deformed areas and regions with fine recrystallized grains was observed. An increase in the annealing temperature to 800 °C and 900 °C led to the further development of the primary recrystallization (Figure 1b, c), characterized by a gradual increase in the recrystallized volume fraction and grain size. The completely recrystallized structure was formed after annealing at 1000 °C (Figure 1d).

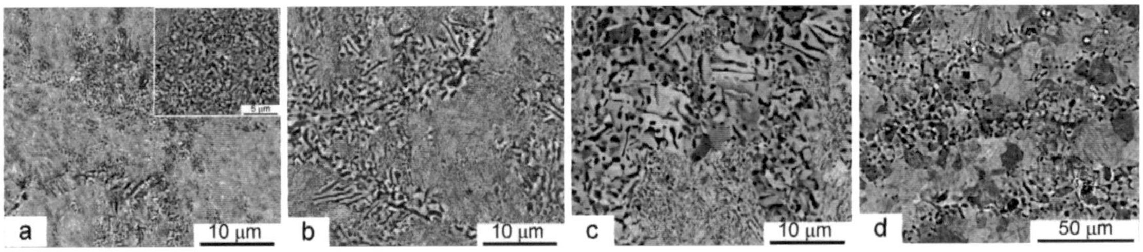

Figure 1. SEM-BSE images of the microstructure of the Al-containing CoCrFeMnNi-type alloy after annealing at 700 °C(a), 800 °C (b), 900 °C (c), 1000 °C (d).

The high magnification image (shown in Figure 1a) showed that the formation of precipitates occurred in the recrystallized regions. It is worth noting that the particles had both lamellar and close to equiaxial morphology. An increase in the annealing temperature was characterized by a rise of the fraction of equiaxed particles. The coarsening of the particles correlated with decreasing their volume fraction in the recrystallized areas during annealing. A gradual increase in the particles' transverse size from 190 nm to 660 nm along with a reduction of the fraction of particles from 33% to 7% were observed with increasing the annealing temperature from 700 °C to 1000 °C (Figure 2a).

The average size of the recrystallized grains of the alloy (Figure 2b) slightly increased from 0.9 μm to 1.5 μm when the annealing temperature changed from 700 °C to 900 °C. At the same time, a further increase in the annealing temperature to 1000 °C resulted in a significant rise of the grain size to 5.5 μm. Most likely, the precipitation of second-phase particles inhibited the grain growth in the Al-containing alloy (compare with the single-phase CoCrFeMnNi-type alloy [11] in Figure 2b) resulting in the formation of an ultrafine-grained matrix. Note that no secondary phases were found in the CoCrFeMnNi-type alloy without aluminum after the same annealing temperatures 700-1000 °C [11].

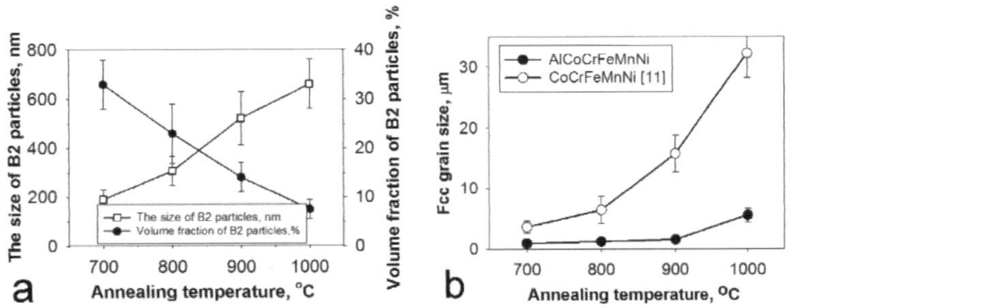

Figure 2. Dependencies of the size and volume fraction of B2 particles (a) and the fcc grain size (b) of the Al-containing CoCrFeMnNi-type alloy on annealing temperature.

A typical EBSD map (Figure 3a,b) of the Al-containing alloy demonstrated a dual-phase microstructure, consisting of the fcc matrix and particles with a bcc crystal structure. In addition, two peaks were found in the distribution of misorientation angles (insert on Figure 3b). The results of orientation image microscopy showed orientation relationship between the matrix grains and the particles corresponding to the Kurdjumov-Sachs OR: {111}fcc // {110}bcc and <110>fcc // <111>bcc (fcc/bcc misorientation of ~45°). The ∑3 boundaries (twin/matrix misorientation of 60°) indicated the presence of numerous annealing twins in the structure. The TEM microstructure of the alloy is presented in Figure 3c. The superlattice reflections {100} in the SAED pattern from the [001] bcc zone axis (Figure 3d) indicate the ordering of the BCC lattice, i.e., the particles had a B2 type structure. The EDS chemical analysis revealed that the B2 intermetallic phase was rich in Al and Ni.

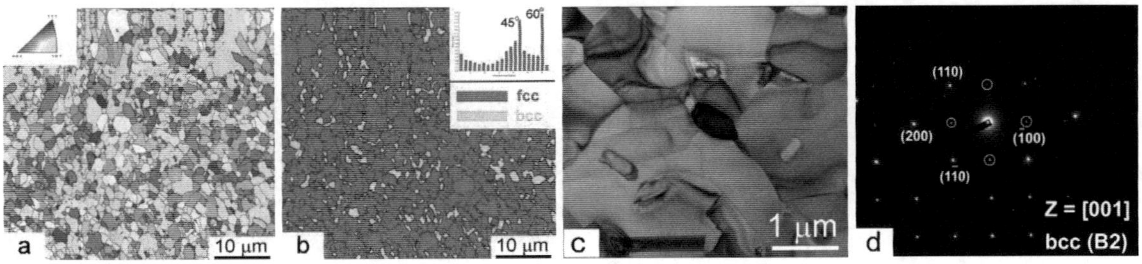

Figure 3. IPF map (a) and corresponding phase map (b); TEM bright-field image (c) and corresponding SAED pattern obtained from [001] B2 zone axis (d) of the Al-containing CoCrFeMnNi-type alloy after annealing at 900 °C.

The formation of a dual-phase precipitation-hardened fcc + B2 structure due to the addition of Al to the CoCrFeMnNi alloy provided a considerable increment in strength. The Al-containing CoCrFeMnNi-type alloy demonstrated higher values of microhardness compared to the undoped single-phase HEA [11] after the same thermomechanical treatment (Figure 4a). The Al-containing alloy in the as-cast condition with a single-phase coarse-grained structure exhibited a rather low yield strength of 220 MPa along with a high ductility of 68% (Figure 4b). Cold rolling followed by annealing resulted in a significant strengthening and some reduction of elongation to fracture of the alloy. For example, annealing at 700 °C provided a high yield strength of 1160 MPa and an ultimate tensile strength of 1500 MPa, but the hardening capacity of the alloy was quite limited. An acceptable ductility of 22% was achieved at the expense of the strength of the alloy: yield strength and ultimate tensile strength of 900 MPa and 1200 MPa, respectively, were obtained after annealing at 800 °C. A further increase in the annealing temperature led to softening and enhancement of the ductility of the alloy.

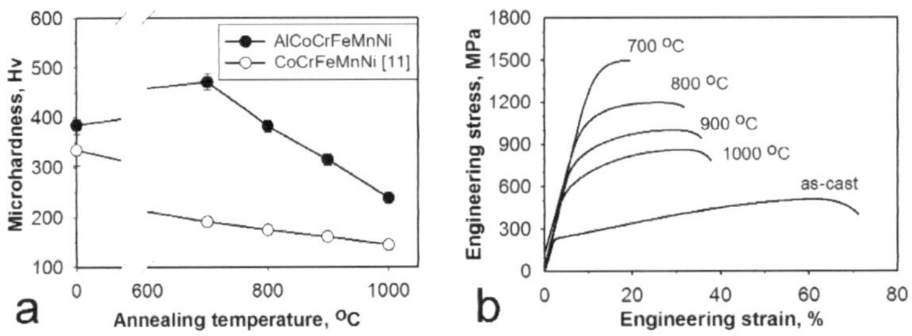

Figure 4. Microhardness (a) and tensile stress-strain curves (b) of the Al-containing CoCrFeMnNi-type alloy after cold rolling and annealing at different temperatures.

4. Conclusions

In summary, the microstructure evolution and mechanical properties of the Al-containing CoCrFeNiMn-type high-entropy alloy subjected to cold rolling followed by annealing at 700-1000 °C were studied. The presented results demonstrated that the addition of 7.5 at.% of Al to the CoCrFeMnNi alloy resulted in the fcc + B2 microstructure formation via thermomechanical processing, which provided a very attractive combination of strength and ductility.

Acknowledgements

The authors gratefully acknowledge the financial support from the Russian Science Foundation Grant No. 18-19-00003. The authors are grateful to the personnel of the Joint Research Center, «Technology and Materials», Belgorod State University, for their assistance with instrumental analysis.

References

[1] Yeh J-W, Chen S-K, Lin S-J, Gan J-Y, Chin T-S, Shun T-T, Tsau C-H and Chang S-Y 2004 *Adv. Eng. Mater.* **6** 299
[2] Gludovatz B, Hohenwarter A, Catoor D, Chang E H, George E P and Ritchie R O 2014 *Science* **345** 1153
[3] Gali A and George E P 2013 *Intermetallics* **39** 74
[4] He J Y, Wang H, Huang H L, Xu X D, Chen M W, Wu Y, Liu X J, Nieh T G, An K and Lu Z P 2016 *Acta Mater.* **102** 187
[5] Wang W R, Wang W L, Wang S C, Tsai Y C, Lai C H and Yeh J W 2012 *Intermetallics* **26** 44
[6] Yang T, Xia S, Liu S, Wang C, Liu S, Zhang Y, Xue J, Yan S and Wang Y 2015 *Mater. Sci. Eng. A* **648** 15
[7] He J Y, Liu W H, Wang H, Wu Y, Liu X J, Nieh T G and Lu Z P 2014 *Acta Mater.* **62** 105
[8] Sathiyamoorthi P and Kim H S 2020 *Prog. Mater. Sci.* 100709
[9] Dasari S, Jagetia A, Chang Y J, Soni V, Gwalani B, Gorsse S, Yeh A C and Banerjee R 2020 *J. Alloys Compd.* **830** 154707
[10] Lozinko A, Zhang Y, Mishin O V, Klement U and Guo S 2020 *J. Alloys Compd.* **822** 153558
[11] Klimova M, Shaysultanov D, Semenyuk A, Zherebtsov S and Stepanov N 2021 *J. Alloys Compd.* **851** 156839

Effect of multi-axial deformation on the structure and properties of a nickel-enriched Ti-Ni shape memory alloy

V Komarov[1,2,3*], I Khmelevskaya[2], R Karelin[1,2], V Yusupov[1],
R Kawalla[3], G Korpala[3], U Prahl[3] and S Prokoshkin[2]

[1] Baikov Institute of Metallurgy and Materials Science RAS, Moscow, Russia

[2] National University of Science and Technology MISIS, Moscow, Russia

[3] TU Bergakademie Freiberg, Germany

*Corresponding author: vickomarov@gmail.com

Abstract. In the present study, multi-axial severe plastic deformation of a nickel-enriched Ti-Ni shape memory alloy for medical applications was performed using the *MaxStrain* module of the *Gleeble* system at 350 °C with a high accumulated true strain of $e = 6.5$. The martensitic transformations and structure were studied using DSC and TEM analysis. The maximum completely recoverable strain was determined by a thermomechanical method. An ultrafine-grained structure with a grain/subgrain size bellow 100 nm was formed in a bulk nickel-enriched Ti-Ni alloy. The resulting structure provides a completely recoverable strain of 10.5%, which is much higher in comparison with the recrystallized structure (4 %).

1. Introduction

Ti-Ni-based shape memory alloys (SMA) are widely used functional materials. Nickel-enriched Ti-Ni SMA, due to the unique combination of their functional and mechanical properties, is one of the most suitable materials for medical applications [1-4].

Severe plastic deformation (SPD) is one of the most effective methods of thermomechanical treatment for the formation of an ultrafine-grained (UFG) structure [4-5]. The best combination of functional and mechanical properties of Ti-Ni SMA can be achieved after formation of a UFG structure. However, this structure was obtained only in thin Ti-Ni samples with a thickness of about 0.1-0.2 mm [6-7]. The development of SPD methods is aimed at searching for new deformation modes, which would allow manufacturing bulk samples containing the UFG structure. The use of equal-channel angular pressing and warm forging allows obtaining only a submicrocrystalline structure in a Ti-Ni SMA [8-9].

It was shown in previous studies that SPD of an equiatomic Ti-Ni SMA in isothermal conditions using the *MaxStrain* module of the *Gleeble* system with multi-axial deformation results in the formation of the UFG structure with the average grain/subgrain size down to 55 nm, and in the realization of high functional properties [10-12].

The present work continues previous investigations and is a preliminary study aimed at investigation of the *MaxStrain* deformation effect on the structure and properties of bulk nickel-enriched Ti-Ni SMA for medical applications.

2. Experimental

In the present work, the nickel-enriched Ti-Ni SMA plate (about 55.9 mass.% Ni) was hot rolled at 900 °C and defined as a reference treatment (RT). Then the specimens were cut and their central part (1 cm^3) was deformed by compression with two strikers, then the sample was rotated by 90° around its longitudinal axis and compressed again in the direction perpendicular to the previous compression direction [10]. This cycle was repeated 10 times at an anvil speed of 0.5 mm/s and a constant temperature of 350 °C. The accumulated true strain was $e = 6.5$.

The characteristic temperatures of martensitic transformations (MT) M_s, M_f (start and finish of forward MT), A_s, A_f (start and finish of reverse MT) were estimated by a *Mettler-Toledo* DSC at a cooling-heating rate of 10 K/min. The structure was studied using a *JEM-2100* transmission electron microscope. The completely recoverable strain ε_r was determined by a thermomechanical method using a bending mode for strain inducing in cold water.

3. Results and Discussion

After RT, the structure can be characterized as a low-defect recrystallized structure with an average grain size of 30-40 microns (Figure 1). The rather large grain size can be explained by the high heating temperature of 900 °C. This structure is homogeneous and mainly consists of uniaxial grains. Black inclusions in Figure 1 are the Ti_2Ni phase, which always remains after smelting in an amount of up to 5% and does not participate in the MT.

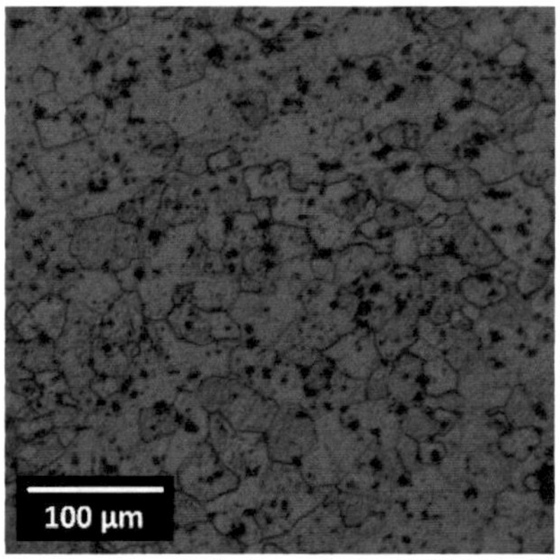

Figure 1. Recrystallized structure of a nickel-enriched Ti-Ni alloy after reference treatment.

The study of the structure and phase composition by transmission electron microscopy was performed after *MaxStrain* deformation at 350 °C with the accumulated strain $e = 6.5$. According to the previous experience [10-12], at these low temperatures in such an alloy the process of dynamic polygonization of B2-aystenite should develop. Analysis of the bright field of TEM images in Figure 2 reveals that after *MaxStrain* deformation at 350 °C, the structure consists of equiaxed and elongated nano-sized (70-120 nm) grains and subgrains with a high density of free dislocations (not less than 10^{11} cm^{-2}). At the same time, a cellular substructure and submicroscale areas with a uniform distribution of dislocations are observed as well.

It should be noted that the processes of dynamic recovery and polygonization in nickel-enriched the Ti-Ni SMA are less developed as compared to the equiatomic Ti-Ni alloy under the same thermomechanical conditions [10]. This fact is a consequence of the retarding effect of finely-dispersed Ti_3Ni_4 phase particles, which appeared as a result of dynamic strain aging in the nickel-enriched alloy [13].

In the diffraction pattern, reflexes of B2-austenite in combination with numerous weak reflexes of the R-phase are identified. A part of the R-phase reflexes can hardly be separated from the B2-austenite once due to low resolution. Specific reflexes of phases are identified in the diffraction pattern by arrows. The observed martensite-like parallel plates are crystals of the R-phase, formed upon cooling to room temperature after *MaxStrain* deformation. The TEM study does not reliably reveal reflexes or images of the Ti_3Ni_4 phase due to the very high defectness of the lattice and the extremely small size of the Ti_3Ni_4 particles. Thus, under this deformation condition as well as at 350 °C, *e* = 6.5, the formation of an UFG structure of B2-austenite is possible.

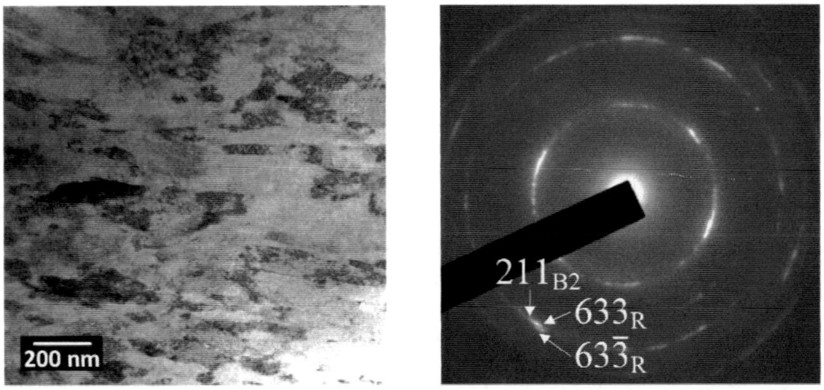

Figure 2 – Structure of the nickel-enriched Ti-Ni alloy after *MaxStrain* deformation at 350 °C, *e* = 6.5. Bright field image and electron diffraction pattern.

The presence of the R-phase after *MaxStrain* deformation is confirmed by calorimetry (Figure 3). After the RT, only single calorimetric peaks are observed, indicating one-stage of forward and reverse B2↔B19′ MT. *MaxStrain* deformation leads to broadening and multiplication of peaks, indicating a two-stage B2↔R↔B19′ transformation. The deformation also shifts the peaks of forward and reverse MT to lower temperatures: from -17 to -70 °C and 13 to -10 °C, respectively, the MT hysteresis becomes larger. The finish temperature of the forward MT is not measured as it is below minus 100 °C.

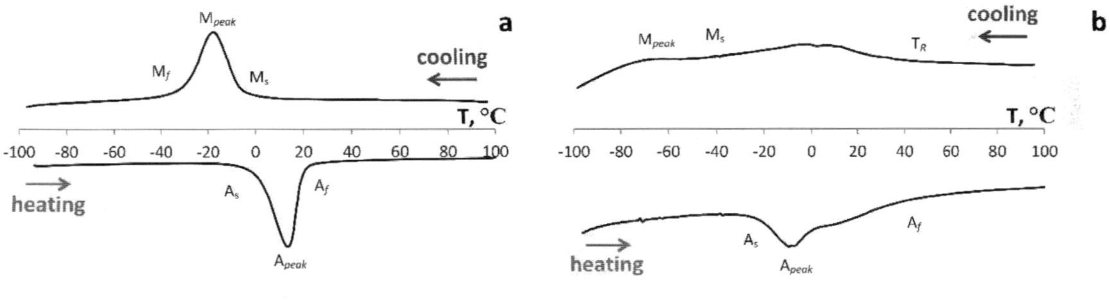

Figure 3. DSC curves of nickel-enriched Ti-Ni alloy after RT (a) and *MaxStrain* deformation (b).

The effect of *MaxStrain* deformation at 350 °C on the MT sequence and temperatures is defined by two main factors: strong strain hardening and strain aging. Strain hardening suppresses the formation of B19′-martensite while facilitates the formation of the R-phase from B2-austenite. The formation of the Ti_3Ni_4 phase as a result of strain aging is also favorable for the formation of the R-phase.

The characteristic temperatures can be regulated over a wide range by aging at different temperatures and heating times. The effect of post-deformation annealing after *MaxStrain* deformation on the structure and properties of a nickel-enriched Ti-Ni alloy needs to be studied in the future.

The completely recoverable strain is an important functional property of Ti-Ni SMA. A calorimetric study reveals that *MaxStrain* deformation leads to the appearance of B2→R transformation and significant widening of the range of existence of the R-phase (Figure 3). The strain inducing has been started in the R-phase and has an advantage as compared to strain inducing in the B19′-martensite state, because the critical stress for B19′-martensite formation from R-phase (transformation yield stress) is lower than the reorientation stress of thermal B19′-martensite [2].

The results show that after the RT, the value of completely recoverable strain is about 4%. The *MaxStrain* deformation leads to realization of completely recoverable strain of 10.5%. This result is comparable with the highest values of completely recoverable strain, obtained in previous studies, and much higher than after the RT.

4. Conclusions

The possibility of conduction of the *MaxStrain* deformation for the nickel-enriched Ti-Ni shape memory alloy was studied. The *MaxStrain* deformation at a temperature of 350 °C with an accumulated true strain of $e = 6.5$ leads to the formation of an ultrafine-grained structure with an average size of structural elements below 100 nm and provides a significant advantage in completely recoverable strain in comparison with the recrystallized structure: 10.5% against 4 %, respectively. The obtained results of preliminary studies indicate that the use of the *MaxStrain* deformation is promising in terms of the formation of a nanocrystalline structure in bulk samples of a nickel-enriched Ti-Ni shape memory alloy and a corresponding improvement in functional properties.

Acknowledgements

The reported study was funded by RFBR, project number 19-33-60090.

References

[1] Otsuka K, Ren X 1999 *Intermetallics* **7** 5 511-528
[2] Brailovski V, Prokoshkin S, Terriault P, Trochu P 2003 *Shape memory alloys: fundamentals, modeling and applications* (Montreal: ETS Publ.) p 851
[3] Jani J M, Leary M, Subic A, Gibson M A 2014 *Mater. & Des.* **56** 1078-1113
[4] Resnina N, Rubanik V 2015 Shape memory alloys: properties, technologies, opportunities (Pfaffikon: Trans Tech Publications)
[5] Valiev R Z, Aleksandrov I V 2000 *Nanostructural Materials Obtained by Severe Plastic Deformation* (Integratsiya, Moscow)
[6] Brailovski V, Prokoshkin S, Inaekyan K and Demers V 2011 *J. Alloys Compd.* **509** 2066-2075
[7] Sundeev R V, Shalimova A V, Glezer A M, Pechina E A, Gorshenkov M V 2018 *Phys. Solid State* **60** 6 1168-1172
[8] Khmelevskaya I Y, Karelin R D, Prokoshkin S D, Isaenkova M G, Perlovich Y A, Fesenko V A, Komarov V S, Zaripova M M 2019 *IOP Conf. Ser.: Mater. Sci. Eng.* **503** 1 012024
[9] Prokoshkin S, Khmelevskaya I, Andreev V, Karelin R, Komarov V., Kazakbiev A. 2018 *Mater. Sci. Forum.* **918** 71-76.
[10] Khmelevskaya I, Komarov V, Kawalla R, Prokoshkin S, Korpala G 2017 J. Mater. Eng. Perform. 26 8 4011-4019.
[11] Khmelevskaya I, Komarov V, Kawalla R, Prokoshkin S, Korpala G 2017 *Mater. Today: Proc.* **4** 3 4830-4835
[12] Komarov V, Khmelevskaya I, Karelin R, Korpala G, Kawalla R, Prokoshkin S 2019 *J. Alloys Compd.* **797** 842-848
[13] Ryklina E P, Polyakova K A, Tabachkova N Y, Resnina N N, Prokoshkin S D 2018 *J. Alloys Compd.* **764** 626-638

IOP Conf. Series: Materials Science and Engineering 1014 (2021) 012020 doi:10.1088/1757-899X/1014/1/012020

Ultralow-cobalt hard alloys obtained by spark plasma sintering

E A Lantsev[1*], V N Chuvil'deev[1], A V Nokhrin[1], M S Boldin[1], N V Malekhonova[1], Y V Blagoveshchenskii[2], N V Isaeva[2], Y V Tsvetkov[2], P V Andreev[1] and K E Smetanina[1]

[1] Lobachevsky State University of Nizhny Novgorod, 23 Prospect Gagarina (Gagarin Avenue), Nizhny Novgorod, 603950, Russia
[2] Baikov Institute of Metallurgy and Materials Science, RAS, 49 Leninskiy av., Moscow, 119334, Russian Federation

*Corresponding author: elancev@nifti.unn.ru

Abstract. The features of spark plasma sintering (SPS) of plasma-chemical nanopowders WC-(0.3, 0.6, 1) wt.% Co were studied. The SPS process of ultralow-cobalt hard alloys can be sequentially represented as a change of the following mechanisms: rearrangement of particles at lower temperatures (Stage I) → sintering of WC-Co particles due to Coble diffusion creep of cobalt, the intensity of which is determined by the grain boundary diffusion rate (Stage II) → sintering due to diffusion creep, the rate of which is limited by the bulk diffusion in cobalt (Stage III-1) → sintering of tungsten carbide particles along the intergranular boundaries of WC / WC under conditions of intensive grain growth (Stage III-2). Samples with a high density (96.4-98.4%) and high mechanical properties were obtained (for the WC-0.3% Co hard alloy: $H_v \sim$ 20.5 GPa, K_{IC} = 7.1 MPa · $m^{1/2}$).

1. Introduction

Hard alloys WC-Co currently play an important role in industry due to their unique physical and mechanical characteristics: high hardness, especially at elevated temperatures, and, at the same time, high crack resistance, wear resistance and corrosion resistance [1]. The physical properties of pure tungsten carbide [2], as well as carbides with the addition of a large volume fraction of a low-melting metal binder [3], are well studied. Recently, much attention has been paid to the methods of high-speed solid-phase sintering of hard alloys [4-6], which make it possible to ensure the formation of a homogeneous high-density structure with the smallest possible grain size and, as a consequence, with increased physical and mechanical properties. One of the promising materials for machine tool and special machine building are ultralow-cobalt hard alloys, which can combine the high hardness inherent in pure tungsten carbide and the increased crack resistance characteristic of conventional hard alloys with a cobalt content of 8-12 wt.%. Such materials are promising for creating a new generation of metal cutting tools and special applications.

Note, there are few data in the literature on sintering ultralow-cobalt hard alloys. Therefore, the purpose of this work is to study the features of high-speed sintering of ultrafine-grained (UFG) ultralow-cobalt hard alloys and, in particular, to analyze the effect of the cobalt content on the microstructure parameters and physical and mechanical properties of new hard alloys.

2. Methods and materials

The objects of study in this work were tungsten monocarbide nanopowders α-WC, obtained by the plasma-chemical method with subsequent reduction annealing in hydrogen at a temperature of 1050°C (3 h) [7]. Various concentrations of cobalt were added to tungsten carbide nanopowder with an initial particle size of 95 nm by precipitation from an alcohol solution of $CoCl_2 \cdot 6H_2O$ salts: 0.3, 0.6, and 1 wt% Co. Powder of pure tungsten carbide is No.1, powders with 0.3, 0.6 and 1wt% cobalt addition are No.2, 3 and 4, respectively.

The compacting of samples with a diameter of 12 mm and a height of h = 4 mm was carried out by the method of spark plasma sintering (SPS) using the Dr. Sinter model SPS-625 (Japan). Sintering of nanopowders was carried out in vacuum (2-5 Pa). During the experiments, the dependence of shrinkage (L) of powders on the heating temperature and isothermal holding time was monitored. The structure of the samples was investigated using a JEOL JSM-6490 scanning electron microscope with an Oxford Instruments INCA 350 energy-dispersive microanalyzer. Qualitative phase analysis was carried out using the Diffrac.EVA program. Quantitative analysis was determined by the Rietveld method. The accuracy of determining the volume fraction of particles of α-WC, W_2C, and η-phase was ±0.5%. Sample density (ρ) was measured by hydrostatic weighing using a Sartorius CPA 225D analytical laboratory balance. Hardness H_v and fracture toughness K_{IC} on sintered specimens were measured using a QnessA60 + microhardness tester. The measurements were carried out with a load of 2 kg.

3. Results

The SEM images of pure tungsten carbide nanopowders and nanopowder compositions WC-Co with various cobalt content show that nanoparticles are collected in agglomerates, the average size of which is the same for all samples and is ~2-3 μm. The structure of pure tungsten carbide contains rather large agglomerates several tens of microns, which are practically absent in the WC – Co nanopowder composites. Analysis of the XRD results shows that the powder compositions contain hexagonal tungsten monocarbide α-WC and cobalt in two modifications - with a cubic and a hexagonal lattice. Figure 1a shows the dependence of shrinkage L of WC-Co nanopowders on the continuous communication temperature. It is seen that the dependence of shrinkage on the system for nanopowders has a three-stage character.

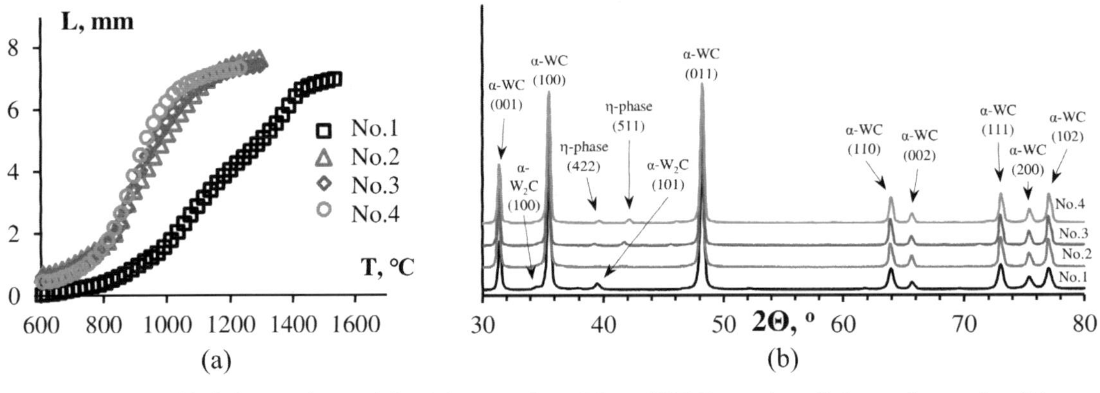

Figure 1. Shrinkage plots L(T) of the powders (a) and XRD results of sintered samples (b)

The microstructure photographs of the surface layers of sintered hard alloys show that in pure tungsten carbide can provide a fairly uniform UFG structure with an average grain size of ~ 0.2–0.3 μm. In the structure of the obtained ceramics, single coarse elongated grains with a size of ~ 1-3 μm are visible. In samples containing cobalt, individual inclusions of cobalt up to 1 μm in size were found.

Figure 1b shows XRD results of the surface layers. It is seen that in the composition of ultralow-cobalt hard alloys, in addition to the α-WC phase, there is a phase of ternary carbide η-Co_3W_3C, the volume fraction of which is less than 0.5 vol.%. The structure of the central layers of ultralow-cobalt hard alloys also contains particles of the η-phase of the composition Co_3W_3C, the amount of which increases from ~1.6 to 3.6% with an increase in the content of cobalt in the composition of the initial nanopowders from 0.3 to 1 wt%, respectively. We emphasize that the bulk η-phase for particles in some samples of ultralow-cobalt hard alloys is several times larger than in the surface layers. The hardness of the surface layer obtained in the continuous mode with an increase in the cobalt content from 0.3 to 1 wt.% decreases from 20.5 GPa to 15.4 GPa. A fairly noticeable spread in hardness is observed, which may indirectly indicate an unequal distribution of cobalt in the structure of sintered hard alloys. Crack resistance of the surface layer of ultralow-cobalt hard alloys with 0.3 and 1% Co is 7-8 MPa·$m^{1/2}$. The hardness of the center hard alloy is slightly different from that of the hard surface layer.

4. Discussion

4.1. Effect of cobalt on the structure and properties of ultralow-cobalt hard alloys

Generalization of the results at figure 2a shows that with an increase in the concentration of cobalt, the absolute and relative values of the density of the sintered specimens decrease. The result obtained, in our opinion, is a consequence of an increase in the volume fraction of the η-phase particles in hard alloys, which have a lower theoretical density (14.48 g / cm^3 for Co_3W_3C). An analysis of the data on the hardness and crack resistance of the sintered specimens is shown in figure 2b. It is seen that there is a decrease in the hardness of the surface layers from 28.8 GPa to 20.5 GPa and an increase in K_{IC} from 4.8 to 7.1 MPa·$m^{1/2}$ with an increase in the cobalt content from 0 to 0.3%. A further increase in the cobalt content leads to a significant decrease in the hardness to 11.4, but does not lead to a noticeable change in the crack resistance of the sintered specimens. This may be due to the non-uniform distribution of cobalt in the process of solid-phase sintering of hard alloys, in which the propagation of cobalt between WC particles occurs only with the help of plastic deformation (viscous flow).

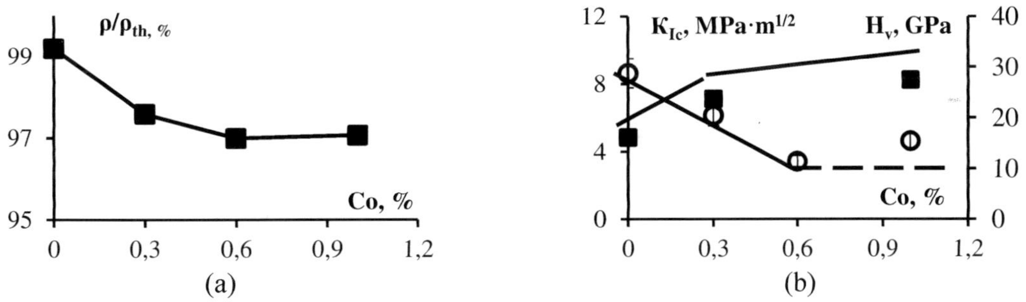

Figure 2. The dependence of the relative density (a) and mechanical properties (b) of hard alloys on the cobalt content

4.2. Analysis of sintering mechanisms

The generalization of presented in figure 1b dependences of shrinkage on heating temperature shows that the dependences L(T) have a traditional three-stage character. To analyze the kinetics of non-isothermal sintering of powders at the stage of intense shrinkage (Stage II), we use the Young – Cutler model [8], which describes the initial stage of nonisothermal sintering of spherical particles under conditions of simultaneous processes of volume and grain boundary diffusion. In accordance with [8], the slope of the shrinkage (ε) dependence on temperature in the coordinates ln ($T\partial\varepsilon/\partial T$) - T_m / T corresponds to the effective activation energy of the non-isothermal sintering process mQ_{s2}. It should be noted that the value of the coefficient m depends on the type of the dominant diffusion mechanism under non-isothermal sintering conditions (m = 1/3 for grain boundary diffusion, m = 1/2 for bulk diffusion, m = 1 for viscous flow [8]).

For ultralow-cobalt hard alloys with 0.3, 0.6, and 1 wt% Co, the effective activation energies at Stage II are mQ_{s2} ~ 11 kTm, 10 kTm, and 9 kTm, respectively. The values obtained in the case m = 1 (see [9])

are quite close to the activation energy of diffusion creep of cobalt, limited by the rate of grain boundary diffusion (Coble creep) [10]. We also note that in [9], the activation energy of SPS for plasma-chemical nanopowders with an increased cobalt content WC-Co, calculated using the Young-Cutler model, was $Q_{s2} \sim$ 7-8 kTm, which is in good agreement with the results obtained in this work. Similar values of the activation energy for diffusion creep were obtained in [10]. The decrease in the activation energy of sintering at Stage II is probably due to the increased volume fraction of the interphase boundaries "tungsten carbide WC – γ-phase (Co)", through which the process of diffusion of carbon and tungsten atoms occurs.

5. Conclusion

The samples obtained by the SPS method have a high density (96.4-98.4% of the theoretical value), a fine-grained structure and high mechanical properties: the hardness of the WC-0.3% Co hard alloy is ~ 20.5 GPa, and the Palmquist fracture toughness coefficient is K_{IC} = 7.1 MPa $m^{1/2}$. The result obtained indicates, in our opinion, the efficiency of the combined use of the SPS technology and the chemical-metallurgical method of applying ultrathin layers of the metal phase on the surface of the synthesized tungsten carbide nanoparticles.

It is shown the process of high-speed sintering of ultrafine-grained ultralow-cobalt hard alloys can be represented as a sequential change of the following mechanisms: rearrangement of particles at low temperatures (Stage I) → sintering of WC-Co particles due to diffusion creep of cobalt according to Coble, the intensity of which is determined by the rate of grain boundary diffusion (Stage II) → sintering due to diffusion creep, the rate of which is limited by the rate of bulk diffusion in cobalt (Stage III-1) → sintering of tungsten carbide particles along the WC / WC grain boundaries under conditions of intensive grain growth (Stage III-2).

Acknowledgments

The authors wish to thank the Russian Science Foundation (grant No. 18-73-10177)

References

[1] Panov V S and Chuvilin A M 2001 *Technology and properties of sintered hard alloys and products from them* [*Tehnologiya i svoistva spechennih tverdih splavov i izdelii iz nih*] (Moscow: MISIS) p 428 (in Russian)
[2] Kurlov A S and Gusev A I 2013 *Tungsten Carbides (Springer Series in Materials Science vol 184)* (Cham: Springer International Publishing) p 242
[3] Upadhyaya G S 1998 *Cemented tungsten carbides: production, properties, and testing* (New Jersey: Noyes Publications) p 420
[4] Cha S I and Hong S H 2003 *Materials Science and Engineering* **356** 381
[5] Sivaprahasam D, Chandrasekar S B and Sundaresan R 2007 *International Journal of Refractory Metals and Hard Materials* **25** 144
[6] Eriksson M, Radwan M and Shen Zh 2013 *International Journal of Refractory Metals and Hard Materials* **36** 31
[7] Isaeva N V, Blagoveshchenskii Y V, Blagoveshchenskaya N V, Mel'nik Y I, Samokhin A V, Alekseev N V and Astashov A G 2013 *Russian Journal of Non-Ferrous Metals* **55** 585
[8] Young W S and Cutler I B 1970 *Journal of American Ceramic Society* **53 [12]** 659
[9] Lantsev E A, Chuvil'deev V N, Nokhrin A V, Boldin M S, Tsvetkov Yu V, Blagoveshchenskiy Yu V, Isaeva N V, Andreev P V and Smetanina K E 2020 *Inorganic Materials: Applied Research* **11** 586
[10] Sargent P M, Malakondaiah G and Ashby M F 1983 *Scripta Metall* **17** 625

IOP Conf. Series: Materials Science and Engineering 1014 (2021) 012021 doi:10.1088/1757-899X/1014/1/012021

Method of molecular dynamics investigation of diffusion in solid solutions with consideration of ordering effects on the example of V50W50 and V90W10 alloys

A G Lipnitskii[1*], V N Maksimenko[1] and I V Nelasov[2]

[1]Belgorod State National Research University, Belgorod, Russia
[2]Institute of Problems of Chemical Physics of RAS, Russia

[*]Corresponding author: lipnitskii@bsu.edu.ru

Abstract. This paper presents the development of methods for the theoretical study of diffusion processes in solid solutions based on molecular dynamics (MD), taking into account the effects of ordering. The influence of concentration and temperature on the features of diffusion characteristics in substitution alloys is studied using the example of solid solutions V50W50 and V90W10. Modeling has shown that the V-W ordering is associated with the separation of plates enriched with tungsten in the planes (100). Vanadium atoms are more mobile and the difference in diffusion coefficients decreases with the equalization of component concentrations and temperature increases.

1. Introduction

At present, considerable attention is paid to establishing the diffusion characteristics of solid solutions (recent works [1,2] and references therein) due to their significant influence on the properties of such materials. In some cases, diffusion characteristics are obtained from experimental studies [1,2], which, however, are often limited by methodological complexity and insufficient resolution of the instrument base [3]. The missing information about the diffusion characteristics of solutions can be compensated by the predictions from the results of atomistic simulation. In this case, it is important to take into account the ordering effects for the correct setting of the atomic structure, which depends on the temperature of the solid solution.

A well-known approach to accounting for ordering is the MC/MD method [4] based on a combination of Monte Carlo (MC) and molecular dynamics (MD) methods. In the MC/MD method [4], the calculation of the change in potential energy when rearranging atoms of different types for the implementation of MC steps is carried out without taking into account the relaxation in the position of the atoms caused by the permutation. Therefore, it remains relevant to develop a new practical implementation of the MC method, which, along with using the advantages of the MD method, would take into account local relaxation to reduce the number of MC steps necessary to achieve equilibrium. Also, the MC/MD method [4] does not use direct modeling of the NPT ensemble, which does not guarantee the achievement of an equilibrium of the atomic system corresponding to the minimum Gibbs energy at a given number of particles N, pressure P, and temperature T.

In this paper, we develop a new method for obtaining the equilibrium atomic structure of a solid solution based on a combination of MD and MC (MD+MC) methods. The new method correctly implements the achievement of the equilibrium of the atomic system as a result of modeling the NPT ensemble. Model samples obtained by the MD+MC method are used to study diffusion in solid solutions by the MD method. A new method for studying diffusion in solid solutions, taking into account the effects of ordering, is tested on the example of two alloys of the V-W system, which forms the basis of many modern metal materials for use at high operating temperatures. To define interatomic interactions of V-W, we use potentials [5-7].

2. Methods

The MD+MC method is based on the equivalence of the MD and MC methods in obtaining the microstates of the NPT ensemble.The microstate of the solution is defined by two constituents: 1) the position of the atoms in local equilibrium, which change as a result of diffusion jumps over significant times in comparison with the periods of vibrations of the atoms; 2) the displacement of the atoms from these positions due to thermal vibrations. The MD+MC method is implemented as an algorithm for sequential actions.

Action 1. In the superlattice model, we set the calculation cell of a random solution containing N_i atoms of each type i in accordance with the given C_i concentrations. We perform MD simulation during 20 ps at a given temperature T and pressure P (=0) until the ensemble's NPT equilibrium is reached.We use the velocity Verlet algorithm for the solution of the equations of motion, the Nose-Hoover thermostat and the Berendsen barostat.

Action 2. Do n_{MC} (=20) steps of MC. To do this, we apply a uniform grid of n_{point} points to the calculation cell with a random position of the starting point of the grid in the cell within periodic boundary conditions as illustrated in figure 1.

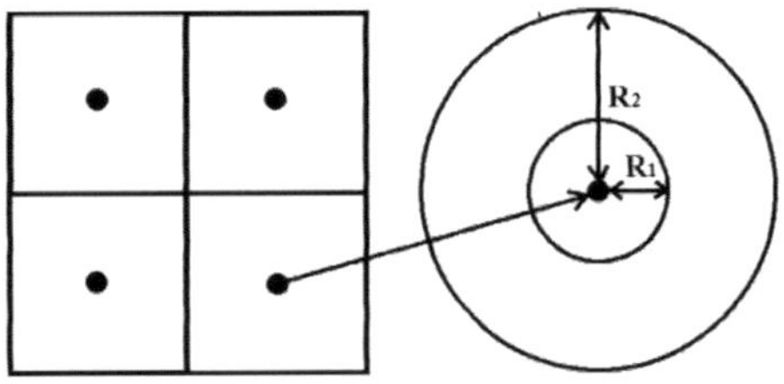

Figure 1. A schematic representation of a uniform grid of points in the calculation cell and the selection of a sphere of atoms near each point with radius R_2. Within a sphere of a smaller radius R_1, mobile atoms are marked, which will participate in a permutation with an atom in the center of the sphere.

Next, select the atom closest to each point and cut a sphere from the cell containing all the atoms within the distance R_2 from the selected atom. Inside the cut-out sphere, we mark about 100 atoms located within the distance R_1 from the atom in the center of the sphere, which will be mobile during MD simulation of the sphere of radius R_2. The coordinates of atoms located at distances ranging from R_1 to R_2 from the central atom are fixed in the MD modeling process. The requirement $R_2 - R_1 > R_c$ is imposed, where R_c is the radius of interatomic interaction for the possibility of parallel modeling of each sphere without their mutual influence. The radius R_c of interatomic interactions in the V-W system made it possible to select a 2×2×2 point grid in the calculation cell of 2000 nodes. MD simulation of each sphere is performed during n_{MD} steps to perform one MC step in it to exchange places of the atom i in the center with an atom of another type j, randomly selected among the atoms in the sphere of radius R_1.The MC step is performed with the Metropolis algorithm according to the change in potential energy $\Delta U = U - U_0$, where U_0 is the average potential energy of the simulated sphere with the initial position of the atoms and U is the average potential energy of the same sphere, but after the rearrangement of atoms i and j.

Action 3. Modeling the NPT ensemble of the entire cell as a whole with the accepted permutations of atoms by the MD method during N_{MD} (=2000) steps to achieve equilibrium. Calculation of the average enthalpy, volume, and other current characteristics of the system using the MD method.

Calculation of the value of the order parameter, configuration entropy, and other values associated with configurations of atoms of different types by cluster variation method. Sequential repetition of steps 2 and 3 is one step of the MD+MC method. The values of MC and MD steps are selected after test calculations with a number of values as a variant with the smallest values of MD steps, which slightly change the result of modeling. To achieve equilibrium in V-W alloys, 100 steps of the MD+MC method are sufficient (the enthalpy, order parameters, and other values change near the average values during further MD+MC modeling).

Model diffusion experiments were performed using MD simulation (*NPT*ensemble) for 32 ns. In the initial calculation cell containing 2000 nodes, one atom was removed to create a vacancy. We used 400 points of mean square displacement of atoms in time for the temperature range 300-2100K. To smooth out random fluctuations, the obtained points were averaged over 20 sets of configurations, the starting points of which are 80 ps apart in time.

3. Results and discussions

The feature of the atomic structure associated with the formation of short-range order in V-W is illustrated in figure 2, which shows the projections of the positions of vanadium and tungsten atoms on the crystallographic planes (100) and (010) in the calculation cell of a V90W10 solid solution at a temperature of 300 K. As can be seen from figure 2, the ordered V90W10 alloy contains areas rich in tungsten. A characteristic feature of these regions is the formation of tungsten plates oriented in the

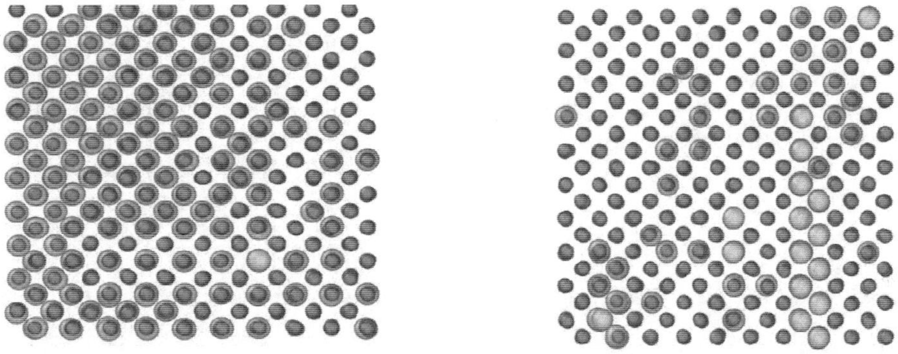

Figure 2. Projections of positions of vanadium atoms (red) and tungsten atoms (green) on the crystallographic plane (100) on the left and the plane (010) on the right in the calculation cellofV90W10 at a temperature of 300K.

crystallographic plane (100) of the BCC lattice. We also observed similar regions at higher temperatures. However, as the temperature increases, the size of areas rich in tungsten decreases. Similar areas were also observed in the V50W50 alloy, but the size of the tungsten plates decreases with increasing tungsten concentration. Here we note that it is possible in principle to observe the formation of tungsten plates in vanadium by atom probe tomography. During the MD+MC simulation, a decrease in the Gibbs energy calculated as the contributions of enthalpy and configuration entropy was always observed from the initial state of a random solution.

As a result of smoothing the calculated values of the diffusion coefficients for a number of temperatures by the Arrhenius dependence, we found that the activation energy of the diffusion process in the V90W10 alloy leaves 1.08±0.04 eV for vanadium atoms and 1.12±0.04 eV for tungsten atoms. The calculated activation energies in the V50W50 alloy were 1.2±0.2 eV for vanadium diffusion and 1.3±0.1 for tungsten diffusion. Thus, within the margin of error, the effective activation energy of vanadium diffusion in the V90W10 alloy corresponds to that of the V50W50 alloy with an equiatomic composition. A lower value of the activation energy of the tungsten diffusion in the V90W10 alloy is due to the diffusion of tungsten atoms along the plate surfaces that are released in the V90W10 solid solution (see Figure 2).

Figure 3 shows the calculated ratio of vanadium and tungsten diffusion coefficients as a function of temperature. As can be seen from figure 3, the diffusive mobility of vanadium atoms significantly exceeds that of tungsten in the V90W10 alloy. At the same time, this ratio of the diffusion coefficients in V50W50 is an order of magnitude less.

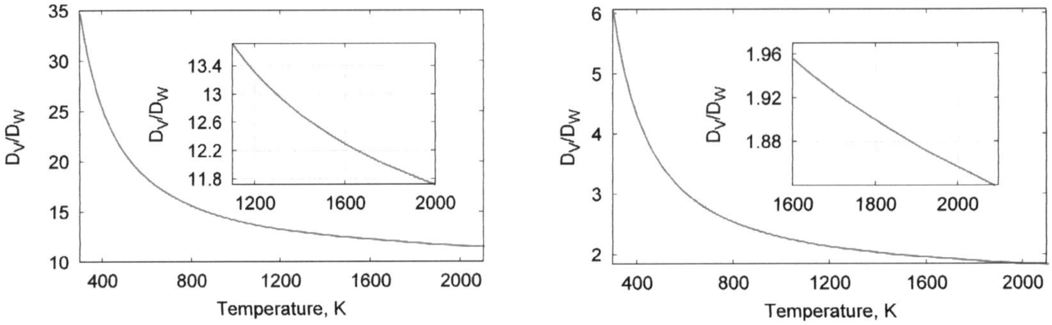

Figure 3. Calculated ratio of vanadium diffusion coefficient to tungsten diffusion coefficient in V90W10 (left) and V50W50 (right) as a function of temperature.

4. Conclusion

The MD+MC method makes a significant contribution to theoretical predictions of the results of diffusion processes in solid solutions and can be used for the study of multicomponent systems, including high-entropy alloys. A decrease in the specific Gibbs energy in the MD+MC simulation from the initial state of a random solution was found, which justifies the reliability of the MD+MC simulation results in achieving thermodynamic equilibrium of alloys at a given temperature and pressure. It is shown that in the considered V-W alloys, ordering is associated with the selection of areas enriched with tungsten. The structure of these regions is characterized by the predominant formation of plates located in the (100) BCC lattice planes. This indicates the possibility of dispersed hardening of vanadium alloys due to alloying with tungsten. The results of MD simulation of diffusion showed that there is a tendency to a faster diffusion of vanadium atoms in comparison with tungsten atoms. With equalization of concentration and an increase in temperature, the difference in the diffusion coefficients of V-W alloy components decreases.

Acknowledgments
The work was supported by RFBR grant No.18-02-00585.

References

[1] Glienke Met al2020 *Acta Materialia* **195** 304
[2] Yang Zet al 2020 *Calphad* **70** 101805
[3] Mehrer H 2007 *Springer Science & Business Media***1 55**
[4] Widom M et al 2014 *Metallurgical and Materials Transactions* **45** 196
[5] Lipnitskii A G and SavelievV N 2016 *Computational Materials Science* **121** 67
[6] Lipnitskii A G et al 2019 *AIP Conference Proceedings* **2167** 020197
[7] Maksimenko V N et al 2019*AIP Conference Proceedings* **216 7**020213

Development of n-body interatomic potentials for calculating the thermodynamic characteristics of V-Nb-Mo-W alloys

V N Maksimenko*, A G Lipnitskii

Belgorod State National Research University, Belgorod, Russia

*Corresponding author: maksimenko.vya.nik@gmail.com

Abstract. New interatomic potentials for Mo and binary systems V-Nb, V-Mo, Nb-Mo, Nb-W and Mo-W are constructed taking into account the angular dependencies of interatomic interactions. Together with the potentials constructed earlier in the same approach, a complete set of potentials for modeling alloys of the V-Nb-Mo-W system at any component concentrations is obtained. A tendency to narrow the range of values of the calculated enthalpy of formation and the volume of formation in equiatomic alloys of the V-Nb-Mo-W system with an increase in the number of components was found. Deviations from the Vegard's rule in the volume of formation of these alloys are shifted to the negative range of values when the number of components is more than two.

1. Introduction

The enthalpy of formation of a solid solution is one of the key values that determine the thermodynamic stability of high-entropy alloys, along with the configuration entropy (recent review [1]). To date, the experimental values of the enthalpy of formation have been determined for many binary systems, including alloys of refractory BCC metals (database [2] and references therein). However, there is still a challenge in establishing the enthalpy of formation of multicomponent alloys. The problem is related to the complexity of experimental work in this direction and the increase in CALPHAD errors with an increase in the number of components due to neglect of the details of the atomic structure. The missing information can be obtained by atomistic modeling. However, the use of such modeling for high-entropy alloys remains limited due to the insufficient development of potentials for multicomponent systems. One of the problems is the need to take into account the angular dependencies when the chemical bond is significantly covalent, which limits the use of a number of well-tested methods for constructing potentials, in which angular dependencies are not taken into account.

In this paper, we develop the V-Nb-Mo-W potentials in the framework of the recently developed n-body approach to constructing interatomic potentials for systems with metallic and covalent characters of chemical bonds, which correctly takes into account angular dependencies [3-5]. The reliability of the theoretical predictions of thermodynamic characteristics in this approach is achieved by including experimental and CALPHAD data on the formation enthalpy and volume of solid solutions at high homological temperatures in the construction of potentials [5]. The advantage of the approach [3-5] is its focus on modeling multicomponent systems at any concentration of components. For this, it is sufficient to construct interatomic potentials only for each pair of components of the system, since the potential functions in this approach depend only on one or two types of atoms [5].

Using the developed potentials, we further calculate the enthalpy of formation and the volume for a number of equiatomic alloys of the V-Nb-Mo-W system to determine the features of the dependence of these thermodynamic characteristics on the number of components in the alloys on the example of the considered system.

2. Methods

The potential energy E_{pot} of the atomic system of N particles depends on the interatomic distances and the angles between bonds according to the following equation [3,5]

$$E_{pot} = \sum_{i<j}^{N} \Phi_{ji}(R_{ji}) + \sum_{i}^{N} \sum_{k<j\neq i}^{N} \sum_{p,q}^{n_3} g_i^{pq}(\cos\theta_{jik})\, f_{ji}^{p}(R_{ji}) f_{ki}^{q}(R_{ki}) + \sum_{i}^{N} F_i(\bar{\rho}_i), \quad (1)$$

where

$$\bar{\rho}_i = \sum_{j\neq i} \rho_{ji}(R_{ji}) \quad (2)$$

and the indices i and j refer to atoms. The discussion of the terms in this expression, potential functions, and the method for optimizing potential parameters is presented in articles [3-5]. To construct interatomic potentials V-Nb-Mo-W, you must specify potential functions $\Phi_{ji}(R_{ji})$, $g_i^{pq}(\cos\theta_{jik})$, $f_{ji}^{p}(R_{ji})$, $F_i(\bar{\rho}_i)$, and $\rho_{ji}(R_{ji})$ for pure elements and cross-potential functions $\Phi_{ji}(R_{ji})$, $f_{ji}^{p}(R_{ji})$, and $\rho_{ji}(R_{ji})$ for each pair of elements in the V-Nb-Mo-W system. We used the potentials of the monatomic systems V, Nb, Mo, W and the binary system V-W [5] that we constructed earlier. At the same time, we modified the potential Mo for a more accurate reproduction of the experimental equation of state and thermal expansion of the BCC lattice of Mo. In this paper, we have constructed potentials for binary systems V-Nb, V-Mo, Nb-Mo, Nb-W and Mo-W. The enthalpy of formation and the lattice parameters of the alloys were used as target values, as implemented in [5]. As a result, a complete set of potentials was obtained for modeling the V-Nb-Mo-W system at any component concentrations.

The enthalpies of formation of binary systems were calculated by molecular dynamics simulations in the framework of the supercell model containing 2000 atoms in the computational cell at a constant number of particles *N*, pressure $P=0$, and temperature *T* (*NPT* ensemble). We use the periodic boundary conditions, the velocity Verlet algorithm for the solution of the equations of motion, the Nose-Hoover thermostat, and the Berendsen barostat. The temperature was chosen to be equal to the temperatures for which the enthalpies of formation of the considered alloys were determined in known experimental studies or calculations using the CALPHAD method. In the case of studying the dependence of the enthalpy of formation and the volume of equiatomic alloys on the number of components, the method of molecular statics at 0 K was used.

3. Results and discussions

Figure 1 shows the results of our calculations of the equation of state for Mo and the enthalpies of formation for V-Nb, V-Mo, Nb-Mo, Nb-W, and Mo-W alloys as functions of the component concentrations. As can be seen from Figure 1, the new potential reproduces well the experimental equation of state for Mo, and the constructed potentials predict the enthalpies of formation of the considered binary systems in good agreement with the experimental and CALPHAD data at high homological temperatures.

Figure 2 shows the results of our calculations of the thermal expansion of Mo and the dependence of the lattice parameters of V-Nb, V-Mo, Nb-Mo, Nb-W and Mo-W alloys on the component concentrations.

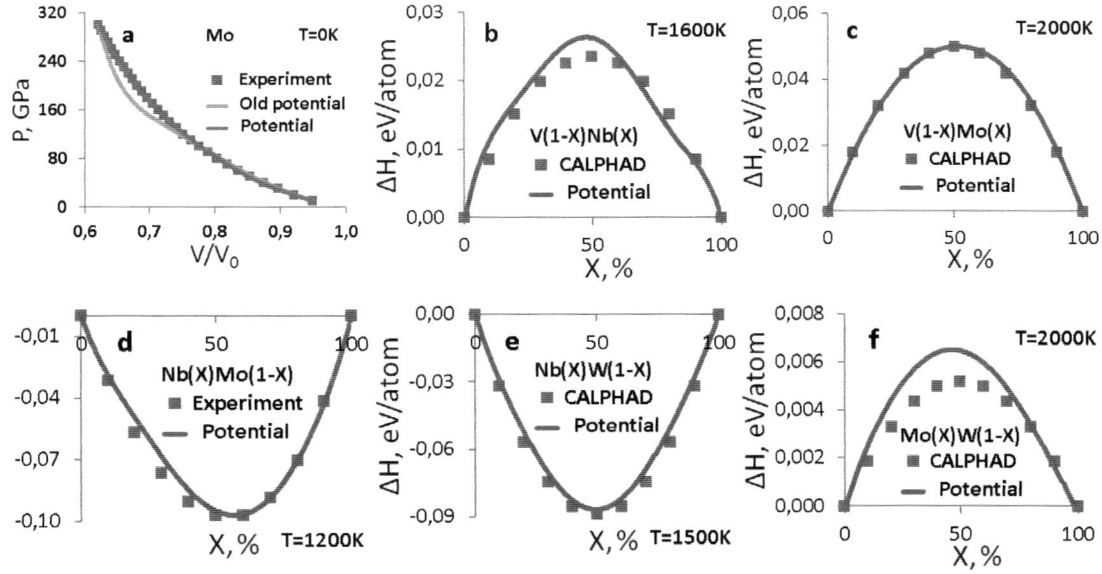

Figure 1. The results of our calculations of the equation of state for Mo in comparison with experimental data [6] and the formation enthalpies of V-Nb, V-Mo, Nb-Mo, Nb-W and Mo-W alloys as functions of the component concentrations.

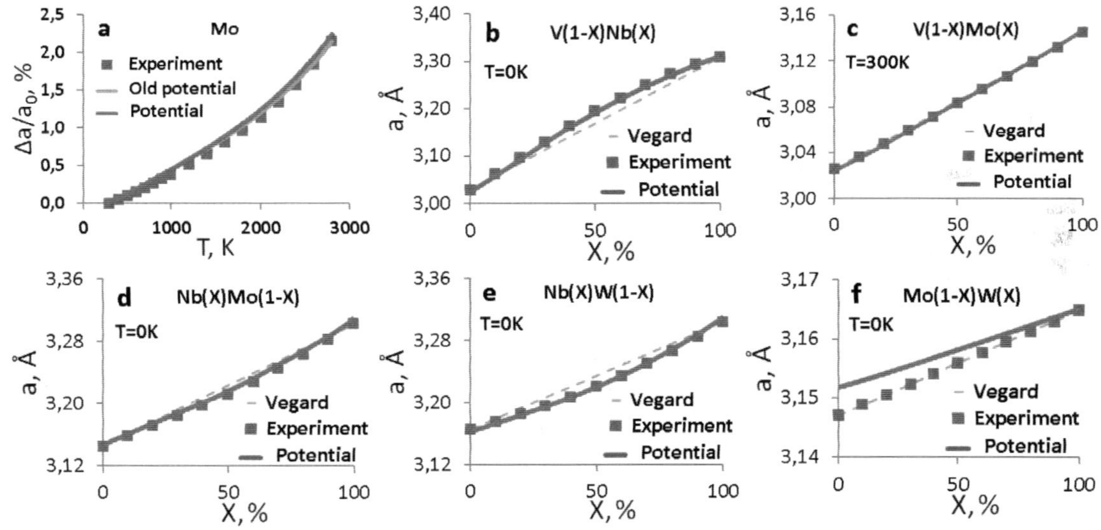

Figure 2. The results of our calculations of the thermal expansion of Mo in comparison with experimental data [7] and the parameters of the BCC lattice of V-Nb, V-Mo, Nb-Mo, Nb-W and Mo-W alloys as functions of the component concentrations.

Figure 2 shows a good agreement between the calculated values of the lattice BCC parameters and their known experimental values for all the systems considered. Deviations from the linear dependence of the lattice parameters on the concentration (from the Vegard's rule) are reproduced in detail. In the case of the Mo-W alloy, the potentials predict the exact execution of the Vegard's rule in accordance with the experimental results, which, however, showed a slight difference (of the order of 0.005 Å) between the calculated equilibrium parameter of the Mo lattice and the experimental value. Thus, the interatomic potentials of V-Nb-Mo-W predict thermodynamic characteristics, the enthalpy of formation and the volume, in good agreement with experimental data and CALPHAD calculations for each binary alloy of the V-Nb-Mo-W system. This justifies the use of the potentials constructed by us for the prediction of thermodynamic characteristics of other alloys of the V-Nb-Mo-W system.

Figure 3 shows the results of our calculations of the volume of formation and the enthalpy for 2 -, 3 -, and 4-component equiatomic alloys of the V-Nb-Mo-W system. As can be seen from Figure 3, there is a slight narrowing of the area of the volume and enthalpy values with an increase in the number of components in the considered alloys. The volume deviation from the Vegard's rule has only negative values for alloys with more than two components.

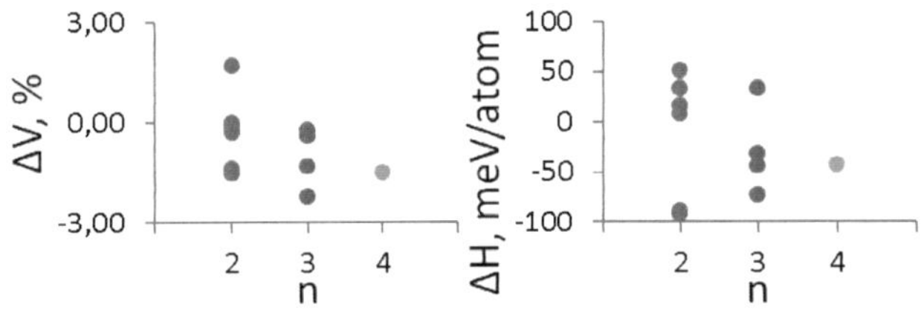

Figure 3. Calculated values of the deviation ΔV of the volume of formation from the Vegard's rule (left) and the enthalpy of formation ΔH (right) for n-component equiatomic alloys of the V-Nb-Mo-W system.

4. Conclusion

In this paper, interatomic potentials for Mo, V-Nb, V-Mo, Nb-Mo, Nb-W and Mo-W are constructed taking into account angular dependencies in interatomic interactions. The potentials reproduce the values of the enthalpy of formation and the lattice parameters of the considered alloys in good agreement with the known experimental data and the results of CALPHAD calculations. A complete set of potentials for modeling the V-Nb-Mo-W system at any component concentrations was obtained. Using these potentials, it is shown that the regions of the volume of formation and the enthalpy of equiatomic alloys of the V-Nb-Mo-W system become narrow with increasing number of components in these alloys. The volumes of formation of V-Nb-Mo-W alloys are characterized by negative deviations from the Vegard's rule for the number of components more than two. The developed new potentials can be used to study alloys of the V-Nb-Mo-W system in a wide temperature range using atomic simulation methods.

Acknowledgments
The work was supported by the Ministry of Education and Science of the Russian Federation under Project No.18-02-00585.

References

[1] Chen J et al 2018 *Journal of Alloys and Compounds* **760** 5
[2] The SGTE Solution Database (SGSOL) www.sgte.net
[3] Lipnitskii A G and Saveliev V N 2016 *Computational Materials Science* **121** 67
[4] Lipnitskii A G et al 2019 *AIP Conference Proceedings* **2167** 020197
[5] Maksimenko V N et al 2019 *AIP Conference Proceedings* **2167** 020213
[6] Hixson R S and Fritz J N 1992 *Journal of Applied Physics* **71** 1721
[7] Touloukian Y S et al 1970 *The TPRC Data Series* **1**

Modelling of the Rolling Process of Titanium Alloy Tube Billets in Laboratory Conditions on a RSP 14-40 Rolling Mill

K G Martynov[1*], D A Negodin[2], D N Klimenko[3], Yu A Gladkov[4] and S V Zherebtsov[3]

[1] The Federal State Unitary Enterprise "Central Research Institute of Structural Materials "Prometey" named by I.V. Gorynin of National Research Center "Kurchatov Institute", St. Petersburg 191015,

[2] SC Chepetsky Mechanical Plant, Glazov 427622, Russia

[3] Laboratory of Bulk Nanostructured Materials, Belgorod National Research University, Belgorod 308015, Russia

[4] Quantorform, Moscow 115088, Russia

*Corresponding author: 4xes@mail.ru

Abstract The development of screw rolling technology for the production of hot-deformed tubes over Ø250 mm on a SVP-500 rolling mill faces a number of challenges that influence the quality of tubes, such as: the screw trace formed on the external surface of tubes and bending of tubes that makes impossible subsequent manufacturing operations. The following experimental and laboratory research was performed to solve these problems: a number of experimental tube billets with and without mandrels were rolled to various strains on a RSP 14-40 laboratory rolling mill to obtain the best ratio of wall thickness to the external diameter. The temperature field distribution and the strain intensity in the deformation area of tube billets rolling were analyzed. A quality assessment of the influence of the section of the calibrating rolls used in industrial technology on the external surface of hot-deformed tubes was also carried out at the modeling stage. The angle changing options of the calibrating rollers section of the screw rolling mill were analyzed for the subsequent tool improvement.

1 Introduction

According to the technology existing at Chepetsky Mechanical Plant (SC ChMP), tubes are manufactured using hot extrusion [1, 2], however this technology imposes constraints on the diameter and length of tubes. One of the possible approaches to obtain proper tubes might be hot rolling on a SVP-500 screw rolling mill enabling to increase the diameter and the length of tubes. However, trial rolling has shown the formation of a "screw" surface, the removal of which produces a large amount of scrap. In addition, rolling on SVP-500 resulted in bending of the tube billet on a mandrel due to which the removal of the tube from the mandrel becomes almost impossible.

2 Materials and methods

An alpha titanium alloy PT-7M (the nominal composition in wt.%: Ti - (1.8-2.5)Al - (2.0-3.0)Zr) was used as the program material. The alloy was supplied by SC ChMP in the form of forged tube billets 39 mm in diameter with drilled internal holes of various diameters (Figure 1). The billets had a uniform microstructure with a grain size of 3-5 μm in both (longitudinal and transversal) directions (Figure 2). A series of experimental rolling was carried out using a RSP 14-40 laboratory rolling mill.

The program of experimental rolling consisted of a series of trial rolling of the billets to various total strain of 31%, 39%, 45%, 46% and 78%. The first rolling was performed on a mandrel, while the remained rollings were done without the mandrel until the final size was obtained. The optimal ratio of wall thickness to diameter did not require additional fixtures in the actual rolling process.

Conditions of experimental rolling:

The heating temperature of the billets was 920 °C;

The angles between the rolls on RSP 14-40 was 16 – 20°;

The rolls rotation speed was 30 – 50 rpm.

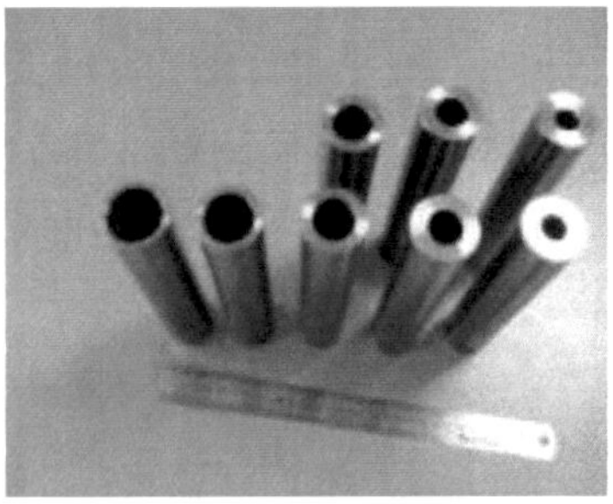

Figure 1. Tube billets for screw rolling.

Figure 2. The initial structure of the tube billet; transversal direction.

Mathematical modeling of the tube billet rolling was performed by the finite element method (FEM) using a QForm VX 8.2.4 software.

3 Results and discussion

3.1 The Evaluation of the Strain Degree Influence on the Structure of Tube Billets made of PT-7M Titanium Alloy

Since the high temperature of strain completely relieves the internal stresses, the rolling temperature of 920 °C can be considered suitable for use in the technological process.

The microstructure in the central part of the samples after 31% strain was found to be recrystallized without notable metallographic texture. However in areas close to surfaces (both internal and external) an unrecrystallized structure was observed. After 39% strain (Figure 3), the microstructure was almost completely recrystallized; preferential structure orientation was not observed similarly to 31% strain.

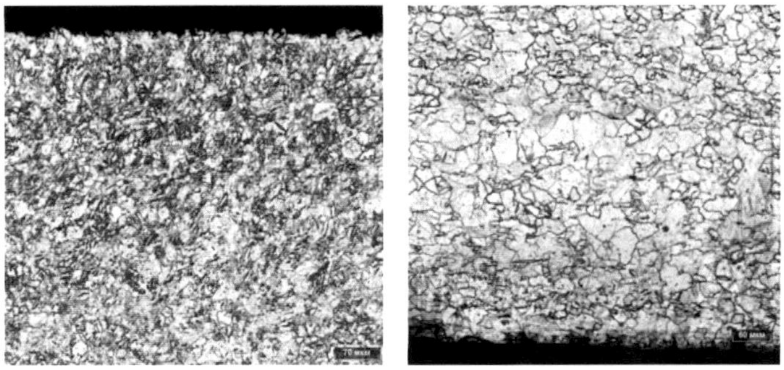

a b

Figure 3 Tube billet structure after 39% deformation (×200); (a) - transversal direction, (b) - longitudinal direction

After 45% - to - 78% strain, traces of plastic deformation in near-surface layers both inside and outside were observed. A noticeable grain orientation in the rolling direction was also detected. Recrystallization in the microstructure of these specimens was not complete again.

The obtained data suggest that the recommended total degree of deformation should be 39%. However the deformed samples were not subjected to recrystallization annealing; static recrystallization most probably could improve the microstructure of the rolled tubes considerably. At the same time work-hardened layer in the near-surface layers can be removed by machining, thereby allowing larger strain to attain required sizes of the tubes.

3.2 Computer mathematical modeling of rolling

Three roll designs were compared using computer simulation in order to study how the geometry of rolls is related to the occurrence of a “screw” mark. Figure 4 shows the considered geometries: industrial version (SVP-500 roll, Figure 4a), with a 2.5° changed angle of the sizing section (Figure 4b) and with a 5° changed angle of the sizing section (Figure 4c).

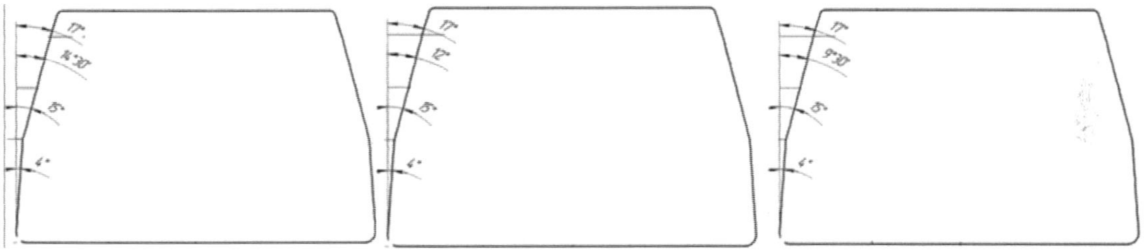

(a) (b) (c)

Figure 4. Roll geometry considered during modeling: (a) industrial version, (b) with a 2.5° changed angle of the sizing section, (c) with a 5° changed angle of the sizing section.

The results of modeling have shown that a change in the SVP-500 roll sizing section results in an increase in the contact mark area on the tube billet. In the industrial version of tool geometry (currently used at SC ChMP), there is no practically reduction in the calibrating section of the roll.

The correction of the computer model with regard to the geometry of the rolls has shown that the metal behaviour during rolling remains unchanged. Figure 5 shows the roll contact point images at the tube billet surface.

(a) (b) (c)

Figure 5. Roll contact point images at the tube billet surface: (a) the sizing section angle is unchanged, (b) the angle change is 2.5° and (c) the angle change is 5°.

It can be seen that for the industrial roll version, no reduction is observed in the sizing section. The 2.5° angle change in the sizing section results in some enlarging the contact point, the reduction occurs on a small area of the sizing section. With increasing the angle change till 5°, a reduction occurs over the entire sizing section range.

The obtained results suggest that the used industrial tool geometry is not perfect and should be significantly improved in order to avoid the appearance of a “screw” mark.

To study the reasons leading to the bending of a tube billet on a mandrel, the processes of heat transfer from the billet to the tool and auxiliary fixtures were investigated using simulation. Figure 6 shows the results of temperature distribution (for better visualization two deforming rolls were hidden).

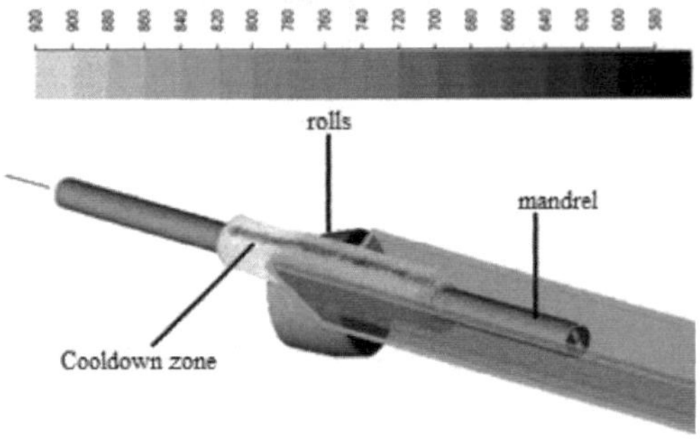

Figure 6. Temperature distribution during screw rolling of tube billets.

It has been found that a prolonged contact of a heated tube billet with cold supply fixtures gives rise to the so-called "cooldown zone" on the tube billet located along its entire length. The minimum temperature in this zone is 750–640°C. The difference in thermal expansion within the bulk of the billet and in the cooldown zone leads to bending of the billet. Deformation-induced heating the stable flow stage of screw rolling has no influence on the cooldown zone and does not lead to heating of the tube billet up to the optimum temperature. The data obtained using computer simulation has shown that reduction of the contact time between the heated tube billet and cold supply fixtures should as much as possible is needed to avoid strong cooldown and the tube billet bennding.

4 Conclusions

Based on the results obtained, the following conclusions were drawn:

i) The total deformation of tube billets to the tube final size on the SVP-500 mill is recommended to be 31 - 39%; the total deformation should be achieved in several passes (3–4 passes).

ii) In order to avoid bending of PT-7M titanium alloy tube billets during screw rolling on the SVP-500 mill, it is required to reduce the time of transferring the tube billet from the furnace to the mill, preventing cooldown.

iii) Before rolling on the SVP-500 mill, the billet should have a well-deformed structure both in the transverse and longitudinal directions.

iv) The temperature of tube billets at the deformation start on the rolls should not be more than 920°C (–20°C).

v) The simulation has shown that the geometry of industrial rolls used in the SVP-500 mill does not allow one to obtain the high-quality outside surface of PT-7M titanium alloy tubes > Ø250 mm. A 5° reduction in the roll sizing section angle is recommended.

References

[1] Leonov V P Mastering and peculiarities of titanium tubes production technology at Russian`s plants. /K G Martynov, L P Rtishcheva, T A Lukianova, M G Shtutsa, U S Karpov // *Titanium* - №4 (50) – 2015 – p. 63.

[2] Leonov V P Development and mastering of production of cold-deformed tubes from titanium alloys at SC ChMP / D A Negodin, D O Hlobystov, A E Moskalev, I A Dubovitskaya, L P Rtishcheva, K G Martynov// *Titanium* - № 4 (78) - 2018 - p.77.

Microstructural aspects of friction stir welding

S Mironov

Laboratory of Mechanical Properties of Nanoscale Materials and Superalloys, Belgorod National Research University, Pobeda 85, Belgorod 308015, Russia

Corresponding author: mironov@bsu.edu.ru

Abstract. Microstructural aspects of friction-stir welding were briefly reviewed. Particular emphasis was given to material flow, crystallographic texture, grain structure development, precipitation phenomena, phase transformations, and thermal stability

1. Introduction

Friction-stir welding (FSW) is an innovative joining technique [1, 2]. Technically, it involves a plunging of a specially designed welding tool, which rotates with a high spindle rate, between two butted sheets and its subsequent traversing along the joint line (Figure 1). The frictional heat and plastic deformation, induced by the rotating tool, rises the local temperature of the welded material so it can be plastically deformed at relatively low stress. The hot material is forced to flow around the rotating tool to fill the cavity at its rear and thus forming a joint in a solid state.

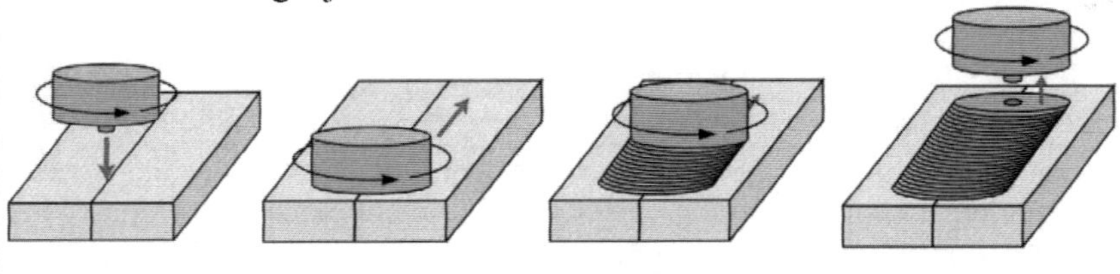

Figure 1. A schematic of FSW process

Due to the solid-state nature of the welding process, FSW avoids solidification defects inherent to conventional fusion techniques and thus routinely produce high-quality welds even in materials which are traditionally considered to be "unweldable", e.g. aluminum alloys. Accordingly, FSW has considerable industrial potential and widely used in transportation industry.

It is important to emphasize that the welded material undergone very large strains at high strain rate and elevated temperature. Material performance at such extreme deformation conditions is not studied well. Hence, FSW represents also a substantial academic interest, and significant research effects have been undertaken during the last two decades to explore the microstructural behavior of friction-stir welded materials. The present work attempts to provide a brief but comprehensive overview of the current state-of-the-art in this area.

2. Material flow and crystallographic texture

Material flow is one of the key issues in FSW. Due to the specific character of FSW, the welded material normally experiences a combination of rotation- and translation motions (Figure 1). Additionally, the material may also undergo a vertical flow along the tool height. Thus, the material actually follows a complex helical trajectory. On the other hand, the rotation component is well accepted to predominate and therefore the FSW straining mode is close to the simple shear [3].

Considering a specific mushroom-shaped design of the welding tool, the character of the material flow (as well as a degree of the imposed strain and even the welding temperature) is principally depended on the depth location. Specifically, the material flow at the upper surface of the welded sheets is governed by the tool shoulder whereas that in the sheets interior is driven by the tool probe (Figure 2).

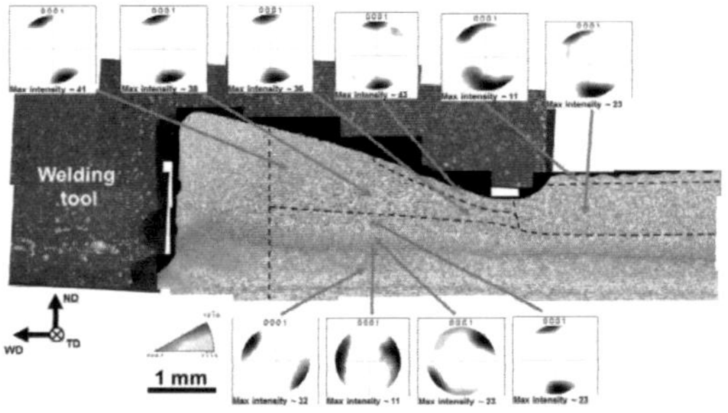

Figure 2. EBSD orientation map showing inhomogeneous texture distribution in thickness direction of friction-stir welded AZ31 magnesium alloy. WD, ND and TD are welding direction, normal direction and transverse direction, respectively. After Suhuddin et al. [9].

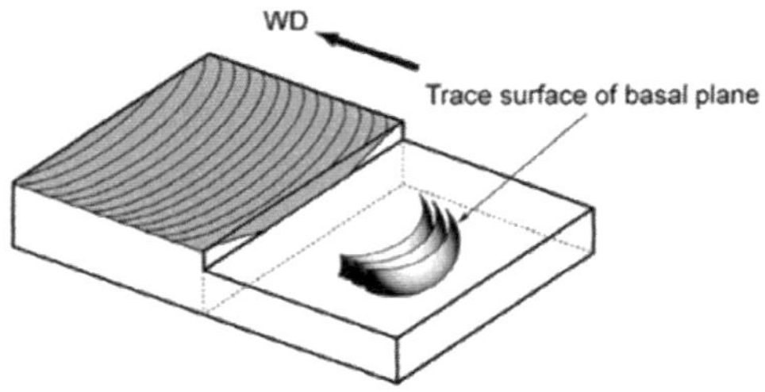

Figure 3. A schematic showing spatial distribution of shear planes within stir zone. WD is welding direction. After Park et al. [12].

FSW typically results in a distinct texture which is typically close to the typical simple-shear orientations [3]. In some cases, however, a formation of recrystallization textures has been reported [4]. In face-centered cubic (FCC) metals, the produced texture was found to be sensitive to the stacking fault energy (SFE). Specifically, a reduction in SFE leads to a transition from $B/\bar{B}\{112\}<110>$ [e.g. 5, 6] to $A/\bar{A}\{111\}<110>$ orientations [e.g. 7]. In body-centered cubic (BCC) crystals, $D_2(11\bar{2})[111]$ texture forms [8]. In hexagonal metals, the evolved texture depends on the c/a ratio. Specifically, magnesium alloys are characterized by a predominance of the *B*-fiber {0001}<uvtw> texture associated with the basal slip [e.g. 9]. In alpha-titanium, on the other hand, the texture is typically dominated by the $P_1\{1\bar{1}00\}<11\bar{2}0>$ attributable to the prism slip [10]. It is also worth noting that the FSW-induced texture in the wrought magnesium alloys is typically very sharp with the peak intensity sometimes achieving ~100 times random [11].

It is important to emphasize that the rotation of the welding tool during FSW gives rise to a variable orientation of the shear direction and the shear plane across the weld zone, shown in the schematic in Figure 3 Accordingly, the texture distribution in the friction-stirred materials is normally not uniform [e.g. 12].

3. Grain structure development

The FSW-induced recrystallization mechanism has been long time one of the research hotspots. The extensive research has demonstrated that grain structure development during FSW is a relatively complex phenomenon which normally includes several stages and is often driven by several mechanisms. These typically include geometrical effect of strain, grain subdivision and discontinuous recrystallization (Figure 4 & 5) but, sometimes, an activation of mechanical- or annealing twinning as well as the grain convergence is also possible [5-11, 13, 14].

The mutual contribution of these of these mechanisms depends primarily on crystal structure and stacking fault energy (SFE) of a particular material but may be also influenced by the temperature of the FSW process. Specifically, the grain structure development during FSW of cubic metals with high SFE is usually governed by the subdivision mechanism [e.g. 5, 6, 8]. In FSW metals with low SFE, in contrast, the key microstructural mechanism is the discontinuous recrystallization [e.g. 7]. In metals with intermediate SFE, e.g. in pure copper, a transition between these two mechanisms has been observed depending on the welding temperature [15]. In hexagonal metals, the microstructural evolution was found to be substantially influenced by crystallographic texture. Specifically, a development of a very sharp texture in stir zone may promote a partial grain convergence [9, 14].

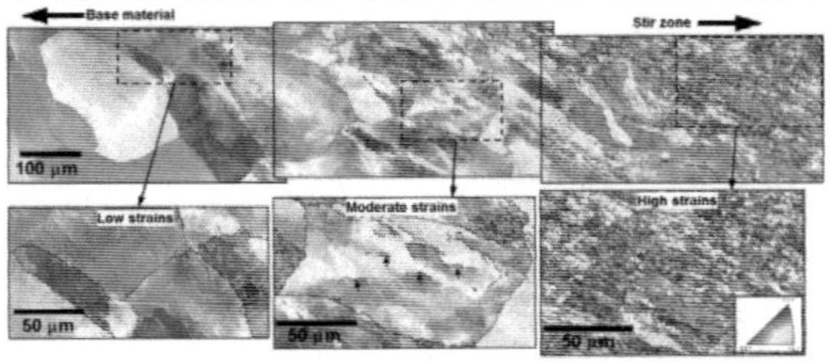

Figure 4. EBSD orientation map showing grain subdivision during FSW of pure iron. Arrows exemplify the transformations of low-angle boundaries to high-angle boundaries. After Mironov et al. [8]

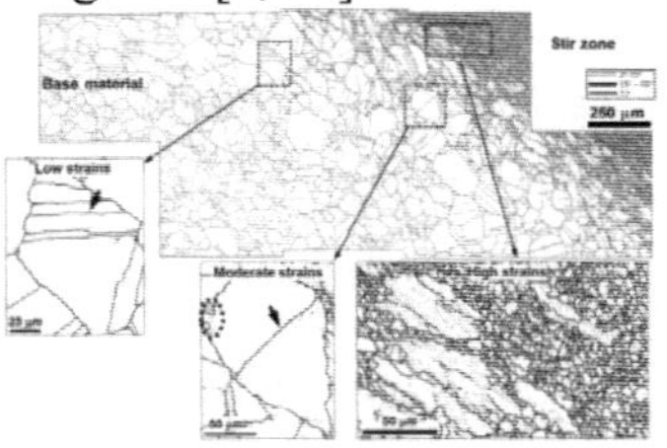

Figure 5. EBSD grain-boundary map showing discontinuous recrystallization during FSW of superaustenitic stainless steel. Arrows show prior-annealing-twin boundaries. Circle indicate grain-boundary bulging. After Mironov et al. [7]

4. Precipitation phenomena

Since commercial alloys often contain second-phase particles, the investigation of precipitation phenomena occurring during FSW attracts significant interest. The extensive research over the last two decades has conclusively demonstrated that a significant temperature gradient inherent to the FSW process gives rise to the complex particle behavior [2, 16]. In the microstructural regions exposed at relatively low temperatures, an essential coarsening of the constituent dispersoids is usually found; this normally leads to the loss of their coherent relationship with matrix. In the microstructural areas influenced by the relatively high heat input, the precipitates often dissolve. Depending on the cooling rate, the dissolved dispersoids may partially re-precipitate back (sometimes, in a form of solute clusters) during weld cooling cycle. Since the welded material normally contains high density of crystal defects (dislocations, subgrain- and grain boundaries), the re-precipitation may occur in a heterogeneous manner.

5. Phase transformations

During FSW of Ti-6Al-4V alloy and carbon steels, material flow may occur in the high-temperature phase filed and the final weld structure develops via the phase transformations occurring during weld cooling cycle. It has been found that these are essentially influenced by the large deformation experienced by the high-temperature phase [17, 18]. Specifically, a significant grain refinement as was as a formation of a distinct simple-shear texture in this phase may result in a pronounced variant selection occurring during the subsequent phase transformation. This, in turn, may give rise to the transformation temperature in the low-temperature phase. Moreover, a formation of a developed substructure in the high-temperature phase leads to local deviations from a characteristic orientation relationships between the phases.

6. Thermal stability

A characteristic of friction-stirred materials is a relatively low stability against abnormal grain growth [19]. This process involves a catastrophic growth of few grains which eventually consume the entire weld zone (Figure 6) and thus may essentially degrade weld properties. This undesirable process has been observed in various alloys thus presumably being a more or less common property of the FSWed materials. It is often reported that the abnormal grain growth often initiates from peripheral regions of stir zone, typically from its upper surface or the weld root, as shown in Figure 6. The origin of this phenomenon is still unclear, however.

Acknowledgment

The author acknowledges the Russian Science Foundation (grant # 19-49-02001) for financial support.

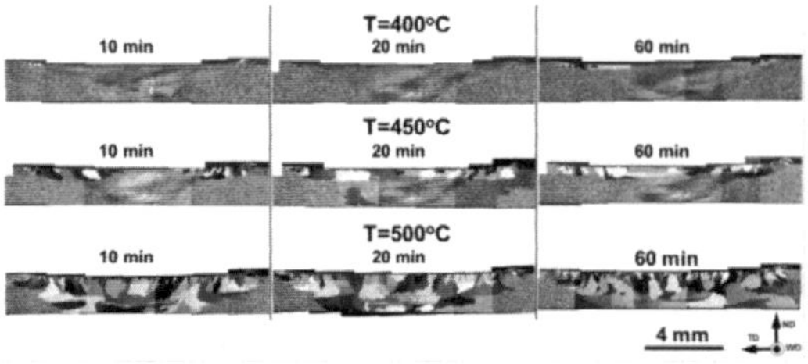

Figure 6. Optical micrographs illustrating abnormal grain growth in friction-stir welded aluminum. WD, ND and TD are welding direction, normal direction and transverse direction, respectively. After Mironov et al. [20]

References

[1] Mishra R S and Ma Z Y, *Friction stir welding and processing*, Mater. Sci. Eng. R 50 (2005) 1-78.
[2] Threadgill P L, Leonard A J, Shercliff H R and Withers P J, *Friction stir welding of aluminium alloys*, Int. Mater. Rev. 54 (2009) 49-93.
[3] Fonda R W and Knipling K E, *Texture development in friction stir welds*, Sci. Tech. Weld. Join. 16 (2011) 288-294
[4] Suhuddin U F H R, Mironov S, Sato Y S, Kokawa H, *Grain structure and texture evolution during friction stir welding of thin 6016 aluminum alloy sheets*, Mater. Sci. Eng. A 527 (2010) 1962-1969.
[5] Prangnell P B, Heason C P, *Grain structure formation during friction stir welding observed by the "stop action technique"*, Acta Materialia 53 (2005) 3179–3192.
[6] Fonda R W, Bingert J F, Colligan K J, *Development of grain structure during friction stir welding*, Scripta Mater. 51 (2004) 243–248.
[7] Mironov S, Sato Y S, Kokawa H, Inoue H, Tsuge S, *Structural response of superaustenitic stainless steel to friction stir welding*, Acta Mater. 59 (2011) 5472-5481.
[8] Mironov S, Sato Y S, Kokawa H, *Microstructural evolution during friction stir-processing of pure iron*, Acta Mater. 56 (2008) 2602–2614.
[9] Suhuddin U F H R, Mironov S, Sato Y S, Kokawa H, Lee C W, *Grain structure evolution during friction stir welding of AZ31 magnesium alloy*, Acta Mater. 57 (2009) 5406-5418.
[10] Mironov S, Sato Y S, Kokawa H, *Development of grain structure during friction stir welding of pure titanium*, Acta Mater. 57 (2009) 4519-4528.
[11] Mironov S, Yang Q, Takahashi H, Takahashi I, Okamoto K, Sato Y S, Kokawa H, *Specific character of material flow in near-surface layer during friction stir processing of AZ31 magnesium alloy*, Metal. Mater. Trans. A. 41 (2010) 1016-1024.
[12] Park S H C, Sato Y S, Kokawa H, *Basal plane texture and flow pattern in friction stir weld of a magnesium alloy*, Metal. Mater. Trans. A 34 (2003) 987-997.
[13] Mironov S, Inagaki K, Sato Y S, Kokawa H, *Development of grain structure during friction-stir welding of Cu–30Zn brass,* Phil. Mag. 94 (2014) 3137-3147
[14] Mironov S, Onuma T, Sato Y S, Kokawa H, *Microstructure evolution during friction-stir welding of AZ31 magnesium alloy*, Acta Mater. 100 (2015) 301-312.
[15] Mironov S, Inagaki K, Sato Y S and Kokawa H, *Microstructural evolution of pure copper during friction-stir welding,* Phil. Mag. 95 (2015) 367–381
[16] Sato Y S, Kokawa H, Enomoto M, Jogan S, Hashimoto T, *Precipitation sequence in friction stir weld of 6063 aluminum during aging*, Metal. Mater. Trans. A 30 (1999) 3125–3130.
[17] Mironov S, Zhang Y, Sato Y S and Kokawa H, Crystallography of transformed β microstructure in friction stir welded Ti-6Al-4V alloy, Scripta Mater. 59 (2008) 511-514.
[18] Hua P, Mironov S, Sato Y S, Kokawa H, Park S H C, Hirano S, *Crystallography of martensite in friction-stir-welded 12Cr heat-resistant steel*, Metall. Mater. Trans. A 50 (2019) 3158-3163
[19] Charita I, Mishra R S, Abnormal grain growth in friction stir processed alloys, Scripta Mater. 58 (2008) 367-371
[20] Mironov S, Masaki K, Sato Y S, Kokawa H, *Relationship between material flow and abnormal grain growth in friction-stir welds*, Scripta Mater. 67 (2012) 983-986.

IOP Conf. Series: Materials Science and Engineering 1014 (2021) 012025 doi:10.1088/1757-899X/1014/1/012025

Investigation of an early stage of the abnormal grain growth in the friction-stir welded 6061-T6 aluminum alloy

I Vysotskiy, S Malopheyev, S Mironov*

Laboratory of Mechanical Properties of Nanoscale Materials and Superalloys, Belgorod National Research University, Pobeda 85, Belgorod 308015, Russia

*Corresponding author: mironov@bsu.edu.ru

Abstract. In this work, quasi in-situ electron backscatter diffraction (EBSD) was applied to study an early stage of the abnormal grain growth occurring in the friction-stir welded 6061 aluminium alloy during post-weld solution heat treatment. It was found that the catastrophic grain growth completed during the heating stage of the annealing treatment, and further static storage at the solutionsizing temperature provided almost no microstructural changes.

1. Introduction

Friction-stir welding (FSW) is often considered as a promising technique for joining of heat-treatable aluminum alloys. Due to the solid-state character of the welding process, FSW avoids numerous solidification defects intrinsic to conventional fusion technologies and thus produces sound joints with excellent mechanical properties [1, 2]. An essential drawback of FSW, however, is the relatively low stability of the welded material against the abnormal grain growth [e.g. 3-6]. This undesirable phenomenon typically occurs during solution heat treatment of welded joints and may essentially degrade their service properties. Currently, this process is usually interpreted in terms of Humphrey's cellular theory [7], being usually attributed to the coarsening- and/or partial dissolution of second-phase particles occurring during FSW [e.g. 4]. In this context, it is worth noting that the constituent precipitates should *completely dissolve* at the solutionizing temperature and therefore, *they cannot exert any influence on the grain growth behavior*.

This apparent controversy may be explained assuming that the abnormal grain growth occurs during the *heating stage* of the solution heat treatment. In this case, the secondary particles still retain in the material and thereby they may affect the microstructural processes. The present work was undertaken in order to check this simple idea. To this end, the early stage of the abnormal grain growth in a friction-stir welded heat-treatable aluminum alloy was studied by using a quasi in-situ electron backscatter diffraction (EBSD) technique.

2. Experimental

The program material used in the present study was a commercial 6061 aluminum alloy. This is a typical heat-treatable alloy which is widely applied in industry and whose FSW behavior is studied relatively well. The material was produced by semi-continuous casting, homogenized at 380°C for 1 hour and then extruded at the same temperature to a cross-area reduction 75%. To provide the peak-hardened condition, the hot extruded material was undergone to the T6 tempering treatment, i.e., solutionized at 550°C for 1 hour, water quenched and then aged at 160°C for 8 hours.

The tempered material was friction-stir welded at the spindle- (rotation) rate of 1100 rpm and the feed rate of 760 mm/min. This particular welding regime was selected on the basis of previous works [8, 9] which showed that this combination of FSW variables promotes excellent mechanical characteristics under both the static- and cyclic loading conditions. The welding tool was fabricated from a tool steel and consisted of a concave-shaped shoulder (with a diameter of 12.5 mm) and an M5 cylindrical probe of 1.9 mm in length. To provide a full-thickness joining of 3-mm-thick welding sheets, double-side FSW was applied in the same welding direction.

Microstructural observations were focused on the transverse cross-section of the welded material. These were conducted by EBSD using a FEI Quanta 600 field-emission-gun scanning electron microscope equipped with TSL OIM software. A suitable surface finish for EBSD was obtained by mechanical polishing in a conventional fashion followed by electro-polishing in a 25% HNO_3 solution in methanol.

To investigate the early stage of the abnormal grain growth, the same welded specimen was placed into a 550°C preheated furnace and sequentially kept at this temperature for 1, 4, and 10 minutes. For comparison, the *heating time* for achieving the annealing temperature (i.e., 550°C) was measured to be ~6 min. Therefore, two first annealing steps represented the heating stage of the solution heat treatment. After each annealing step, the microstructural specimen was immediately quenched in water, and EBSD maps were taken from the same microstructural region within the stir zone without any surface polishing. This provided a quasi in-situ condition for the EBSD experiment. For the aid of comparison, appropriate EBSD maps were also obtained from the as-welded material condition as well as from the fully-annealed sample subjected to the standard solution heat treatment (550°C, 1 hour).

3. Results and Discussion

3.1. *Preliminary analysis*

The typical microstructure found in the stir zone in the *as-welded* material condition is shown in Figure 1a. It is seen, the microstructure was dominated by fine- nearly-equiaxed grains with a mean boundary intercept of ~5 µm. Such microstructures are typically observed in friction-stirred aluminum alloys, and are usually attributed to the continuous recrystallization operating during FSW [e.g. 10-12].

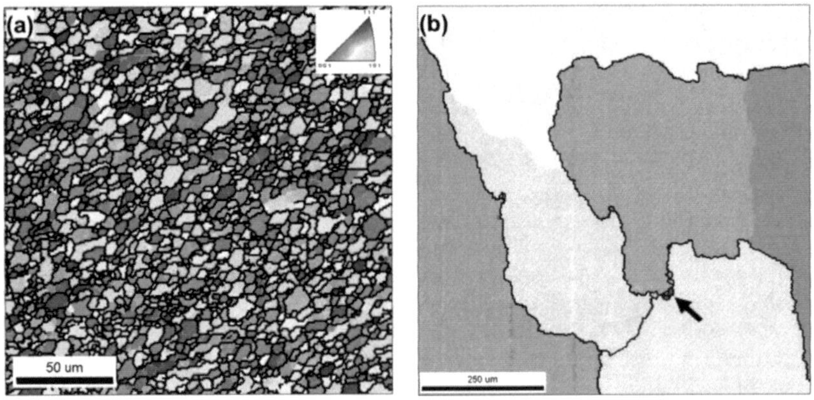

Figure 1. EBSD orientation maps taken from the central section of stir zone (a) in the as-welded material condition and (b) after subsequent solution heat treatment (550°C, 1 hour). In the maps, individual grains are colored according to their crystallographic orientations relative to welding direction; low- and high-angle boundaries are depicted as white and black lines, respectively. In (b), arrow exemplifies the survived remnant of fine-grained structure. Note difference in scales

Subsequent solution heat treatment resulted in the development of mm-scale irregular-shaped grains (Figure 1b). Importantly, the material also contained a minor fraction of the survived fine-grained structure (an example is arrowed in Figure 1b). This microstructural bimodality evidenced the obvious abnormal character of the grain growth, in close agreement with the literature data [3-6].

3.2. *Quasi in-situ EBSD observations*

A series of sequential EBSD maps taken during the quasi in-situ experiment is shown in Figure 2. It is seen that the abnormal grain growth initiated at the early stage of the heating period (Figure 2a) and virtually completed after achieving the solitionizing temperature (Figure 2b). On the other hand, the subsequent static storage at the annealing temperature provided almost no further microstructural changes (compare Figures 2b & 2c). Thus, the present study confirmed the proposed idea on the preferential development of the abnormal grain growth during heating stage of the solution heat treatment.

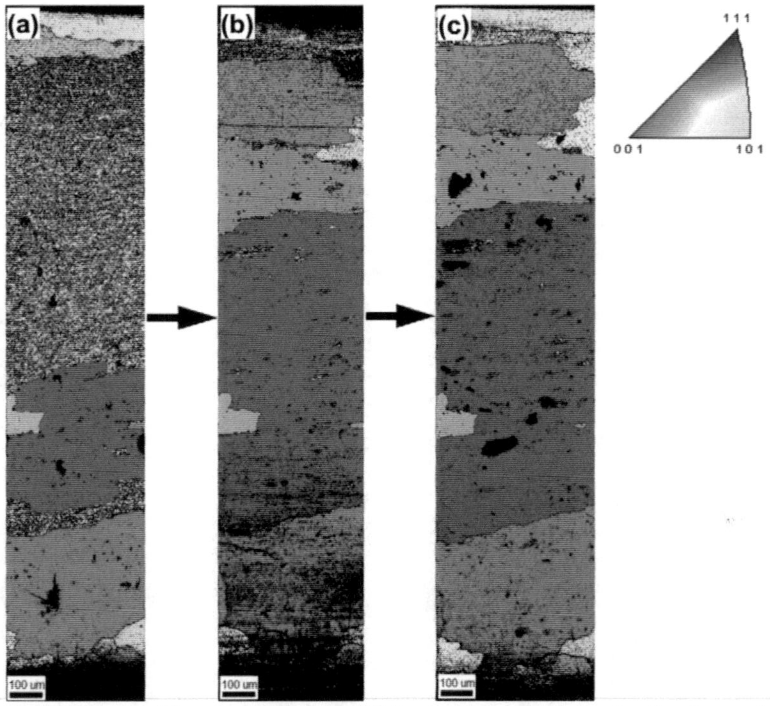

Figure 2. A series of the cross-thickness EBSD orientation maps showing the abnormal grain growth in the stir zone as a function of annealing time at 550°C: (a) 1 min (i.e., the initial step of the heating stage), (b) 1+4 min (i.e., the nearly-completion of the heating stage), and (c) 1+4+10 min (i.e. a static storage at the annealing temperature). In the maps, individual grains are colored according to their crystallographic orientations relative to welding direction. The black pixels are the EBSD data with confidence index below 0.1

This result could be rationalized by considering the characteristic temperatures for the onset of the grain growth and the particle dissolution. In fine-grained aluminum alloys, gross grain boundary migration is typically observed above 250°C [13], i.e. it should normally develop during the heating stage. On the other hand, the constituent β-phase precipitates in the examined 6061 alloy should completely dissolve only at 500°C [14]. Thus, the secondary particles should influence the grain growth and this presumably resulted in the activation of the abnormal mechanism.

It is interesting to note that the abnormal grains nucleate at the peripheral regions of the stir zone, i.e. at its upper surface and near the weld root (Figure 2a). This agrees well with reports in scientific literature [e.g. 15, 16], and is sometimes explained by the very specific microstructure evolved in these areas during FSW [17]. Subsequently, however, the abnormally coarse grains also evolve in the stir zone interior (Figure 2b) thus indicating that the abnormal grain growth was not a surface phenomenon.

4. Conclusion

This work was undertaken in order to investigate an early stage of the abnormal grain growth in the friction-stir welded 6061 aluminum alloy. To this end, a quasi in-situ EBSD technique was applied. It was shown the abnormal grain growth preferentially develops during the heating stage of the solution heat treatment. Moreover, it seems from the experimental results that the abnormal grain growth almost completely ceased after achieving the solutionizing temperature and subsequent static storage provides almost no microstructural changes.

Acknowledgment

The authors acknowledge the Russian Science Foundation (grant # 19-49-02001) for financial support. They also would like to thank the staff of the Joint Research Centre "Technology and Materials" at Belgorod national research University for assistance in experimental works.

References

[1] Mishra R S and Ma Z Y 2005, *Mater. Sci. Eng. R* **50** 1-78.
[2] Threadgill P L, Leonard A J, Shercliff H R and Withers P J 2009 *Int. Mater. Rev.* **54** 49-93.
[3] Charit I, Mishra R S 2008 *Scripta Mater.* **58** 367-371.
[4] Hassan Kh A A, Norman A F, Price D A, Prangnell P B 2003 *Acta Mater.* **51** 1923-1936.
[5] Attallah M M, Salem H G 2005 *Mater. Sci. Eng. A* **391** 51-59.
[6] Ipekoglu G, Erim S, Cam G 2014 *Metall. Mater. Trans. A* **45** 864-877.
[7] Humphreys F J 1997 *Acta Mater.* **45** 5031-5039.
[8] Malopheyev S, Vysotskiy I, Kulitskiy V, Mironov S, Kaibyshev R 2016 *Mater. Sci. Eng. A* **662** 136–143.
[9] Vysotskiy I, Malopheyev S, Rahimi S, Mironov S, Kaibyshev R 2019 *Mater. Sci. Eng. A* **760** 277–286
[10] Prangnell P B, Heason C P 2005 *Acta Materialia* **53** 3179–3192.
[11] Fonda R W, Bingert J F, Colligan K J 2004 *Scripta Mater.* **51** 243–248.
[12] Mironov S, Inagaki K, Sato Y S, and Kokawa H 2015 *Metal. Mater. Trans. A* **46** 783-790.
[13] Humphreys F J, Prangnell P B, Bowen J R, Gholinia A, Harris C 1999 *Phil. Trans. R. Soc. Lond. A* **357** 1663-1681.
[14] Kalinenko K, Kim K, Vysotskiy I, Zuiko I, Malopheyev S, Mironov S, Kaibyshev R 2020 *Mater. Sci. Eng. A* **793** 139858
[15] Mironov S, Masaki K, Sato Y S, Kokawa H 2012 *Scripta. Mater.* **67** 983-986.
[16] Mironov S, Masaki K, Sato Y S, Kokawa H 2013 *Metall. Mater. Trans. A* **44** 1153-1157.
[17] Mironov S, Yang Q, Takahashi H, Takahashi I, Okamoto K, Sato Y S, Kokawa H 2010 *Metal. Mater. Trans. A.* **41** 1016-1024.

IOP Conf. Series: Materials Science and Engineering 1014 (2021) 012026 doi:10.1088/1757-899X/1014/1/012026

On the local variation of mechanical properties in friction-stir welded 6061 aluminum alloy

A Kalinenko, K Kim, I Vysotskiy, S Malopheyev, S Mironov*

Laboratory of Mechanical Properties of Nanoscale Materials and Superalloys, Belgorod National Research University, Pobeda 85, Belgorod 308015, Russia

*Corresponding author: mironov@bsu.edu.ru

Abstract. In this work, digital image correlation technique was applied to investigate local variations of mechanical properties in friction-stir welded 6061 aluminium alloy. A pronounced strain concentration in the bottom section of stir zone was revealed during transverse tensile tests. This effect was suggested to be associated with residual stress generated in this microstructural region during friction-stir welding.

1. Introduction

Friction-stir welding (FSW) is an innovative solid-state joining technique having a very large industrial potential [1, 2]. Due to the specific character of the welding process, FSW is characterized by the sharp gradients of temperature, strain and strain rate. This normally results in drastic variations of microstructure and mechanical properties across the weld zone. So, the typical friction-stir welded joint could be considered as a very heterogeneous material.

So far, the local variations of mechanical properties in such materials were typically estimated by employing the ordinary microhardness measurements. Recently, however, an advanced digital-image-correlation (DIC) technique has been invented for this purpose [3]. The first experience of using of this approach in FSW field showed very encouraging results which enabled a significant progress in understanding of the microstructure-property relationship [e.g. 4-9].

This work is part of the wide-ranging research project intended to explore the microstructural aspects of friction-stir welded of 6061 aluminum alloy. In this particular study, the DIC technique was applied to study the local variations of mechanical characteristics of the produced welded joints.

2. Experimental

The program material used in the present study was a commercial 6061 aluminum alloy with measured chemical composition of 0.88 Mg, 0.66 Si, 0.26 Cu, 0.12 Mn, 0.12 Cr, 0.09 Zn (all in wt.%), and a balance being Al. A cast ingot of the material was homogenized at 380°C for 1 hour and then extruded to 75% of area reduction at the same temperature. To provide the peak-hardened condition, the hot extruded material was undergone to the T6 tempering treatment, i.e., solutionized at 550°C for 1 hour, water quenched and then aged at 160°C for 8 hours.

The tempered material was sliced into 3-mm-thick welding sheets and then friction-stir welded in the wide range of FSW conditions, as described in Ref. [10]. For this particular work, the weld produced at the spindle- (rotation) rate of 500 rpm and a feed rate of 125 mm/min was used. The welding tool was fabricated from a tool steel and consisted of a concave-shaped shoulder (with diameter of 12.5 mm) and an M5 cylindrical probe of 1.9 mm in length. A steel backing plate was used.

Microstructural observations were performed by using optical microscopy. For this purpose, the specimens were cut perpendicular to the welding direction, mechanically polished in a conventional fashion and finally etched with Keller's reagent. To provide a broad view on microstructure distribution within the welded joint, microhardness profiles were measured across the weld zone. Vickers microhardness data were obtained by applying a 200 g load with a dwell time of 10 s by using Wolpert 402MVD microhardness tester.

To provide a more detailed insight into local variation of mechanical properties, the DIC technique was employed. To this end, tensile specimens were machined perpendicular to the welding direction. The specimens were centered relative the weld centerline, had a gauge section of 30 mm in length and 7 mm in width, and included all characteristic microstructural zones generated during FSW. The upper and lower surfaces of the specimens were mechanically polished to achieve a uniform thickness and remove the characteristic kissing bond defect at the weld root inherent to the FSW process. For the DIC measurements, a random ink pattern was applied to the sample surface. The tension tests to failure were conducted at ambient temperature and the nominal strain rate of 10^{-3} s^{-1}. An Instron 5882 universal testing machine equipped with high speed camera was used for the tests. The in-plane Lagrangian strains were measured by using commercial Vic-3D system.

3. Results and Discussion

3.1. *Microstructural observations*

The low-magnification optical image of the cross-section of the produced weld is shown in Figure 1a. A distinct stir zone is seen. A relatively dark optical contrast of this microstructural area suggested significant grain refinement occurred during FSW. A characteristic basin-like shape of the stir zone evidenced a significant influence of the tool shoulder on the material flow in the upper portion of the welded joint. Another remarkable issue was a characteristic kissing bond defect which usually forms during FSW (arrow in Figure 1a). To exclude the possible influence of this defect on tensile behavior, the bottom layer of the tensile specimens was removed by mechanical polishing.

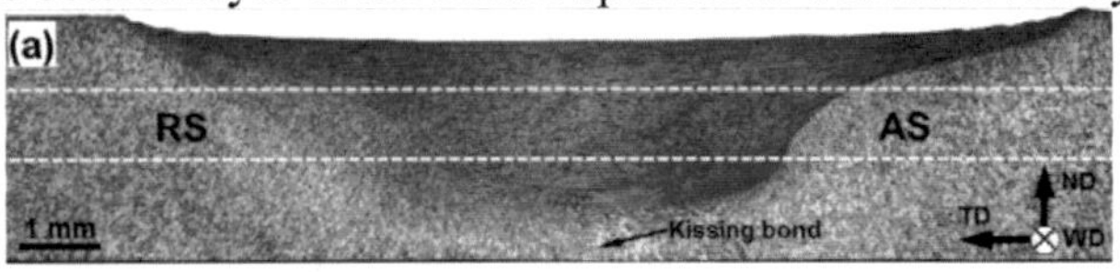

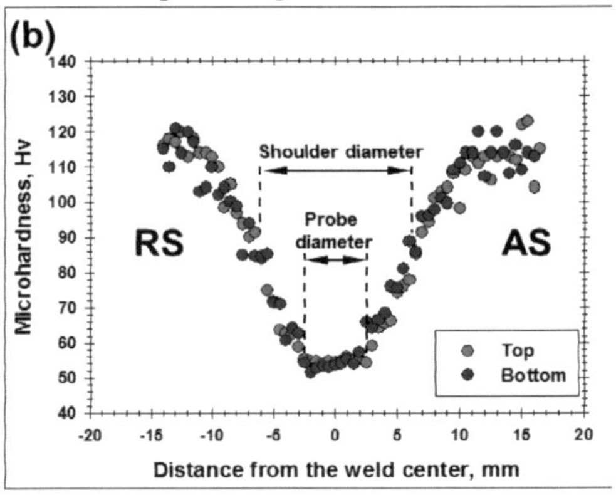

Figure 1. (a) Low-magnification optical image of the weld cross-section and (b) microhardness profiles measured across the dotted lines indicated in (a). For clarity, the probe- and shoulder diameter are indicated in (b). WD, ND and TD are welding direction, normal direction, and transverse direction, respectively; RS and AS are retreating side and advancing side, respectively

3.2. *Microhardness measurements*

With the aim of the preliminary evaluation of the local variation of mechanical properties within the weld zone, microhardness profiles were measured across the welded joint with the results being shown in Figure 1b. To account for the possible variation of the material properties in thickness direction of the weld, the measurements were made in its upper- and bottom sections, as indicated by the dotted lines in Figure 1a. For facilitate interpretation of the microhardness data, the probe- and shoulder diameters were indicated in Figure 1b. To the first approximation, the probe delineates stir zone whereas the shoulder diameter indicates the heat-affected zone.

It is seen that FSW resulted in essential material softening, and this effect was most pronounced in the stir zone (Figure 1b). This phenomenon is typically found during FSW of heat-treatable aluminum alloys being usually attributable to ether coarsening- or dissolution of strengthening precipitates due to the weld thermal cycle [1, 2]. Remarkably, both measured profiles showed broadly similar results (Figure 1b) thus suggesting a relatively homogeneous microstructure distribution within the stir zone.

3.3. Tensile behavior

The deformation diagrams comparing the tensile behavior of the welded joint with that of the base material are shown in Figure 2a. It is seen that the welded material exhibited a substantial degradation in both the strength and ductility characteristics. As follows from Figure 2b, this effect was associated with pronounced strain localization in relatively soft weld zone.

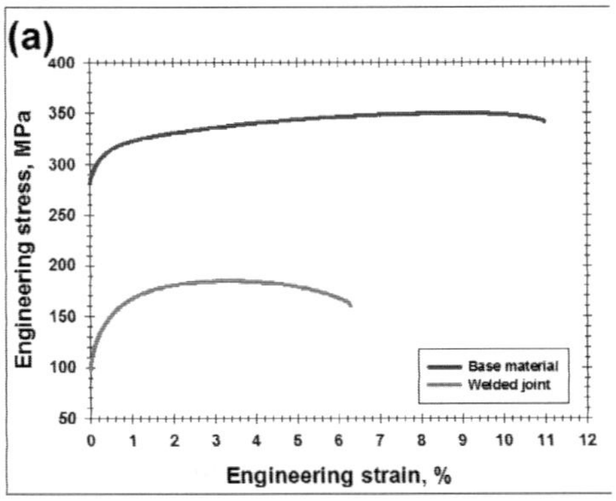

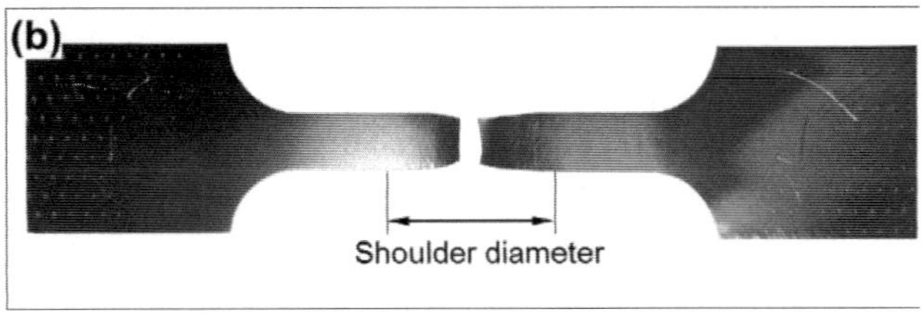

Figure 2. (a) Deformation diagrams comparing tensile behavior of welded joint and base material and (b) appearance of the failed weld specimen. For clarity, the shoulder diameter is indicated in (b)

3.4. Digital-image correlation measurements

The distribution of local longitudinal strains measured on the side surface of the welded specimen as function of global elongation during transverse tensile tests is shown in Figure 2. It seen, that the strain distribution was highly localized being almost completely concentrated in the stir zone, i.e. the softest microstructural region. On the other hand, the base material zone experienced almost no plastic strain.

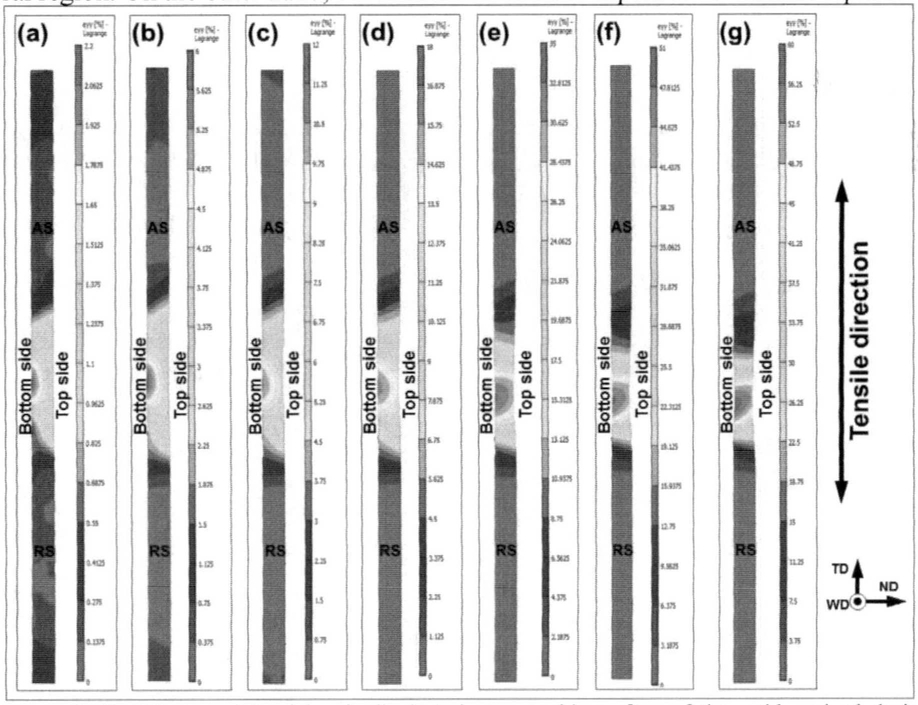

Figure 3. Distribution of local longitudinal strains on a side surface of the weld evolved during transverse tensile test after global elongation of 0,2% (a), 1% (b), 2% (c), 3% (d), 5% (e), 6% (f), and immediately before failure (g). AS and RS abbreviate advancing side and retreating side, respectively

Of particular interest was the observation that the plastic strain was most pronounced in the bottom section of the stir zone, perhaps initiating at the weld root (Figure 2a). With increasing of the global elongation, this local strain spot gradually propagated in the upward direction gradually encompassing the entire stir zone (Figures 2b to 2f). The sample failure also occurred in the stir zone, and likely also initiated from the bottom part of the stir zone (Figure 2g).

Since the "kissing bond" has been carefully polished out prior the tensile tests, it is unlikely that the revealed strain concentration at the weld root was associated with the welding defects. On the other hand, considering the microhardness measurements (Figure 1b), it is hard to believe that the above effect was associated with inhomogeneous microstructure distribution within the stir zone (though this issue still requires an additional check).

It was surmised, therefore, that the preferential strain concentration may be associated with *residual stress*. Due to the local nature of material heating during FSW, the concomitant thermal expansion of the hot material normally gives rise to the *tensile* residual stress in the weld zone [2]. As the welded sheets were place on a steel backing plate during FSW, the weld cooling cycle was presumably governed by the heat transfer into the plate. Accordingly, the highest cooling rate (and thus the largest residual stress) should be presumably generated at the weld root. This should facilitate material flow in this microstructural region during the tensile tests and thus give rise to the strain distribution measured in the present work (Figure 2).

It should be pointed out, however, that the above suggestion is purely speculative and warrants experimental verification.

4. Conclusion

In this work, digital image correlation technique was applied to investigate local variations of mechanical properties in friction-stir welded 6061 aluminum alloy. During transverse tensile tests, a pronounced strain concentration in the bottom section of stir zone was found. This effect was difficult to explain neither in terms of welding defects nor a heterogeneous microstructure distribution. Accordingly, it was suggested to be associated with residual stress generated during FSW.

Acknowledgment

The authors acknowledge the Russian Science Foundation (grant # 19-49-02001) for financial support. They also would like to thank the staff of the Joint Research Centre "Technology and Materials" at Belgorod national research University for assistance in experimental works.

References

[1] Mishra R S, Ma Z Y 2005 *Mater. Sci. Eng. R* **50** 1-78.
[2] Threadgill P L, Leonard A J, Shercliff H R and Withers PJ 2009 *Int. Mater. Rev.* **54** 49-93.
[3] Chu T C, Ranson W F, Sutton M A 1985 *Exp. Mechan.* **25** 232-244.
[4] Genevois C, Deschamps A, Denquin A, Doisneau-Cottignies B 2005 *Acta Mater* **53** 2447-2458
[5] Lockwood W D, Reynolds A P 2003 *Mater. Sci. Eng. A* **339** 35-45.
[6] Genevois C, Deschamps A, Vacher P 2006 *Mater. Sci. Eng. A* **415** 162-170
[7] Sutton M A, Yang B, Reynolds A P, Yan J 2004 *Mater. Sci. Eng. A* **364** 66-74
[8] Mironov S, Onuma T, Sato Y S, Yoneyama S, Kokawa H 2017 *Mater. Sci. Eng. A* **679** 272-281
[9] Mironov S. Sato Y S, Yoneyama S, Kokawa H, Fujii H T, Hirano S 2018 *Mater. Sci. Eng. A* **717** 26-33
[10] Kalinenko A, Kim K, Vysotskiy I, Zuiko I, Malopheyev S, Mironov S, Kaibyshev R 2020 *Mater. Sci. Eng. A* **793** 139858

IOP Conf. Series: Materials Science and Engineering 1014 (2021) 012027 doi:10.1088/1757-899X/1014/1/012027

Effect of tempering temperature on the microstructure and creep resistance of a 10%Cr martensitic steel

R V Mishnev*, N R Dudova, R O Kaibyshev

Belgorod State National Research University, Pobeda, 85, Belgorod, 308015, Russia

*Corresponding author: mishnev91@mail.ru

Abstract. The tempered martensite lath structure of a 10% Cr martensitic steel after normalizing and tempering at 750 °C was studied. The effect of reducing the tempering temperature from 770 by 20 °C on the microstructural parameters in steel was shown. Tempering at a lower temperature of 750 °C provides thinner laths and finer $M_{23}C_6$ particles (55 nm) than tempering at 770 °C. Such structural changes do not affect the creep rupture time at 650 °C and applied stresses of 180 and 160 MPa, whereas the positive effect occurs at 140 MPa, the time to rupture increases by more than two times from 1426 to 3909 hours.

1. Introduction

High-chromium creep-resistant martensitic steels are widely used for critical components of fossil power plants that operate at temperatures up to 620 °C due to their excellent creep resistance [1]. The newest generation of 9–10% Cr steels with 3 wt.% Co addition, an increased B content (≥0.008 wt.%) and a highly reduced N content (<0.003 wt.%) was developed to increase the stability of the tempered martensite lath structure and, therefore, to enhance the creep strength [2,3]. A ten-fold increase in the creep rupture time of these steels as compared with that of the P92-type steels was attained [4]. This fact is associated mainly with improving the coarsening resistance of $M_{23}C_6$ boundary carbides during tempering and creep. Thus, the size of carbides after tempering at 770 °C in this steel is ~70 nm [4]. It can be suggested that steel tempered at a lower temperature due to the smaller $M_{23}C_6$ carbide size could exhibit a higher creep rupture time. The aim of this work is to study the effect of lowering the tempering temperature from 770 to 750 °C on the microstructural and creep characteristics at 650 °C. The results of studying the steel tempered at 750 °C are compared with those for 770 °C published earlier [5].

2. Material and methods

A 10% Cr steel with the chemical composition (in wt.%) 0.1 C, 0.06 Si, 0.1 Mn, 10.0 Cr, 0.17 Ni, 0.7 Mo, 0.05 Nb, 0.2 V, 0.003 N, 0.008 B, 2.0 W, 3.0 Co, 0.002 Ti, 0.006 Cu, 0.01 Al and Fe-balance was examined. The steel was subjected to normalizing at 1060 °C for 30 min and tempering at 750 °C for 3 h. The microstructure was examined using a JEOL JEM-2100 transmission electron microscope (TEM) and a Quanta Nova scanning electron microscope (SEM) equipped with an electron backscatter diffraction (EBSD) pattern analyzer incorporating an orientation imaging microscopy (OIM) system. Flat specimens with a gauge length of 25 mm and cross-sectional dimensions of 7 mm×3 mm were subjected to creep tests until rupture under applied stresses of 140, 160 and 180 MPa at 650 °C.

3. Results and discussion

3.1. Tempered martensite lath structure

The structure of the 10% Cr steel after tempering at 750 °C is characterized as a tempered martensite lath structure with an average size of prior austenite grains (PAG) of about 35 μm as well as after tempering at 770 °C. On the other hand, the lath structural and precipitation parameters of 750 °C tempered steel Figure 1 sufficiently differ from those of 770 °C tempered as can be seen from Table 1. The average width of martensitic laths is smaller and comprises 275 nm (Figure 1a). The dislocation density is slightly higher, comprises about 2.3×10^{14} m^{-2}. Finer $M_{23}C_6$ – type carbides with a mean size of 55 nm are located at the high-angle boundaries (HABs) of PAGs, packets, and blocks and low-angle boundaries (LABs) of laths nearly in the equal proportion. The average size of Nb-enriched M(C,N) carbonitrides is also smaller (20 nm). They are uniformly distributed within the martensitic laths. It should be noted that V-rich M (C,N) particles, as after tempering at 770 °C are not observed [5]. Scarce, fine W-rich M_6C carbides with the same average dimension of 25 nm are found whereas no Laves phase precipitates are observed.

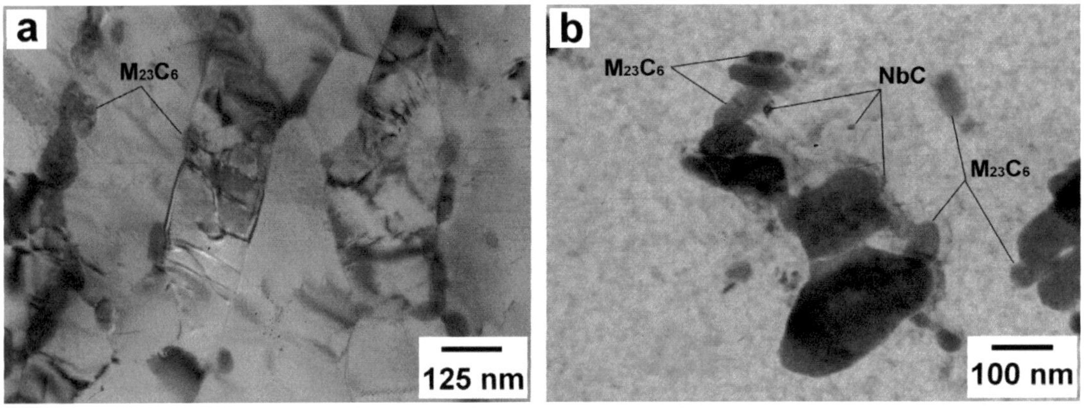

Figure 1. Microstructure of the 10% Cr steel after tempering at 750 °C: TEM micrographs show precipitated phases on foil (a) and replica (b).

Figure 2 demonstrates the data from EBSD analysis. ‘‘Bain circles” appeared on the {001} pole figure (Figure 2d) reconstructed from a separate PAG (Figure 2a), indicating that an austenite→martensite transformation followed the Bain, Kurdjumov – Sachs and Nishiyama – Wassermann orientation relationships [5] took place in the steel. The meso-scale strain distribution indicated by the kernel average misorientation (KAM) map (Figure 1(a)) is non-uniform. Highly-misoriented areas of fine laths alternate with low-misoriented coarse laths within the packets. The KAM value after tempering at 750 °C is ~0.73° that is somewhat higher than after tempering at 770 °C (~0.67°) [5].

Table 1. Effect of tempering temperature on microstructural parameters.

Tempering temperature, °C	Mean size of PAGs, μm	Lath width, nm	Dislocation density, $\times10^{14}$ m^{-2}	Mean size of $M_{23}C_6$, nm	Mean size of Nb(C,N), nm	Mean size of M_6C, nm
750	35±5	275±30	2.3	55±5	20±5	25±5
770	35±5	380±30	1.7	70±7	30±5	25±5

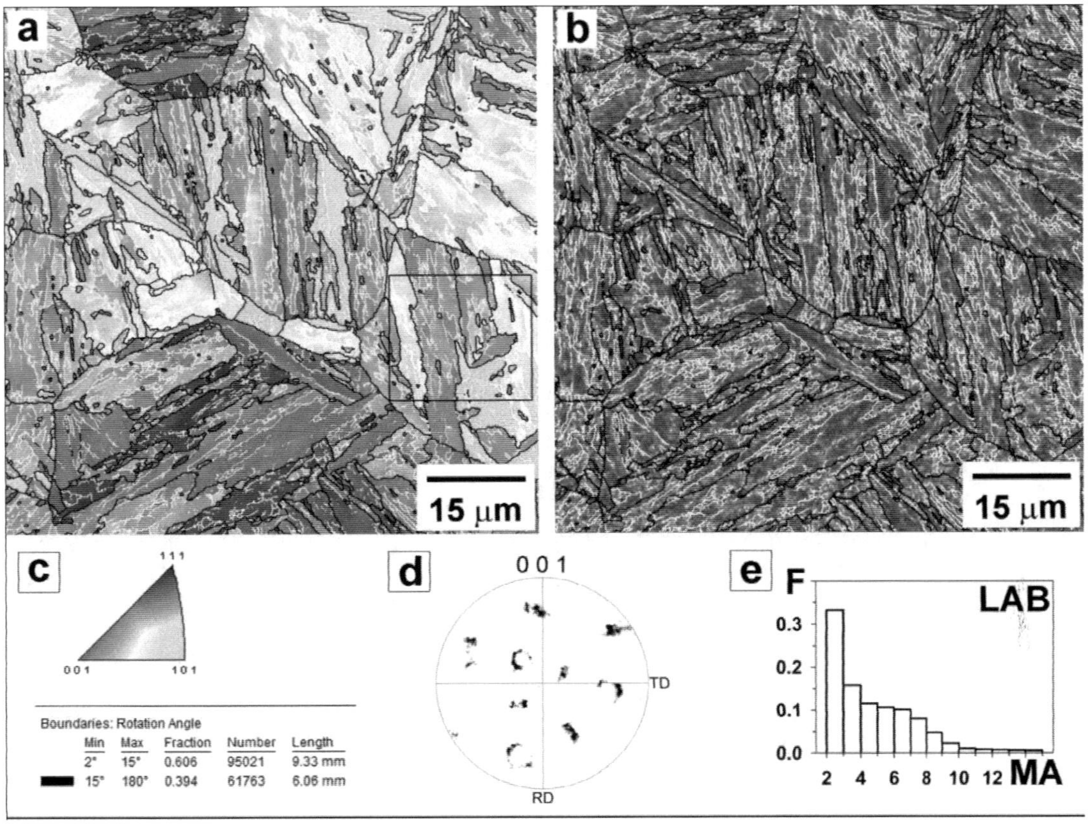

Figure 2. Microstructure of the 10% Cr steel after tempering at 750 °C: OIM image with {001} pole figure and distribution of LAB misorientations (F – fraction, (%), MA – misorientations angle, (degree)) (a,c,d,e); kernel average misorientation map (b).

3.2. Creep behavior

As can be seen from Table 2, structural changes associated with a decrease in the tempering temperature had a positive effect on the creep resistance of the 10% Cr martensitic steel. In the short-term region (180-160 MPa), the time to rupture does not depend on the studied tempering temperatures. This fact can be associated with an insignificant contribution of the structure stability to the creep resistance, while the localization of deformation has a large effect on the creep behavior at high applied stresses [6].

A noticeable positive effect of tempering at 750 °C was revealed at an applied stress of 140 MPa. The time to rupture is about 2.7 times higher after tempering at 750 °C as compared to tempering at 770 °C.

Table 2. Effect of tempering temperature on creep rupture time (h) at 650°C.

Tempering temperature, °C	180 MPa	160 MPa	140 MPa	120 MPa
750	32.95	203.6	3 909	-
770	18.7	211	1 426	39 437

The time and strain dependencies of the creep rate at different stresses of the 10% Cr steel after tempering at 750°C are shown in Figure 3(a) and (b), respectively. The minimum creep rate is reduced by approximately one order of magnitude with a decrease in the applied stress from 180 to 160 MPa. It is interesting to note that a more significant reduction in the minimum creep rate by approximately 2 orders of magnitude occurs when the applied stress decreases from 160 to 140 MPa.

Taking into account that the 10% Cr steel shows a satisfactory impact toughness of 70 J/cm^2 [7] at ambient temperature after tempering at 750 °C, then temperatures of 750…770 °C can be recommended as the feasible tempering interval for this heat-resistant steel, which provides high creep rupture strength during 100,000 h at 650 °C.

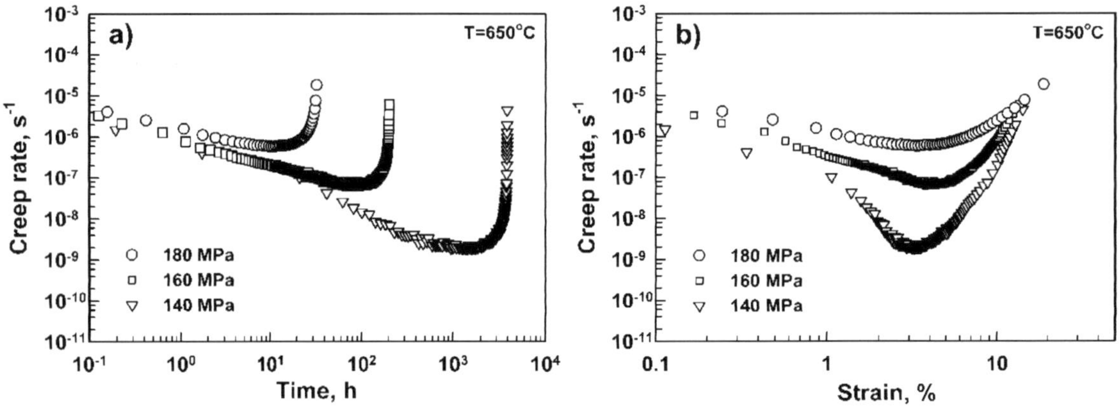

Figure 3. Creep rate vs. time (a) and strain (b) curves at 650 °C and different applied stresses from 140 to 180 MPa of the 10% Cr steel tempered at 750 °C.

4. Summary

A decrease in the tempering temperature from 770 to 750 °C ensures the formation of a tempered martensite lath structure with thinner laths and finer $M_{23}C_6$ carbides with a mean size of 55 nm in the 10% Cr steel with high B and low N contents. Such structural changes do not affect the creep rupture time at 650 °C and applied stresses of 180 and 160 MPa, whereas at 140 MPa the positive effect on increasing creep rupture time by more than two times from 1426 to 3909 hours occurs as compared to steel tempered at 770 °C.

Acknowledgments

The study was financially supported by the Russian Science Foundation, under grant No. 19-79-00195. The authors are grateful to the staff of the Joint Research Center, “Technology and Materials”, Belgorod State University, for providing the equipment for instrumental analysis.

References

[1] Abe F, Kern U T and Viswanathan R 2008 *Creep-Resistant Steels* (Cambridge: Woodhead Publishing)
[2] Semba H and Abe F 2006 *Energy Materials* **1** 238
[3] Tabuchi M, Hongo H and Abe F 2014 *Metall. Mater. Trans. A* **45** 5068
[4] Kaibyshev R, Mishnev R, Fedoseeva A and Dudova N 2016 *Mater. Sci. Forum* **879** 36
[5] Dudova N, Mishnev R and Kaibyshev R 2019 *Mater. Sci. Eng. A* **766** No 138353
[6] Abe F 2015 *Metall. Mater. Trans. A* **46** 5610
[7] Dudova N, Kaibyshev R 2012 *Mater. Sci. Forum* **706-709** 847

Mechanical behavior of titanium alloy Ti-6Al-4V with vacuum-plasma protective coating V+(Ti-V)N at elevated temperatures

Iu M Modina[1*], R R Valiev[1,2], Yu M Dyblenko[1], M K Smyslova[1] and I P Semenova[1]

[1] Institute of Physics of Advanced Materials, Ufa State Aviation Technical University, Ufa, 450008, Russia
[2] Saint Petersburg State University, St. Petersburg State University, 7/9 Universitetskaya nab., St. Petersburg, 199034, Russia

*Corresponding author: modina_yulia@mail.ru

Abstract. This work is devoted to improving the performance properties of the titanium alloy Ti-6Al-4V due to hardening by forming an ultrafine-grained structure (UFG) using severe plastic deformation (SPD) by equal channel angular pressing (ECAP) and then applying a protective coating V+(Ti-V)N on its surface by vacuum-plasma deposition. The effect of the coating on the increase in the strength characteristics of the UFG material is demonstrated in comparison with the CG and UFG state without coating.

1. Introduction

The titanium alloy Ti-6Al-4V is widely used as a structural material in aerospace and aircraft engine building due to its high specific strength and corrosion resistance. However, more and more strict requirements regarding the structural strength are imposed on products and parts of gas turbine engines due to their operation under high static and dynamic loads, temperature, erosion and corrosive effects. One of the most promising approaches to dramatically improve the mechanical and fatigue characteristics of metals and alloys is the formation of an ultrafine-grained (UFG) structure in them by the methods of severe plastic deformation (SPD) [1]. While the use of ionic surface modification by the vacuum plasma deposition of a protective coating provides resistance to wear, erosion and corrosion [2,3].

It is known that the most favorable combination of physico-chemical and mechanical properties is possessed by coatings based on titanium nitrides [2,3]. In our recent works [4, 5], a comprehensive analysis of the Ti+(Ti-V)N protective coating, which was deposited on the surface of the UFG titanium alloy T-6Al-4V, was carried out, where commercially pure Ti was used as the first sublayer. The results of this analysis showed that the use of such a combined approach increases the wear resistance and operational stability of two-phase titanium alloys. However, the use of V as the first sublayer can give an additional increase in erosion resistance, since this element has good viscosity [6]. Therefore, the purpose of this work is to study the mechanical properties of the V+(Ti-V)N vacuum-plasma protective coating at operating temperatures.

2. Material and experimental procedure

The titanium alloy Ti-6Al-4V was used as the material for investigation. The chemical composition of the alloy was (in wt%): 6.2% Al, 4.3% V, 0.02% Zr, 0.039% Si, 0.16% Fe, 0.06% C, 0.168% O, 0.015% N, 0.003% H, Ti - base. To ensure the conditions for plastic deformation of hard-to-deform materials, workpieces with a length of 20 mm were subjected to preliminary heat treatment (HT) by quenching in water at a temperature of 960 °C for 1 h and subsequent annealing at 600 °C for 4 h with air cooling to form a duplex $\alpha+\beta$ structure with a fraction of the primary α-phase no more than 25% [7].

The ultrafine-grained state was obtained by the SPD method of equal-channel angular pressing (ECAP), 4 passes according to the route Bc with an angle of intersection of the channels of 120° at a temperature of 750 °C (e=2.7 with a strain rate of 4 mm s^{-1}).

Prior to the coating deposition, the samples were subjected to electrolytic-plasma polishing (EPP), which is used to polish complex-profile parts from titanium alloys and other structural materials [8]. Then, the vacuum-plasma coating V+(Ti-V)N was deposited on the surface of all samples according to the technique described in [4, 5].

The microstructure of the samples was investigated using scanning and transmission electron microscopy (SEM and TEM, respectively). Investigations of the coating on substrates with different microstructures included the measurement of hardness at a load of 300 g for 15 s on a Struers Duramin instrument; the measurement of the coating thickness was carried out on a Calotest instrument, which made it possible to prepare a spherical cut of the coating to the base material (substrate). Cylindrical specimens with a diameter of 3 mm and a gauge length of 15 mm were used for the tensile tests. The tensile testing was performed at an initial strain rate of 10^{-3} s^{-1} at room and elevated temperatures on an Instron testing machine.

3. Results and their discussion

3.1. Microstructure of Ti-6Al-4V

The microstructure of the coarse-grained titanium alloy Ti-6Al-4V was a duplex structure (Figure 1a) with a primary α-phase size of ~ 7 µm and its volume fraction 65%, as well as a lamellar $\alpha+\beta$ phase.

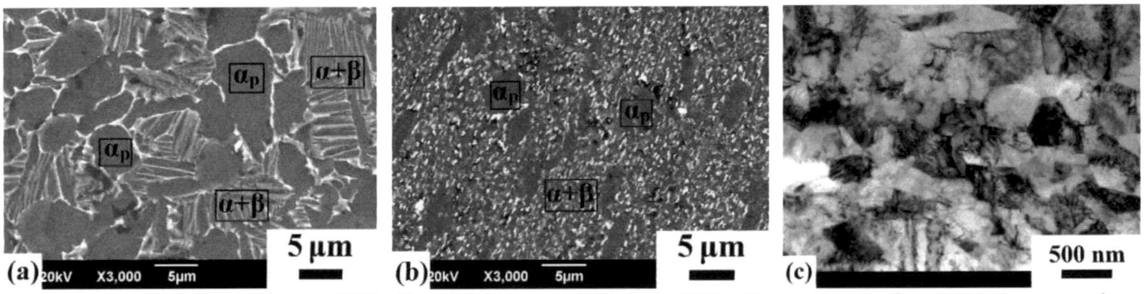

Figure 1. Microstructure of Ti-alloy (a) in-received state, SEM-image; and after ECAP-processing: (b) SEM; (c) TEM.

SPD by ECAP completely transformed the lamellar component of the structure into a globular one (Figure 1b), with a grain / subgrain size of ~ 0.48 µm (Figure 1c).

3.2. Architecture and mechanical behaviour of vacuum-plasma protective coating V+(Ti-V)N

The architecture of the deposited coating with a total thickness of ~ 6 µm (Figure 2) consisted of the first sublayer V deposited on the UFG substrate of the Ti-6Al-4V alloy, the first functional layer (Ti-V)N, the second sublayer (Ti-V), and the second functional layer (Ti-V)N. Sublayers are not visible in the image due to their small thickness ~ 0.2 µm.

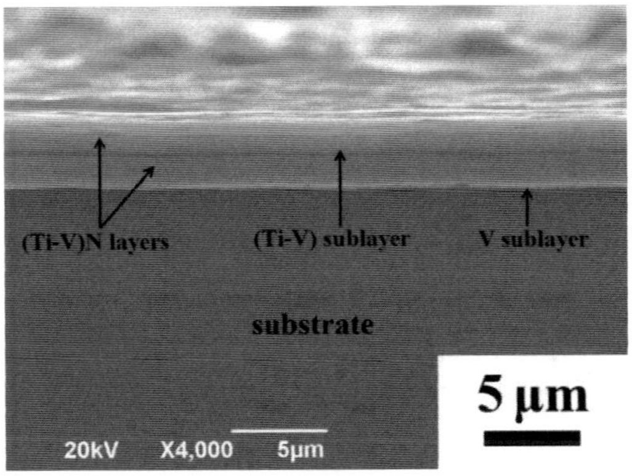

Figure 2. SEM-image of the Ti-alloy with the deposited vacuum-plasma coating V+(Ti-V)N.

Figure 3 shows the results of mechanical tensile tests at room and elevated temperatures, the scale on the left shows the dependence of the ultimate strength (UTS), and on the right, the dependence of the total elongation (δ). It is noted that the formation of the UFG state increases the ultimate strength of the titanium alloy Ti-6Al-4V by 30%, and the deposition of a protective coating gives an additional increase in strength by ~ 5% for UFG specimens in this temperature range. With an increase in temperature, a decrease in UTS is observed for all states, however, the V+(Ti-V)N coating restrains the softening of the UFG alloy in the temperature range of 350-400 °C, the total elongation of which was 25% with the coating and 35% without the coating.

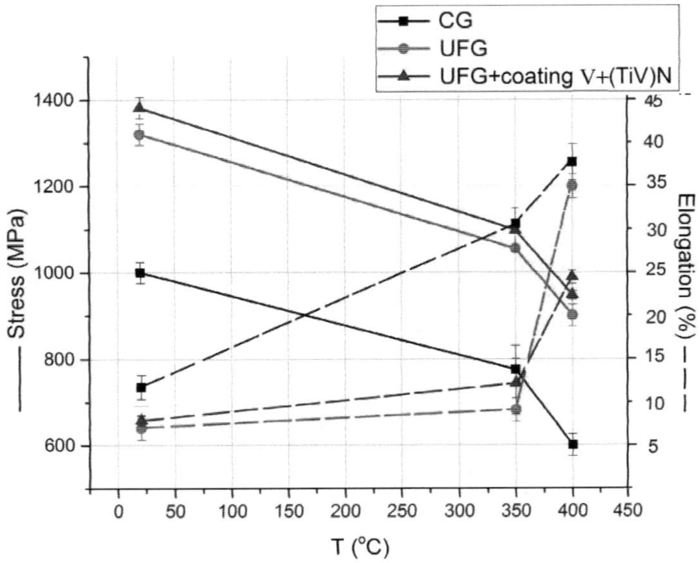

Figure 3. Dependence of UTS and Total elongation on tensile test temperature of Ti-6Al-4V titanium alloy specimens in the CG, UFG states and V+(Ti-V)N coated.

Thus, it was found that the V+(Ti-V)N coating has a positive effect on the strength of the UFG alloy at room temperature and elevated temperatures up to 400 °C. This behavior is typical for vacuum-plasma coatings deposited on titanium alloys [4, 9], which is due to a change in the mechanism of crack initiation in the surface and the creation of a barrier effect in a coating with a higher modulus of elasticity than a substrate [4, 9].

The presented experimental data demonstrate the principal possibility of increasing the strength properties of the Ti-6Al-4V alloy with an ultrafine-grained structure at operating temperatures (up to

400 °C) by deposition of a high-strength and wear-resistant coating of the V+(Ti-V)N system by the plasma-vacuum method.

Acknowledgements

Investigations of the microstructure in section 3.1 were carried out with the financial support of the Ministry of Science and Higher Education of the Russian Federation, project 0838-2020-0006 "Fundamental study of new principles for the creation of promising electromechanical energy converters with characteristics above the world level, with increased efficiency and minimum specific indicators, using new highly efficient electrotechnical materials". Preparation of section 3.2 related to the study of deposited coating became possible through the funding of the Russian Science Foundation, project No. 19-79-10108. Authors are grateful to the personnel of the research and technology Joint Research Center "Nanotech" of the Ufa State Aviation Technical University for their assistance with instrumental analysis.

References

[1] Valiev R Z, Estrin Y, Horita Z, Langdon T G, Zehetbauer M J and Zhu Y T 2015 *Mater Res. Lett.* **4** 1

[2] Styazhkin V A, Kopylov A A and Paleeva S Y 2000 *Protect of Metals* **36(3)** 292

[3] Ahmad Z, Khan A U, Farooq R, Saif T and Mastoi N R 2016 Mechanism of corrosion and erosion resistance of plasma-sprayed nanostructured coatings *High temperature corrosion* ed Z Ahmad (London: InTechOpen) chapter 7 pp 123-146

[4] Semenova I P, Valiev R R, Selivanov K S, Modina Y M, Polyakov A V, Smyslova M K and Valiev R. Z 2017 *Rev on Adv Mater Sci* **48(1)** 62

[5] Valiev R R, Selivanov K S, Modina I M, Dyblenko Y M, Semenova I P and Valiev R Z 2020 *Adv Eng Mater* 2000121

[6] Latellaa B A, Ganb B K, Daviesb K E, McKenzieb D R and McCulloch D G 2006 *Surf and Coatings Technol* **200(11)** 3605

[7] Semenova I P, Saitova L R, Raab G I, Korshunov A I, Zhu Y T, Lowe T C and Valiev R Z 2006 *Mater. Sci. Forum* **503–504** 757

[8] Nestler K, Böttger-Hiller F, Adamitzki W, Glowa G, Zeidler H and Schubert A 2016 *Procedia CIRP* **42(1)** 503

[9] Semenova I P, Selivanov K S, Valiev R R, Modina I M, Smyslova M K, Polyakov A V and Langdon T G 2020 *Adv. Eng. Mater.* 1901219

IOP Conf. Series: Materials Science and Engineering 1014 (2021) 012029 doi:10.1088/1757-899X/1014/1/012029

Microstructure of a low alloyed Cu-Cr-Zr alloy after ECAP-Conform

A Morozova[1,2*], A Lugovskaya[1], A Pilipenko[1], M Tkachev[1], G Raab[3], A Belyakov[1] and R Kaibyshev[1]

[1]Laboratory of Mechanical Properties of Nanostructured Materials and Superalloys, Belgorod National Research University, 308015, Belgorod, Pobeda str. 85, Russia
[2]The KINETICA Engineering Center, National University of Science and Technology MISiS, Moscow, 119049, Moscow, Leninskiy ave 4, Russia
[3]Ufa State Aviation Technical University, 450000 Ufa, K. Marksa st. 12, Russia

[*]Corresponding author: morozova_ai@bsu.edu.ru

Abstract. The microstructure of a Cu-0.1%Cr-0.1%Zr alloy after annealing or aging and equal channel angular pressing with conform (ECAP-C) process was investigated. ECAP-C led to formation of strain induced low-angle boundaries (LAB) which transformed in high-angle boundaries (HAB) during deformation. Deformation microbands developed after 1 ECAP-C pass, leading to new ultrafine grain formation in the microbands upon further processing. Average grain size gradually decreased with strain while dislocation density increased. After 8 ECAP-C passes average grain size achieved 1 μm. Maximum of LAB density occurred after 4 ECAP passes and then LAB density decreased with strain while HAB density gradually increased during ECAP-C. Relationships between dislocation density, grain size, and density of crystallite boundaries were discussed.

1. Introduction.

Cu-Cr-Zr alloys are perspective candidates for electrical application due to combination of high electrical conductivity and strength after appropriate treatment [1-3]. The most investigated and prevalent treatment of Cu-Cr-Zr alloys is severe plastic deformation (SPD) with annealing [4-5]. Ultrafine grained structure with fine particles providing high strength and conductivity can be formed by different deformation methods and heat temperatures. The main disadvantage of SPD technique is production of small samples with volume less 10^{-4} m^{-2}. New SPD technique combining equal channel angular pressing and conform process can help to solve this problem [6]. ECAP-C can produce long rods and wires with unique properties, such as electrodes for point welding, automobile wire, contact wire for high speed trains, etc. [7]. However, there is a deficiency of experimental data about microstructure evolution of Cu-Cr-Zr alloys during ECAP-Conform and effect of microstructure on mechanical properties and electrical conductivity. The aim of the present work is to study the microstructure evolution of a Cu-Cr-Zr alloy during ECAP-Conform.

2. Experiment

A Cu-0.1%Cr-0.1%Zr alloy was chosen as the starting material. The billets were annealed at 920 °C during 1 h and water quenched (ST). Then, a part of the samples was aged at temperature of 550 °C during 4 h (AT). Both types of samples were subjected to ECAP-C at room temperature via route B_C to 1, 2, 4, 8 passes. The intersection angle of matrix channels was 120 °. The microstructures were investigated using a Nova NanoSEM 450 scanning electron microscope with an electron backscattered diffraction (EBSD) analyzer. The average grain and subgrain size were estimated as average values between transversal and longitudinal boundary and subboundary intercept. The dislocation density was determinated as function of the Kernel Average Misorientation [8]. The fraction of ultrafine grains (with a size below 2 micron), the density of crystallite boundaries, average misorientation angle, Kernel average misorientation (for scan step of 200 nm) were estimated using the OIM Software. The specimens for microstructure investigations were polished by 25% HNO_3 and 75% CH_3OH electrolyte at temperature of -20 °C and voltage of 10 V by a Tenupol 5 machine.

3. Results and discussion

Microstructure in initial state

After solution treatment at 920 °C, the Cu-Cr-Zr alloy contained low fraction of course particles that crystallized during cooling of ingot. The average grain size was about 140 μm, and the dislocation density was about 10^{13} m^{-2}. Aging at 450 °C led to precipitation of fine ellipsoid bcc Cr-rich particles. The size of particles in the longitudinal direction and transverse direction was 10-15 nm and 5-7 nm, respectively. Numerical annealing twins were observed .HAB fractions were above 0.98 and the average misorientation angles comprised 45-48° in both initial states. Kernel Average misorientations over a distance of 3 μm were about 0.5°.

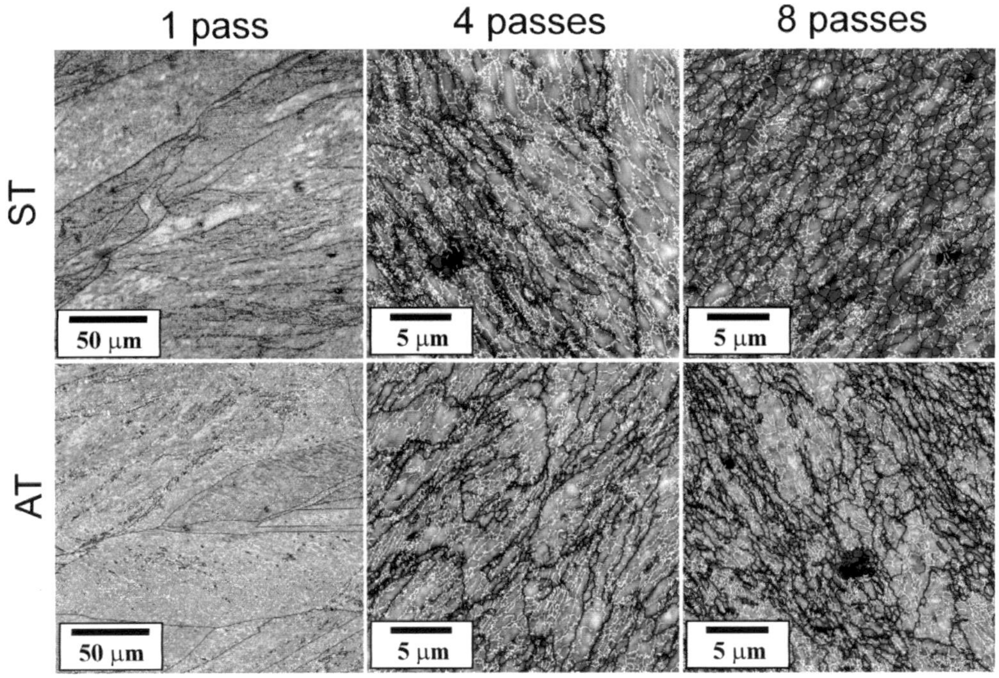

Figure 1. Microstructure evolution during ECAP-C in the Cu-Cr-Zr alloy. Black and white lines represent HAB and LAB, respectively.

Microstructure after ECAP-C

ECAP-C was accompanied by the formation of strain-induced LAB and elongation of initial grains. The deformation microbands consisting of long parallel HABs splitted initial grains. Chains of crystallites with irregular LAB-HAB were observed in AT samples after 1 ECAP-C pass. Further deformation led to increase in misorientations of LABs with subsequent transformation of LABs into HABs. New

IOP Conf. Series: Materials Science and Engineering 1014 (2021) 012029 doi:10.1088/1757-899X/1014/1/012029

ultrafine grains formed predominantly in deformation microbands. Fraction of ultrafine grains increased significantly after 4 ECAP-C passes.8 ECAP-C passes led to formation of equiaxed crystallites with a size about 0.5 μm and high fraction of HAB.

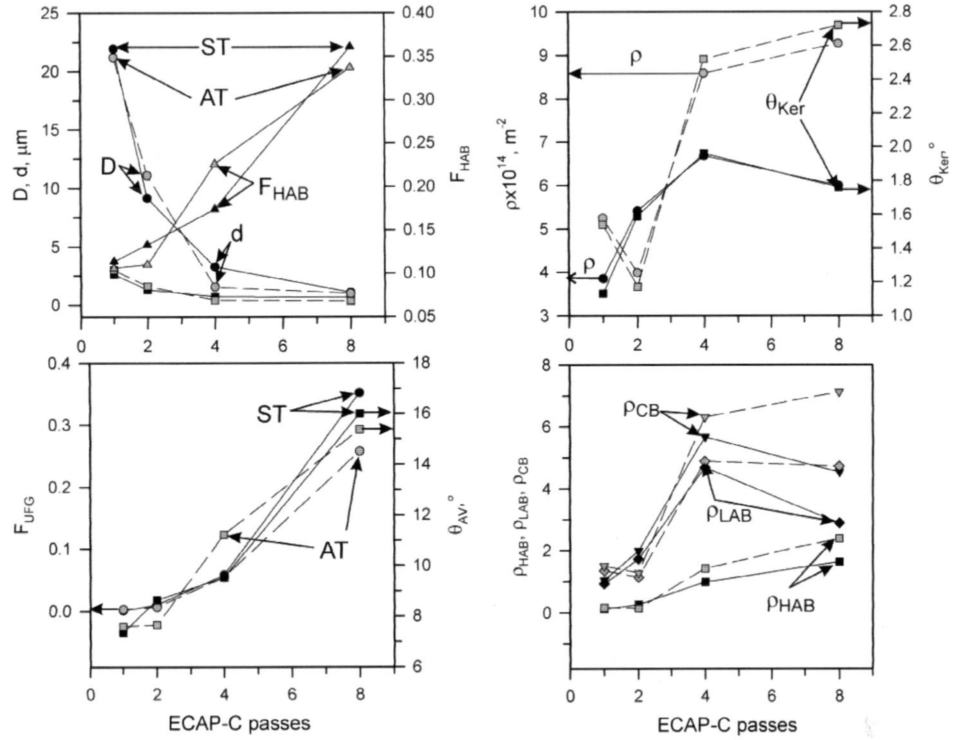

Figure 2. Effect of ECAP-C on grain (D) and subgrain size (d), fraction of HAB (F_{HAB}), dislocation density (ρ), Kernel average misorientation (θ_{Ker}), fraction of ultrafine grains (F_{UFG}), average misorientation angle of grain boundaries (θ_{av}), density of HAB (ρ_{HAB}), LAB (ρ_{LAB}) and all crystallite(ρ_{CB}).

ECAP-C at room temperature led to significant grain refinement along with an increase in dislocation density and microstrain. Gradual transformation of strain-induced LAB into HAB during deformation decreased the mean grain size from 120-140 μm to 1 μm. HAB fraction slowly changed after 1-2 ECAP-C passes followed by significant increase after 4-8 passes. Deformation was accompanied by an increase in HAB fraction to 0.36 and 0.34 in ST and AT condition after 8 ECAP-C passes, respectively. Dislocation densities in ST and AT conditions approached an apparent saturation at 66.5*10^{14} m^{-2} and 9*10^{14} m^{-2} after 4 ECAP passes, respectively. Tendencies of change in average boundary misorientation angle and UFG fraction were the same, i.e., a kind of incubation period during 1-4 passes and a drastic increase after 8 ECAP-C passes. ECAP-C promoted an increase in the boundary density. The density of LAB achieved maximum about 4.7-4.8 μm^{-1} after 4 ECAP-C passes and then decreased to 2.9 μm^{-1} and 4.7 μm^{-1} for ST and AT condition, respectively. The density of HAB in AT samples was higher than in ST condition and gradually increased with deformation that testify to the development of LAB to HAB transformation.

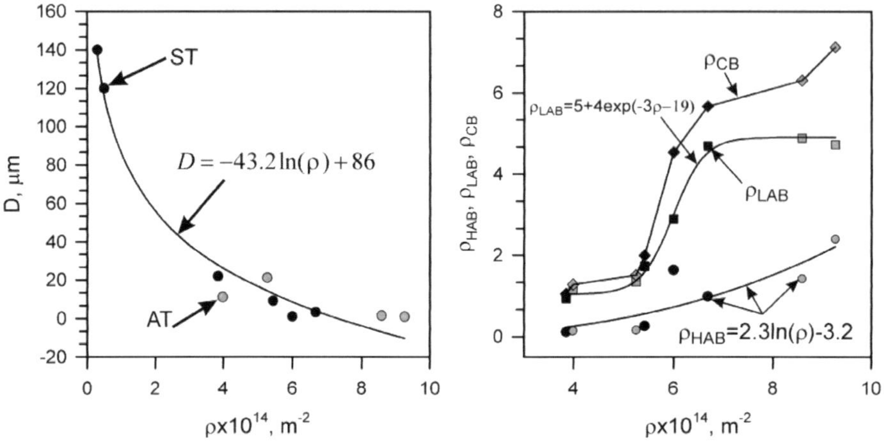

Figure 3. Relationship between grain size (D) and dislocation density (ρ); the density of HAB (ρ_{HAB}), LAB (ρ_{LAB}), and crystallites boundaries (ρ_{CB}) vs the dislocation density.

It is clearly seen that the relationship between the grain size D and the dislocation density ρ can be expressed by logarithmic law:

$$D = -43.2\ln(\rho) + 86 \quad (1)$$

An increase in the dislocation density during ECAP-C led to the formation smaller grains. The dislocation density affected the development of strain-induced boundaries. The change in LAB density ρ_{LAB} can be described by sigmoidal law:

$$\rho_{LAB} = 5 + 4\exp(-3\rho - 19) \quad (2)$$

In contrast, the HAB density (ρ_{HAB}) increased with dislocation density according to logarithmic law:

$$\rho_{HAB} = 2.3\ln(\rho) - 3.2 \quad (3)$$

Equitations (1-3) good described experimental correlation of grain size and grain boundaries density with dislocation density. An increase in dislocation density promoted their rearrangement in strain-induced boundaries, which quickly increased their misorientation, transformed in HAB, leading to new ultrafine grains.

Acknowledges

The financial support received from the Ministry of Science and Education, Russia, under President grant No. 075-15-2020-407 18.03.2020 (MK-68.2020.2) is gratefully acknowledged.

References

[1] Wang Q, Liu D, Tong L, Zhou Y, Wang W, Zhou H and Fan R 2019 *Metals* **9(8)** 824.
[2] Purcek G, Yanar H, Shangina D V, Demirtas M, Bochvar N R and Dobatkin S V 2018 *J. Alloys Compd.* **742** 325.
[3] Borodin E N, Morozova A, Bratov V, Belyakov A and Jivkov A P 2019 *Mater. Charact.* **156** 109849.
[4] Murashkin M Y, Sabirov I, Sauvage X and Valiev R Z 2016 *J. Mater. Sci.* **51(1)** 33.
[5] Morozova A, Mishnev R, Belyakov A and Kaibyshev R 2018 *Rev. Adv. Mater. Sci.* **54(1)** 56.
[6] Raab G J, Valiev R Z, Lowe T C and Zhu Y T 2004 *Mater. Sci. Eng.* A **382(1-2)** 30.
[7] Islamgaliev R K, Sitdikov V D, Nesterov K M and Pankratov D L 2014 *Rev. Adv. Mater. Sci.* **39** 61.
[8] Zhilyaev A P, Shakhova I, Morozova A, Belyakov A and Kaibyshev R 2016 *Mater. Sci. Eng. A* **654** 131.

IOP Conf. Series: Materials Science and Engineering 1014 (2021) 012030 doi:10.1088/1757-899X/1014/1/012030

Effect of deformation techniques on the microstructure and mechanical properties of a copper alloy

A Morozova[1,2*], A Pilipenko[1], M Tkachev[1], A Lugovskaya[1], A Belyakov[1] and R Kaibyshev[1]

[1]Laboratory of Mechanical Properties of Nanostructured Materials and Superalloys, Belgorod National Research University, Pobeda Str. 85, Belgorod, 308015, Russia
[2]The KINETICA Engineering Center, National University of Science and Technology MISiS, Leninskiy Ave. 4, Moscow, 119049, Russia

*Corresponding author: morozova_ai@bsu.edu.ru

Abstract. The microstructure and mechanical properties of a Cu-0.25%Mg alloy subjected to either equal channel angular pressing (ECAP) or combined cold working, including ECAP followed by rolling and then drawing at room temperature, were investigated. ECAP led to the formation of strain-induced boundaries and the development of ultrafine equiaxed grains with an average size of about 0.6 μm after 4 passes. The microstructure after the combined cold working included fibrous grains elongated in the rolling/drawing direction with a sharp texture containing <001> and <111> fibers. The transversal grain size after combined cold working to a total strain of 5.7 achieved 0.43 μm. The yield strength and ultimate tensile strength after ECAP to a total strain of 9.6 were 570 and 600 MPa; whereas those after combined cold working to a total strain of 8.7 were 745 and 780 MPa, respectively. The reason of the difference in mechanical properties was discussed.

1. Introduction

Cu-Mg alloys are widely used materials for electrical application due to their high electrical conductivity and significant work hardening [1-2]. Adaptation of severe plastic deformation (SPD) techniques as appropriate treatment can improve the combination of strength and electrical conductivity [3-5]. SPD techniques, for example, equal channel angular pressing (ECAP), high pressure torsion, multiple forging, etc. lead to an increase in the dislocation density, the formation of new strain-induced grain and subgrain boundaries [6-7]. The grain refinement and high dislocation density caused by SPD should improve strength without reduction of electrical conductivity in Cu-Mg alloys. Therefore, the combination of the SPD technique with common deformation methods such as rolling or drawing could be of practical significance and deserves special scientific investigation. The aim of the present work is to study the microstructure and mechanical properties of a Cu-Mg alloy after ECAP or combined cold working including ECAP followed by cold rolling and cold drawing.

2. Experiment

A Cu-0.25%Mg alloy was investigated. Original ingots were forged at 800 °C to a total strain of 1.4 and rolled at 450 °C to a total strain of 1. Then, a part of the samples was subjected to ECAP at room temperature to 1, 2, 4 and 8 passes via route B_C with an intersection angle of the matrix channels 90°.

The total strain after 1 ECAP pass was about 1.2. The other part of the samples was cold worked by 1 pass of ECAP followed by cold rolling (CR) to a total strain of 1.5 and then cold drawing (CD) to total strains of 1.5-5.8.

Microstructure investigations were carried out using a Nova NanoSEM 450 scanning electron microscope equipped with an electron backscattered diffraction (EBSD) device. The samples for EBSD analysis were electro-polished in 25% HNO_3 and 75% CH_3OH electrolyte at a temperature of -20 °C and a voltage of 10 V using a Tenupol 5 equipment. The average grain and subgrain size, fraction of high-angle boundaries (HAB), ultrafine grained structure (grain size below 2 micron), average misorientation angle, dislocation density, Kernel average misorientation (for scan step of 200 nm) were calculated using OIM Software. The mechanical properties were studied using an Instron 5882 machine with specimens having the following relationship between the gauge length l vs cross-section area F: $l=5.65F^{0.5}$. The initial strain rate was $2x10^{-3}$ c^{-1}.

3. Results and discussion

Microstructure evolution

The original microstructure of the copper alloy after forging at 800 °C and rolling at 450 °C consisted of equiaxed grains with a size of about 3.3 μm and numerous annealing twins. The HAB fraction was about 0.32. 1 pass of ECAP at room temperature led to the formation of new strain induced low-angle boundaries (LAB) and to the elongation of initial grains (Figure 1). Naew ultrafine equiaxed grains appeared, forming chains along initial grain boundaries. Further ECAP resulted in transformation of LAB into HAB and promoted the development of an ultrafine grained microstructure. A homogeneous microstructure with a large HAB fraction and a crystallite size of about 0.5 μm was formed after 4 ECAP passes. The Taylor factor distribution was characterized by two maximums, i.e., relatively small maximum for 2.5 and huge one for 3.3, which did not change significantly with increasing the number of ECAP passes.

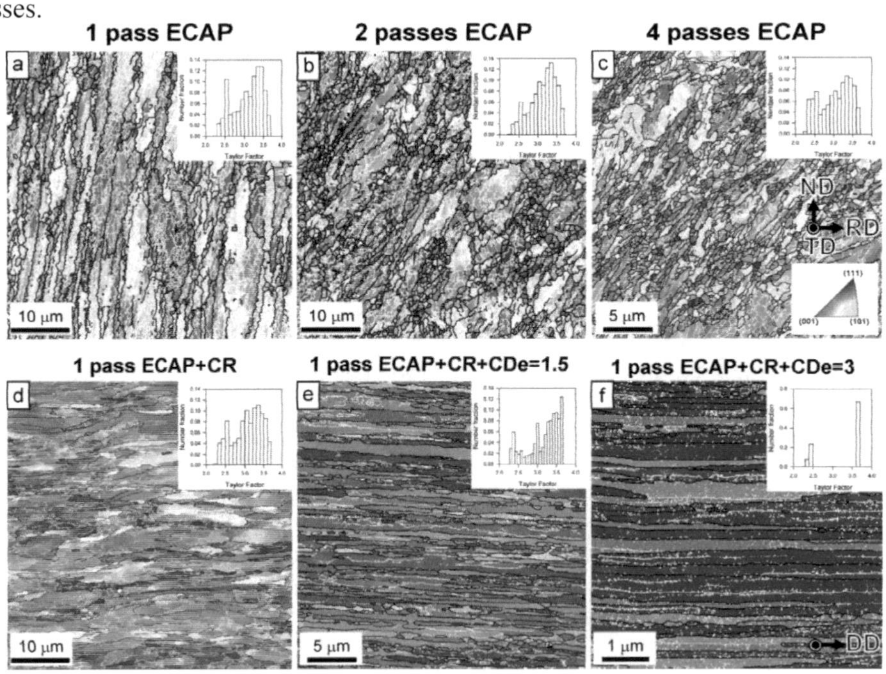

Figure 1. Microstructure evolution during 1 (a), 2 (b) and 4 (c) ECAP passes and combined cold working by 1 pass of ECAP with cold rolling (CR) to total strain of 1.5 (d) and cold drawing (CD) to total strains of 1.5 (e) and 3 (f) of the Cu-Mg alloy. Black and white lines represent HAB and

LAB, respectively. Colors indicate the crystallographic direction along the drawing direction (DD) or transversal direction (TD).

Cold rolling after 1 ECAP pass lengthened the grains along the rolling direction and flattened them in the normal one. Further cold drawing was accompanied by a decrease in the transversal grain size and an increase in the HAB fraction. A sharp crystallographic texture contained <001> and <111> fibers was observed. A fibrous microstructure with a transversal grain size of about 0.3 μm was formed after cold drawing to a strain of 3.0. Distribution of Taylor factor after cold rolling had a huge maximum for 3.2-3.5, which increased with drawing. Bimodal distribution of Taylor factor with maximums at 2.4-2.5 and 3.6 was observed after drawing to a strain of 3.0. Grains with strong fibers of <001> and <111> corresponded to Taylor factors of 2.4-2.5 and 3.6, respectively.

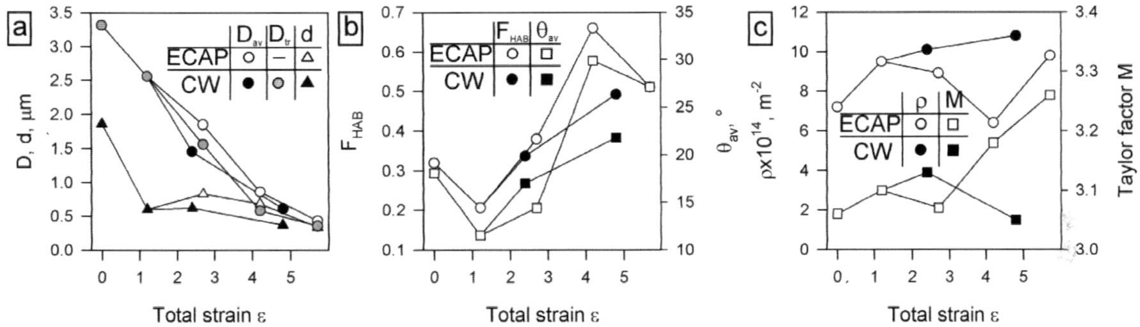

Figure 2. Effect of strain on the average (D_{av}) and transversal (D_{tr}) grain size, subgrain size (d) (a), fraction of HAB (F_{HAB}), average boundary misorientation angle (θ_{av}) (b), dislocation density (ρ) and Taylor factor (M) (c) of the Cu-Mg alloy.

Plastic deformation refined the microstructure (Figure 2). The difference between grain size and subgrain size decreased with strain. The grain and subgrain sizes were almost the same regardless of the deformation methods after total strains of about 4. The average grain size after ECAP (0.6 μm) was larger than after drawing (0.43 μm). Probably this difference was associated with the smaller total strain for ECAP as compared to combined cold working. The HAB fraction decreased after 1 ECAP pass to 0.21 and then increased with ECAP to 0.5-0.66. Drawing promoted a quick increase in the HAB fraction and average boundary misorientation angles. The dislocation density increased with strain during ECAP from $7x10^{14}$ m^{-2} to $11x10^{14}$ m^{-2} and almost unchanged during drawing $(7\text{-}9)x10^{14}$ m^{-2}). The average Taylor factor increased after 1-2 ECAP passes to 3.1 and then decreased to 3.05 as in the initial state after 4 ECAP passes. The combined cold working was accompanied by a gradual increase in Taylor factor with straining that attained 3.26 after a total strain of 5.7.

Mechanical properties after deformation

The microstructure refinement improved substantially the strength properties. After cold deformation the yield strength (YS) tended to approach the ultimate tensile strength (UTS) irrespective of the processing technique. ECAP was accompanied by gradual strengthening of about 20-30 MPa per unit strain, while drawing provided strengthening of about 40-60 MPa for the same strain increment. So, the difference in UTS after a total strain of about 8 for ECAP and drawing was almost 180 MPa. The maximum YS/UTS of ECAPed and drawn samples achieved 570/600 MPa and 745/780 MPa, respectively. Elongation dramatically decreased with strain to 1% and 5 % for drawing and ECAP, respectively.

The difference in strength after ECAP and drawing could be attributed to different deformation microstructures. A fibrous microstructure with small transverse grain size and sharp texture provided high strength. The strength (σ) should correlate with the grain size (D) according to the Hall-Petch relationship [8]:

$$\sigma = \sigma_0 + k_y D^{-0.5} \quad (1)$$

where σ_0 and k_y are constants of material. The texture should lead to strengthening due to the change of the Taylor factor (M) in dislocation strengthening by the Taylor equation [9]:

$$\sigma = \sigma_0 + \alpha M G b \rho^{0.5} \quad (2)$$

where G is Gibson modules, b is Burger vector, ρ is dislocation density. The grain reorientation with <111> along the drawing direction corresponded to an increase in the Taylor factor and was accompanied by drastic strengthening. Figure 3 shows that both grain refinement and texture amplification resulted in high strength of the Cu-Mg alloy after drawing.

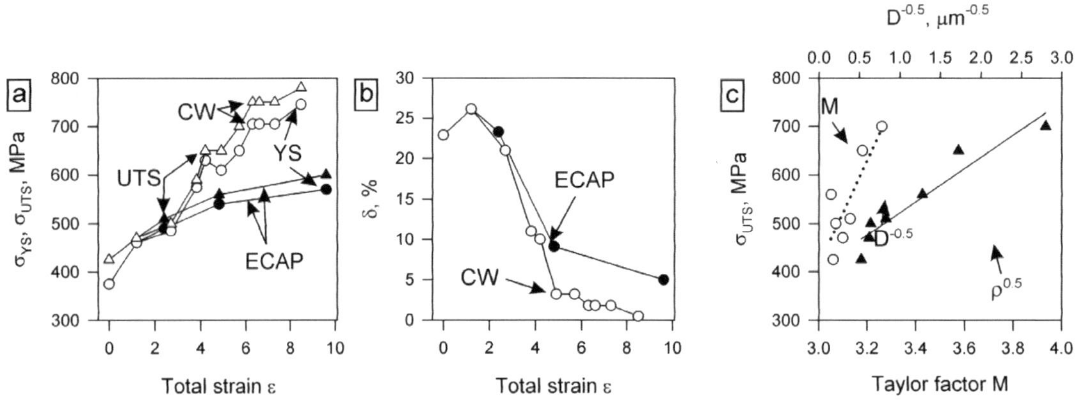

Figure 3. Effect of total strain on the yield strength (σ_{YS}), ultimate tensile strength (σ_{UTS}) (a), and elongation (δ) (b) of the Cu-Mg alloys along with relationships between σ_{UTS} and D or M (c).

Acknowledges

The financial supports received from the Ministry of Science and Higher Education of the Russian Federation, under President grant No. 075-15-2020-407 for microstructure investigation, and from the Russian Science Foundation, Russia, under grant No. 19-79-30025 for mechanical property analysis are gratefully acknowledged. The authors are grateful to the personnel of the Joint Research Center, "Technology and Materials", Belgorod National Research University, for their assistance with instrumental analysis.

References

[1] Gorsse S, Ouvrard B, Goune M and Poulon-Quintin A 2015 *J. Alloys Compd.* **633** 42.
[2] Kim Y, Lee K, Cho Y H, Guo Z, Koo J M and Seok C S 2016 *Int. J. Precis. Eng. Manufac.-Gr.Tech.* **3(4)** 353.
[3] Yun X B, Wei Y O U, Ying Z H A O, Bing L I and Fan Z X 2013 *Trans. Nonfer. Met. Soc. China* **23(4)** 1108.
[4] Morozova A, Mishnev R, Belyakov A and Kaibyshev R 2018 *Rev. Adv. Mater. Sci.* **54(1)** 56.
[5] Zhu C, Ma A, Jiang J, Li X, Song D, Yang D and Chen J 2014 *J. Alloys Compd.* **582** 135.
[6] Borodin E N, Morozova A, Bratov V, Belyakov A and Jivkov A P 2019 *Mater. Charact.* **156** 109849.
[7] Vinogradov A, Patlan V, Suzuki Y, Kitagawa K and Kopylov V I 2002 *Acta Mater.* **50** 1639.
[8] Petch N J 1953 *J. Iron Steel Inst.* **174** 2528.
[9] Taylor G I 1934 *Proc. R. Soc. London Ser. A.* **145(855)** 362.

IOP Conf. Series: Materials Science and Engineering 1014 (2021) 012031 doi:10.1088/1757-899X/1014/1/012031

Laser fusion of aluminum-alumina composites obtained in the vortex layer apparatus

A Nalivaiko[1*], D Ozherelkov[1], A Morozova[1,2], A Shinkaryov[1], A Gromov[1]

[1]MISIS Catalysis Lab, National University of Science and Technology MISIS, Moscow, 119991, Russia
[2]Laboratory of Mechanical Properties of Nanostructured Materials and Superalloys, Belgorod National Research University, Belgorod, 308015, Russia

*Corresponding author: nalivaiko@misis.ru

Abstract. The current study is devoted to laser fusion of aluminum-matrix composites with 3 wt. % of alumina obtained in the vortex layer apparatus. The mixing parameters and the distribution of alumina in the composite structure were studied. Three parameters of laser fusion for plate synthesis were used. The finest results were achieved using a volumetric energy density of 24 J/mm^3 with 370 W laser power. The maximum mechanical properties of the samples synthesized by the optimal parameters were: UTS = 384 ± 5 MPa; E_f = 0.54 ± 0.1 %.

1. Introduction

Metal matrix composites are widely used in aerospace, defense, and automotive industries due to their high strength and excellent wear resistance. Recently, aluminum-based composites have been increasingly used among composites with a metal matrix due to their combination of lightweight and high mechanical characteristics [1]. To date, many scientific groups have investigated a wide range of reinforcements for aluminum-matrix composites, such as SiC, AlN, TiC, TiB_2, and others. However, in comparison with other reinforcing additives, Al_2O_3 particles can improve both the mechanical and high-temperature properties of aluminum products without forming any undesirable phases [2,3]. An optimal set of mechanical properties can be achieved with evenly dispersed small Al_2O_3 particles in the metal matrix. Based on this, this study was aimed to form Al–Al_2O_3 composite materials in a vortex layer apparatus with efficient mixing of powders. The prepared powders were subjected to laser fusion followed by the characterization of the resulting composites.

2. Materials and Methods

Silumin and alumina powders were the initial materials for producing aluminum–alumina composites. The silumin contained ~ 86.5 % Al and ~ 11.0 % Si and included small additives of Mg, Mn, Fe, and Ti [4]. The average particle size of the silumin powder was 43 μm. Alumina obtained from aluminum chloride was used as reinforcement [5]. Alumina contained more than 99.0% Al_2O_3 and corresponded to the α-modification of alumina. The alumina used was pre-grounded in a ball mill and carefully classified using a sieve with a hole diameter of 64 μm. The average particle size of the alumina was 20 μm.

Mixing of powders was performed in a classical vortex layer apparatus, a similar scheme of which is discussed in detail in [6]. Mixing was carried out by ferromagnetic steel bodies managed by a high-power electromagnetic field. The ferromagnetic bodies had a cylindrical shape with a height of 1.3 cm. The ratio of the mass of the mixture to the mass of the ferromagnetic bodies was 2 to 1. The powders were mixed for 60 seconds. The speed of rotation of the magnetic field in the working chamber did not exceed 3000 rpm.

Laser fusion of aluminum-alumina composites was performed on an SLM Solutions 280 HL 3D printer. The synthesized samples were square plates with a thickness of 3 mm and a side length of 25 mm. The thickness of the initial powder layer was 80 μm for each plate. The volumetric energy density (VED, J/mm³) was used as an optimization parameter [7]. The VED was calculated using the equation (1)

$$VED = P / (V \cdot h \cdot t) \quad (1)$$

where P is the laser power (W), V is the scan speed (mm/s), h is the hatch spacing (mm), and t is the layer thickness (mm).

For all modes, the VED was about 24 J/mm³, but the parameters P and V varied: P = 370 W was used for Mode #1, 200 W for Mode #2, and 100 W for Mode #3.

Scanning electron microscope (SEM) images of the powders and the synthesized samples were obtained using a Tescan Vega 3 SEM with an element mapping attachment. The analysis of porosity of the synthesized samples was performed using a Carl Zeiss Axio Observer A1M optical microscope (OM). Tensile tests were performed on an Instron 5969 testing machine at room temperature with a crosshead speed of 0.5 mm/min.

3. Results and Discussion

For the study, a mixture of aluminum powder with 3 wt.% of alumina was prepared in a vortex layer apparatus. As can be seen in Figure 1c, the vortex layer apparatus allowed a thorough mixing of powders, as well as subjected to the grinding of alumina. The initial alumina powder (see Figure 1b) contained particles up to 30 μm in size, while after mixing the powders, the alumina particle size was 1-3 μm on average. However, as can be seen from the elemental distribution map (see Figure 1d), there are areas of alumina accumulation in the powder mixture that reach 20-25 μm in size. The size and shape of silumin particles did not change after mixing in the vortex layer apparatus. The distribution of alumina in the mixture was fairly uniform, which indicates the effectiveness of using the vortex layer apparatus to produce powder composites.

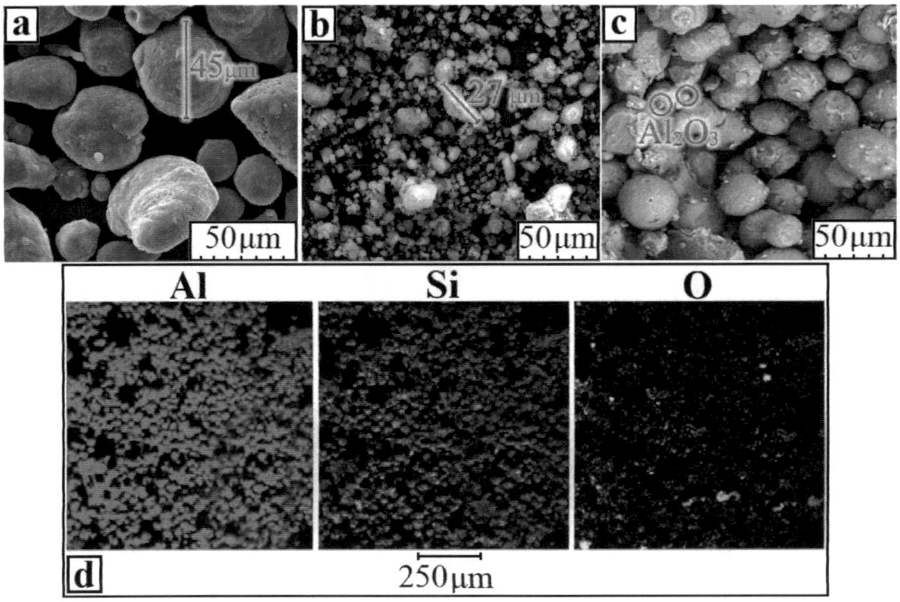

Figure 1. SEM images of powders: silumin powder (a); alumina powder (b); silumin and alumina powder mixture (c); element mapping of silumin and alumina powder mixture (d).

Images of synthesized samples obtained using optical microscopy are shown in Figure 2. All samples had a noticeable pore content, and the samples synthesized in Modes #2 and #3 had distinct areas of insufficient fusion of the composite material. The microstructure of the samples fused in Modes #2 and #3 indicated the choice of an incorrect laser fusion mode. The lowest porosity (0.3±0.02 %) was observed in the plate synthesized according to Mode #1.

The SEM image of the synthesized sample using Mode #1 is shown in Figure 2d. As can be seen from Figure 2, during the laser fusion, alumina was evenly distributed over the surface of the aluminum alloy, and areas of alumina accumulation were no longer observed. Probably, due to the brittleness of the used α-Al_2O_3, it was crushed by mechanical mixing with ferromagnetic steel bodies, and the areas of alumina accumulation in Figure 1d are caused by sticking with partially fused aluminum particles.

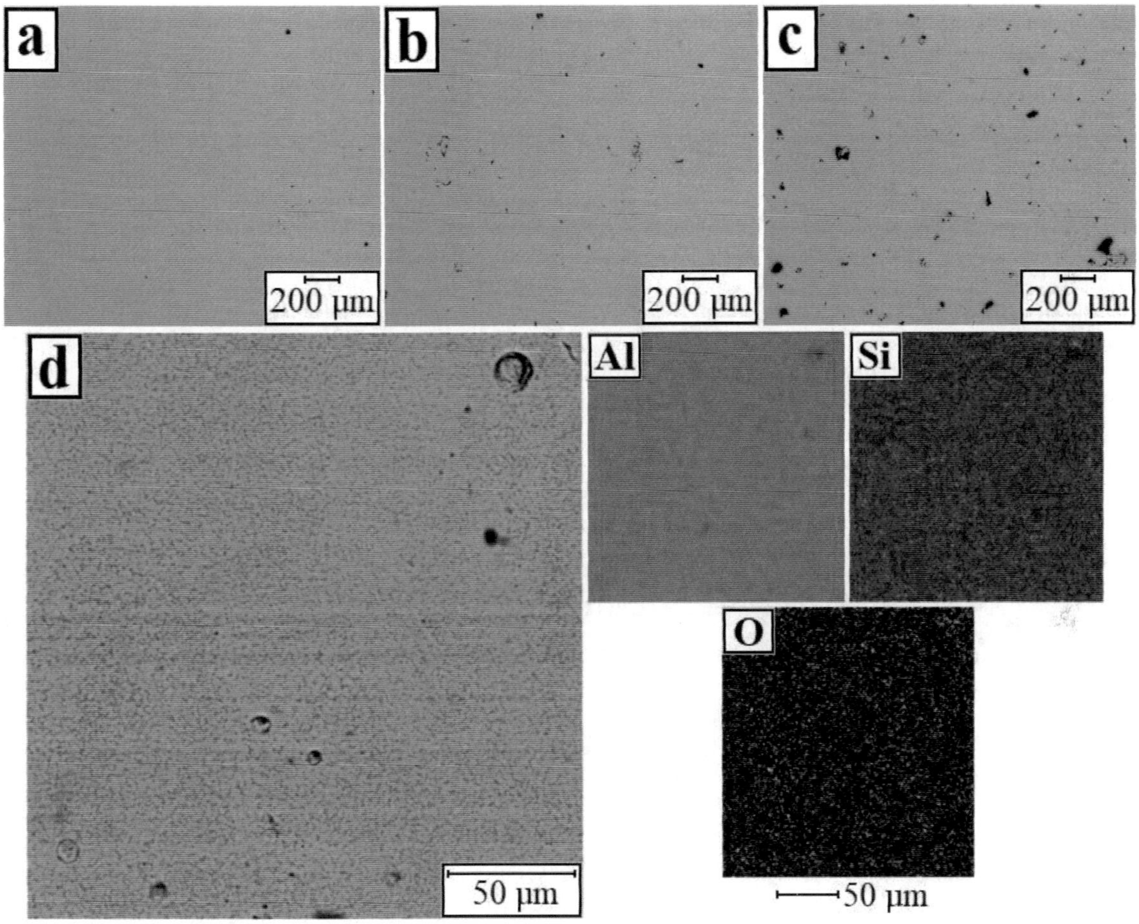

Figure 2. OM and SEM images of the synthesized samples using Mode #1 (a, d), Mode #2 (b) and Mode #3 (c).

The mechanical properties of the synthesized plates are presented in Table 1. The lowest ultimate tensile strength (UTS) was observed with Mode #3, which is explained by insufficient heat input of the laser beam energy and only partial melt of the aluminum matrix.

During the laser fusion process, the composite powder formed a molten pool. α–Al_2O_3 as a reinforcement, practically was not subjected to melting in comparison with silumin. Due to their small size, alumina particles were distributed in the molten pool as a suspended substance, which increased the viscosity of the melt. Despite the same VED value, the formation mechanism of the molten pool in each sample was different. At a low laser power value, the composite material did not have time to completely melt. The melt viscosity was too high for the effective wetting of the composite particles and previous molten layers, which led to the formation of internal defects [8]. As can be seen from Table 1, the strength of the samples increased with an increase in laser power.

Table 1. The mechanical properties of the synthesized objects.

Laser fusion mode	State	UTS (MPa)	E_f (%)
Mode #1 (P = 370 W)	as processed	384 ± 5	0.54 ± 0.1
Mode #2 (P = 200 W)	as processed	241 ± 4	0.25 ± 0.1
Mode #3 (P = 100 W)	as processed	150 ± 9	0.33 ± 0.1

However, for all samples, UTS was noticeably less than for samples synthesized from a classic aluminum alloy for additive manufacturing ($AlSi_{10}Mg$, state: as processed, UTS = 430 MPa) [9]. Also, the elongation of all synthesized samples did not exceed 0.54%. This indicates a high content of Al_2O_3 in the samples, which increased their brittleness. Despite the uniform distribution of alumina in the samples, its high content caused insufficient laser fusion of the composite material, which reduced the bonding strength between the reinforcement and the matrix. According to the obtained results, further research will be concentrated on a lower content of alumina in Al–Al_2O_3 composites obtained by the vortex layer apparatus.

4. Conclusions

In this paper, the process of aluminum–alumina composites formation in a vortex layer apparatus and subsequent laser fusion was considered. Composite materials obtained in the vortex layer apparatus were characterized by a uniform distribution of alumina in the matrix. Three laser fusion modes were tested, the most effective being Mode #1, which was performed at VED = 24 J/mm^3 and P = 370 W. The sample synthesized using Mode #1 had the following mechanical properties: UTS = 384 ± 5 MPa; E_f = 0.54 ± 0.1 %. The obtained results indicated a too high content of alumina for the used laser fusion modes, causing high porosity of the synthesized samples. According to the results, further research using a smaller amount of Al_2O_3 in the composite is planned.

5. Funding

The work is financially supported by the Russian Science Foundation (RSF), grant No. 19–79–30025.

References

[1] Han Q, Setchi R, Evans S L 2016 *Powder Technol.* **297** 183–92

[2] Nalivaiko A Y, Arnautov A N, Zmanovsky S V, Ozherelkov D Y, Shurkin P K, Gromov A A 2020 *J. Alloy Compd.* **825** 154024

[3] Durai T G, Das K, Das S 2007 *Mater. Sci. Eng., A* **15** 445–46

[4] Nalivaiko A Y, Arnautov A N, Zmanovsky S V, Gromov A A 2019 *Mater. Res. Express.* **6(8)** 086536

[5] Pak V I, Kirov S S, Nalivaiko A Y, Ozherelkov D Y, Gromov A A 2019 *Materials* **12(23)** 3938

[6] Adoshev A, Antonov S, Yastrebov S, Melnikov M 2017 *Engineering for Rural Development* **16** 804–10

[7] Aboulkhair N T, Simonelli M, Parry L, Ashcroft I, Tuck C, Hague R 2019 *Prog. Mater Sci.* **106** 100578

[8] Jue J, Gu D, Chang K, Dai D 2017 *Powder Technol.* **310** 80–91

[9] Trevisan F, Calignano F, Lorusso M, Pakkanen J, Aversa A, Ambrosio E P, Lombardi M, Fino P, Manfredi D 2017 *Materials* **10(1)** 76

On the formation of AlSiMgCu/quasicrystal powder composite single tracks by selective laser melting

D Yu Ozherelkov[1,*], **A Yu Nalivaiko**[1], **A A Gromov**[1], **D O Solodov**[2], **A A Komissarov**[2], **A S Shinkaryov**[1], **A I Morozova**[1,3]

[1] MISIS Catalysis Lab, National University of Science and Technology MISIS, Moscow, 119991, Russia
[2] MISIS Laboratory of Hybrid Nanostructured Materials, National University of Science and Technology MISIS, Moscow, 119991, Russia
[3] Laboratory of Mechanical Properties of Nanostructured Materials and Superalloys, Belgorod National Research University, Belgorod, 308015, Russia

*Correspondence: d.ozherelkov@gmail.com

Abstract. The process of selective laser melting (SLM) of an Al-Si-Mg-Cu powder composite with the addition of quasicrystal reinforcements under conditions of a limited powder amount was investigated. The morphology and microhardness of the obtained scan single tracks were discussed, and the optimal sintering parameters were found to be in the Linear Energy Density (LED) range 0.21-0.30 J/mm. The following SLM parameters: P = 370 W and v = 1650 mm/s given the optimal characteristics of scan single tracks geometry with less defects. The obtained composites exhibited an improved hardness with the highest values of 235 HV with the 8% of reinforcements in the powder mixture.

1. Introduction

Quasicrystal (QC) is an intermetallic compound with a long-range order, but without translational periodicity, was first discovered by D. Shechtman et al. in 1984 [1] and more than a hundred kinds of quasicrystals with different industrial applications have been developed up to date. QC are well known for their high strength and hardness, low friction coefficient, high corrosion and wear resistance [2,3] However, due to their high room-temperature brittleness, the applications of QC alloys are limited by their low workability using conventional manufacturing processes. Therefore, manufacturing of any complex shape details from QC materials is an important challenge and issue [4].

One of the possible ways to extend the potential applications of QC materials and benefit from their useful properties is to use them as reinforcement in metal matrix composite materials (MMCs) with the profit from the ductility of the matrix material. As a manufacturing method for MMCs, selective laser melting (SLM) have some obvious advantages. Amongst them are: the production of 3D components with customized geometries, ultra-fine microstructure due to the high cooling rate peculiar for the SLM process, which causes higher mechanical properties. Moreover, the metal-based reinforcements are also possible to be in-situ formed during the preparation of MMCs with strong metallurgical bonding [2,5].

Therefore, developing of Al-based QC-reinforced composite materials via SLM may be interesting for the production of lightweight metallic components with low density/high wear resistance which can be used in automotive and aviation industries [6]. Thus, in this study, Al-based QC reinforced system was selected, due to its low density and potential in-situ reaction between matrix and reinforcement.

Aluminum alloy acted as matrix material in order to obtain a high bonding strength between reinforcement and matrix, due to the high content of Al in reinforcement [3,7]. Current research aims at the investigation of optimal SLM parameters for the experimental composition of a new Al-Si-Mg-Cu powder with different contents of QC reinforcement. The objective of this work is to select the optimal printing mode considering the limited amount of available material based on the study of the formation of the scan single tracks with the investigation of their microstructure and microhardness. This method is a powerful way to rapidly investigate the behavior of new compositions for SLM and to study the effect of the parameters on the geometrical features of the molten pool.

2. Materials and methods

The initial powders for the samples fabrication were Al-Si-Mg-Cu (chemical composition: 87% Al, 10.7% Si, 0.5% Mg, 0.7% Cu, fractional amount of Mn, Ti, Fe and other impurities) alloy (D_{50} = 40 μm) powder obtained by the molten metal spraying method and gas atomized $Al_{45}Cu_{33}Fe_{22}$ QC (D_{50} = 35 μm). Before SLM, the two types of powder were blended with a 1%, 4% and 8% content of QC in the Al alloy matrix in a tumbling mixer for 120 min and dried at 80 °C for 4 h.

The SLM Solutions 280 HL 3D printer was used for selective laser melting. An aluminum–magnesium alloy (AMg5 brand, chemical composition: 94% Al, 5% Mg, 0.5% Mn, 0.2% Si, and no more than 0.1% of other impurities) was used as the substrate material. The laser spot (LP) size was was 80 μm. All samples were synthesized in a nitrogen atmosphere, and the residual oxygen content in the working area was less than 0.2 vol. % for all sample types. A powder layer of approximately 100 μm in thickness and a single pass with a laser beam was used to form scan single tracks with a length of 100 mm. The SLM printing parameters for scan single tracks were selected and varied in the following ranges: P = 250–370 W and V = 850–1650 mm/s.

Scanning electron microscopy SEM and characterization of powders were performed using a Tescan VEGA 3 scanning electron microscope with a NORAN X-ray energy dispersion spectrometer (EDX) and an element mapping attachment. Optical microscopy (OM) of scan single tracks was carried out using a ZEISS Axio Observer Z1 microscope. The microhardness of the samples was measured using a Tukon 1102 microhardness tester. The microhardness was measured at a load of 0.025 N with an indentation time of 10 seconds.

3. Results and discussion

SEM images of the composite obtained by mechanical mixing of the powder with QC particles are demonstrated in Figure 1. It can be noticed that all Al-Si-Mg-Cu particles have a quasi-spherical shape and the method used for the preparation of the composite powder do not damage the initial particles. The EDX mapping analysis showed that the reinforcement of QC particles had a normal distribution in the composite structure. Here it can be pointed out that the introduction of QC particles with non-spherical shape into the molten sprayed powder leads to reduction of its flowability [8], which is one of the key property for the SLM processes. However, in the present work only scan single tracks were considered and the influence of the QC addition on the flowability was not investigated. Based on the results obtained on the size and geometry of the particle, the initial powder was applicable for selective laser melting.

The influence of printing parameters was studied using the Linear Energy Density (LED, J/mm) calculated using equation (1):

$$LED = P / V. \qquad (1)$$

where P is the laser power (W) and V is the scan speed (mm/s).

Cross-section images of the most representative single tracks within different LED ranges are presented in Figure 2. According to the scans geometry, three main process ranges were selected: LED = 0.15-0.20 J/mm, LED = 0.21-0.30 J/mm and LED > 0.35 J/mm. For different powders, the geometrical features have a similar behavior and the total scan width increases with LED values (Figure 3 (a)). The morphology of scan single tracks is dependent on the applied laser parameters by varying the thermodynamic and kinetic characteristics of the molten pool.

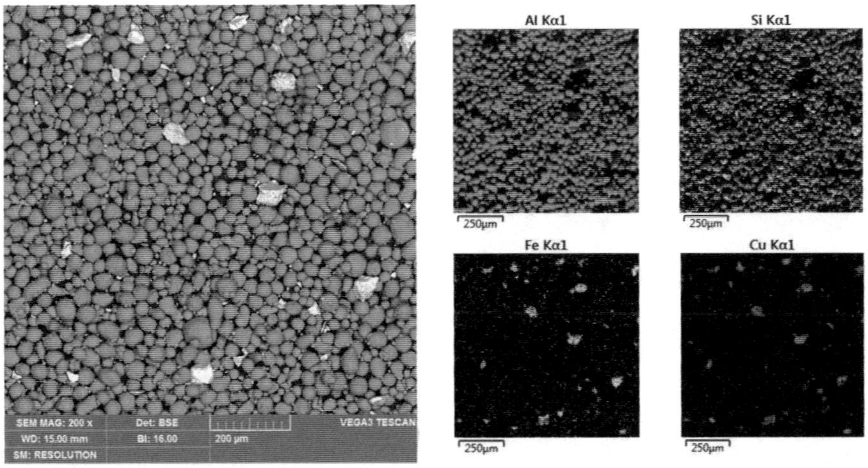

Figure 1. SEM and EDX mapping of initial powder composite after mechanical mixing

Scans in the LED range of 0.15-0.2 J/mm were not completely melted are characterized by a low penetration depth due to the low wettability and high viscosity of the melt due to the limited applied laser energy. On the other hand, deep melt pools were observed with the LED more than 0.35 J/mm with the presence of defects and pores in the structure due to the overheated molten pool and significant thermal accumulation. Study of cross-section images revealed that most suitable SLM parameters to process an Al-Si-Mg-Cu powder reinforced by QC lies in the LED range of 0.21-0.3 J/mm, taking into account the optimal characteristics of the molten pool and wetting characteristics of the melt.

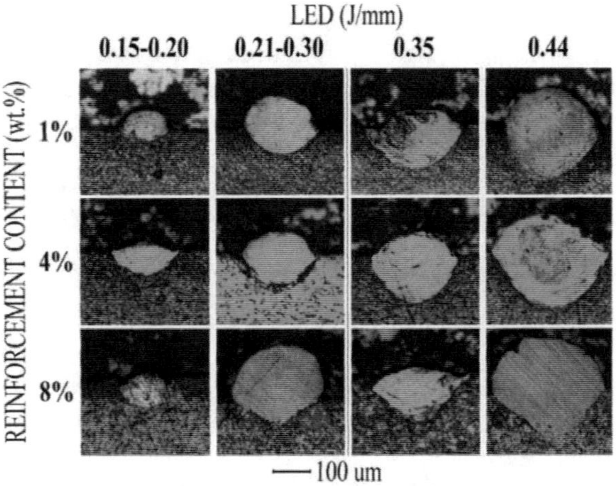

Figure 2. Cross-section images of scan single tracks

The described above was confirmed by the microhardness results, measured on the cross-sections and presented in Figure 3 (b). The microhardness of the composites with different QC content underwent a continuous increase from 159 HV to 235 HV with an increase in LED from 0.15 to 0.23 J/mm and followed by a minor changes up to LED = 0.31 J/mm. These process parameters result in increased microhardness and less defects, such as pores and micro cracks in the structure. A further increase in LED to more than 0.35 J/mm leads to a decrease in microhardness to 123 HV, which indicates the presence of microcracks and pores, which tend to lower hardness performance. But comparing to unreinforced Al-Si-Mg-Cu powder manufactured by SLM with an average microhardness value of 125 HV [7], the studied composite demonstrates an increased microhardness.

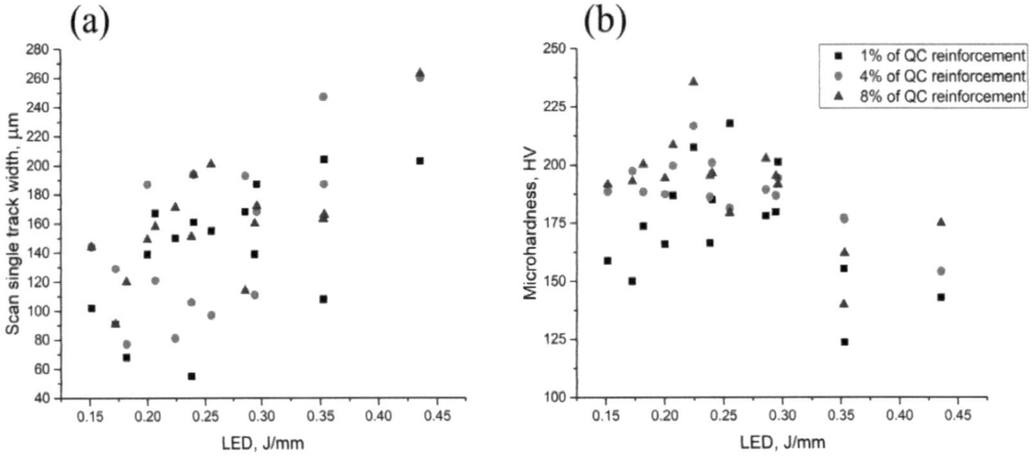

Figure 3. Influence of the LED on the (a) width and (b) microhardness of printed scan single tracks.

4. Conclusions

The formation of scan single tracks of an Al-Si-Mg-Cu powder composite reinforced with different amount of quasicrystal was studied, and the influence of selective laser melting parameters on the geometry and microhardness were investigated. The optimal process parameters were found to be in the LED range 0.21-0.30 J/mm taking into account the optimal scan single tracks geometry. The composites exhibited improvement hardness comparing to the one without QC reinforcement. The highest hardness values of 235 HV were obtained with 8% of QC reinforcements with the following SLM parameters: P = 370 W and v = 1650 mm/s. These process parameters resulted in an increased microhardness and structure with less defects throughout the material.

5. Funding

The work is financially supported by the Russian Science Foundation (RSF), grant No. 19–79–30025.

References

[1] Shechtman D, Blech I, Gratias D, Cahn J W 1984 *Phys. Rev. Lett.* **53** 1951-1953
[2] Lu J L, Lin X, Liao H L, Kang N, Huang W D, Coddet C 2020 *Opt. Laser. Technol.* **129** 106277
[3] Kang N, Fu Y, Coddet P, Guelorget B, Liao H, Coddet C 2017 *Mater. Des..* **132** 105-111
[4] Shadangi Y, Sharma S, Shivam V, Basu J, Chattopadhyay K, Majumdar B, Mukhopadhyay N K 2020 *J. Alloy Compd.* **828** 154258
[5] Jue J, Gua D, Chang K, Dai D 2017 *Powder Technol.* **310** 80-91
[6] Nalivaiko A Y, Arnautov A N, Zmanovsky S V, Ozherelkov D Y, Shurkin P K, Gromov A A 2020 *J. Alloy Compd.* **825** 154024
[7] Nalivaiko A Y, Arnautov A N, Zmanovsky S V, Gromov A A 2019 *Mater. Res. Express.* **6(8)** 086536
[8] Aversa A, Moshiri M, Librera E, Hadi M, Marchese G, et al. 2018 *J. Mater. Process Technol.* **255** 17-25

Effect of the deformation temperature on the deformation behavior of a Cu-Cr-Zr alloy

A Morozova[1,2*], M Tkachev[1], A Pilipenko[1], A Lugovskaya[1], A Belyakov[1] and R Kaibyshev[1]

[1]Laboratory of Mechanical Properties of Nanostructured Materials and Superalloys, Belgorod National Research University, Pobeda Str. 85, Belgorod, 308015, Russia
[2]The KINETICA Engineering Center, National University of Science and Technology MISiS, Leninskiy Ave. 4, Moscow, 119049, Russia

*Corresponding author: morozova_ai@bsu.edu.ru

Abstract. The mechanical properties and deformation behavior of a Cu-Cr-Zr alloy after solution treatment (ST) or aging (ST+AT) and equal channel angular pressing (ECAP) at elevated temperature were investigated. ECAP at elevated temperature led to strengthening and degradation of plasticity compared with the initial state. The deformation behavior of the alloy after 1 and 2 ECAP passes at 200 °C was almost the same. An increase in ECAP temperature promoted uniform strain and plasticity. An increase in the number of ECAP led to a gradual increase in strength. YS and UTS in ST+AT condition were higher than in ST samples regardless of the ECAP temperature and the number of passes. The maximum values of YS and UTS were obtained after 4 ECAP passes at 400 °C and achieved 525 MPa and 600 MPa in ST and 570 MPa and 650 MPa in ST+AT condition, respectively. The effect of deformation strain and ECAP temperature on work hardening rate and recovery were discussed in terms of the Voce equitation.

1. Introduction

Cu-Cr-Zr alloys have been advanced materials for electrical applications due to their high electrical conductivity and strength [1-2]. Fine precipitations formed during heat treatment or warm deformation provide the required combination of these properties [3-4]. Severe plastic deformation, as a promising working technique, may change the microstructure and deformation behavior of Cu-Cr-Zr alloys [5-6]. Strain-induced low- and high-angle boundaries particle precipitation, high dislocation density promote work hardening and recovery [7]. However, the microstructure-property relationship in Cu-Cr-Zr alloys has not been sufficiently investigated. The aim of this work is to study the mechanical behavior of the Cu-Cr-Zr alloy after severe plastic deformation at elevated temperatures and to quantitatively estimate the effect of the deformation temperature on work hardening and recovery of the material.

2. Experiment

The Cu-0.3%Cr-0.5%Zr alloy after solution treatment (ST) or aging (ST+AT) followed by ECAP at elevated temperature was chosen as the object of the present study. Solution treatment was carried out at 920 °C during 0.5 h with subsequent water quenching. A part of the samples was subjected to aging at 450 °C for 1 h. Then, 1, 2 and 4 ECAP passes was performed at temperatures of 200 °C, 300 °C or 400 °C via route B_C. The channel intersection angle of the ECAP die was 90°. The mechanical properties were studied using an Instron 5882 machine using the specimens with a gauge length of 6 mm. The initial strain rate was about 6×10^{-3} s^{-1}.

3. Results and discussion

True stress – true strain curves

Deformation behaviour after ST or ST+AT was characterized by a rather high uniform elongation and significant plasticity (Figure 1). The strength in the ST condition was lower and the plasticity was higher than after the ST+AT. ECAP at elevated temperature led to strengthening and degradation of plasticity. The difference between the yield strength (YS) and the ultimate tensile strength (UTS) was about 10 MPa after ECAP at 200 °C; and the uniform strain was about 0.01. An increase in the ECAP temperature was beneficial for uniform strain and plasticity. The mechanical behaviour of Cu-Cr-Zr alloys after 1 and 2 ECAP passes at 200 °C was almost the same. YS and UTS after 1 pass of ECAP at 300 °C were lower by about 50 MPa than that for the samples ECAPed at 200 °C. 2 and 4 ECAP passes at 300 and 400 °C were characterized by a gradual increase in strength. In the ST+AT condition plasticity achieved its maximum value after 2 ECAP passes at all the ECAP temperatures. YS and UTS in ST+AT samples were higher than that in the ST condition at a different ECAP temperature and number of passes. The maximum YS and UTS have been measured to be 525 MPa and 600 MPa after 4 ECAP passes at 400 °C for ST samples and 570 MPa and 650 MPa for ST+AT samples, respectively.

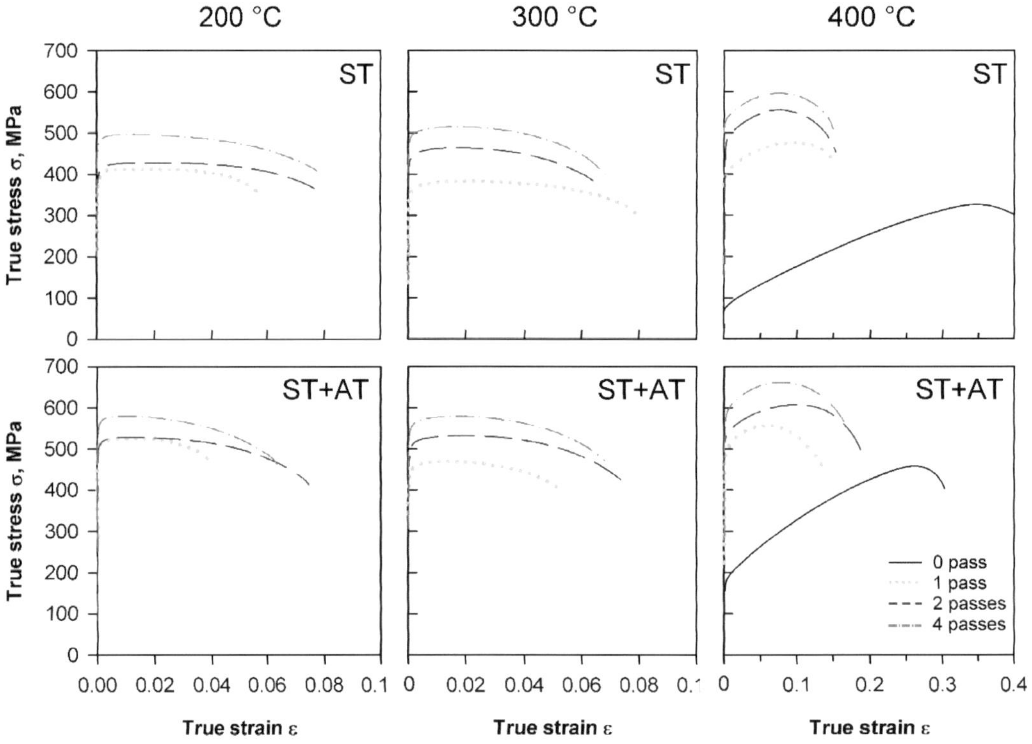

Figure 1. True stress – true strain curves of the Cu-Cr-Zr alloy in the solution treated (ST) and aged (ST+AT) condition after 0 (black line), 1 (green line), 2 (blue line) and 4 (pink line) ECAP passes at temperatures of 200 °C, 300 °C, or 400 °C.

Voce analysis

Analysis of the mechanical behavior can be carried out with the Voce approach. The Voce equation describes the mechanical behavior during uniform deformation as follows [8-9]:

$$\sigma = \sigma_v + (\sigma_0 - \sigma_v)\exp\left(-\frac{\varepsilon}{\varepsilon_v}\right) \qquad (1)$$

where σ is the flow stress, σ_v is the saturation stress, ε_v is a characteristic strain corresponding to σ_v, σ_0 is the back-extrapolated stress to ε=0, ε is a true strain. Then the strain hardening can be expressed by the linear function as:

$$\frac{d\sigma}{d\varepsilon} = \frac{\sigma_v}{\varepsilon_v} - \frac{1}{\varepsilon_v}\sigma \qquad (2)$$

The $\Delta\sigma/\Delta\varepsilon$ - σ curves are shown in Figure 2. The lines in Figure 2 can be approximated by linear function in accordance with equitation (2).

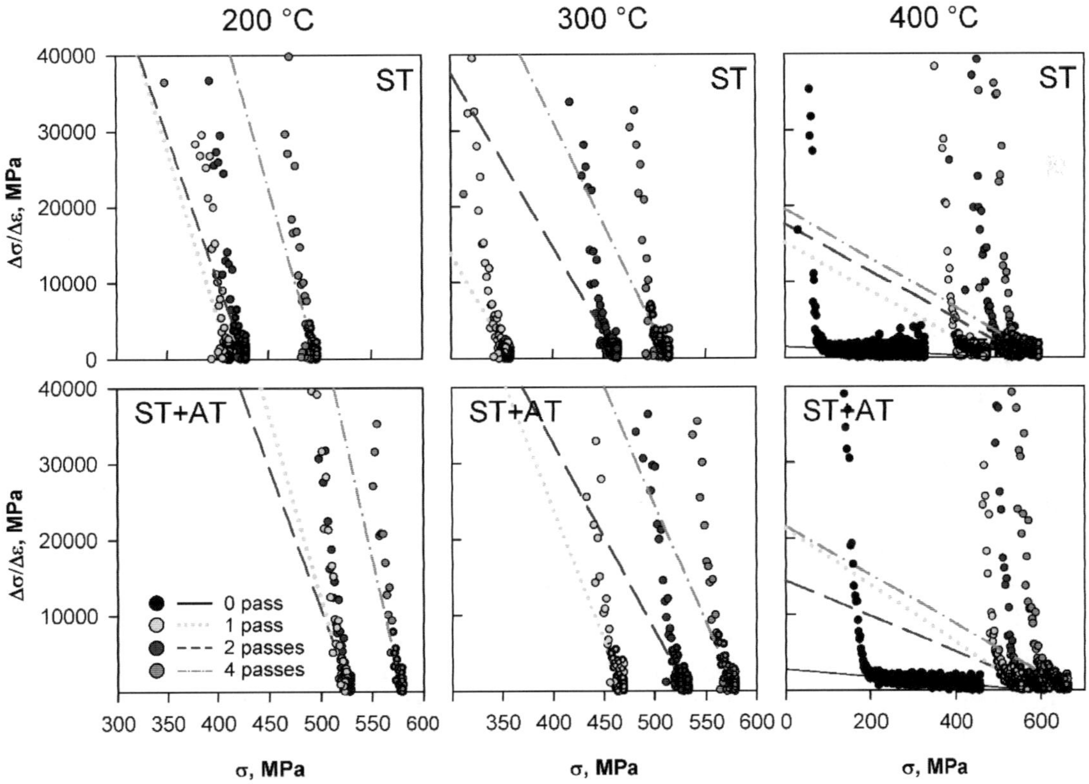

Figure 2. Work hardening rate ($\Delta\sigma/\Delta\varepsilon$) vs. true stress ($\sigma$) curves of the Cu-Cr-Zr alloy in the solution treated (ST) and aged (ST+AT) condition after 0 (black), 1 (green), 2 (blue) and 4 (pink) ECAP passes at temperatures of 200 °C, 300 °C, 400 °. Lines correspond to the Voce equitation (2).

The slope of Voce line $1/\varepsilon_v$ was higher in ST+AT samples regardless of the ECAP temperature and strain. An increase in the number of ECAP passes was accompanied by an increase in Voce line slop $1/\varepsilon_v$ and σ_v/ε_v constant in the all ECAP temperature range in the ST condition (Figure 3). Voce line slop in ST+AT samples had a minimum after 2 ECAP passes at 200-400 °C. An increase in the ECAP temperature led to a decrease in Voce line slop and σ_v/ε_v constant in both conditions.

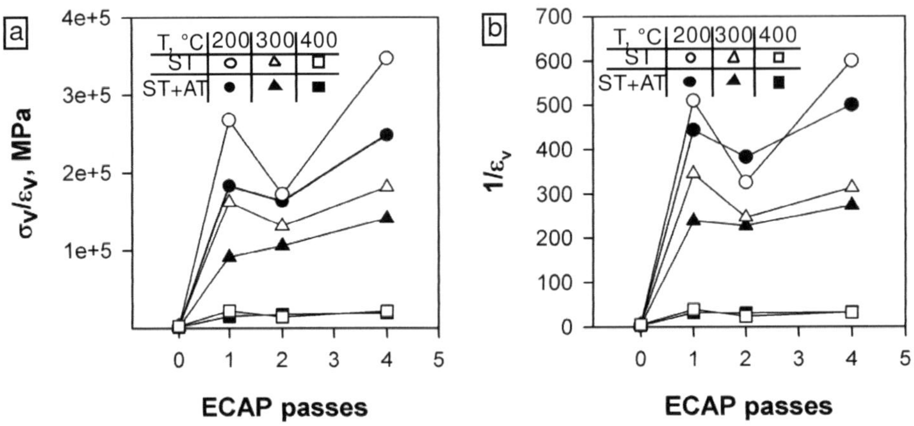

Figure 3. Effect of ECAP temperature and strain on constant, σ_v/ε_v, (a) and a slope of Voce line, $1/\varepsilon_v$, (b) in the Cu-Cr-Zr alloy in the solution treated (ST, black symbols) and aged (ST+AT, white symbols) condition.

The coefficient, σ_v/ε_v, corresponds to the work hardening rate and strongly depends on ε_v. An increase in the ECAP temperature led to a decrease in the work hardening rate due to a significant change of ε_v from 0.001-0.005 for ECAP at 200 °C to 0.01-0.05 for ECAP at 400 °C. The Voce line slop of $1/\varepsilon_v$ is reciprocal of characteristic strain that defines the velocity of σ_v attaining. An increase in ε_v corresponds to an increase in uniform deformation and prolongation of dislocation accommodation and their rearrangement that is associated with recovery of the material.

So, ECAP at elevated temperature led to strengthening and degradation of plasticity. An increase in the ECAP temperature promoted uniform strain and plasticity. An increase in ECAP numbers led to a gradual increase in strength. YS and UTS in ST+AT condition were higher than in ST samples regardless of ECAP temperature and number of passes. An increase in the ECAP temperature led to a decrease in the work hardening rate and promoted recovery significantly.

Acknowledges

The financial support received from the Ministry of Science and Education, Russia, under President grant No. 075-15-2020-407 is gratefully acknowledged. The authors are grateful to the personnel of the Joint Research Center, "Technology and Materials", Belgorod National Research University, for their assistance with instrumental analysis.

References

[1] Murashkin M Y, Sabirov I, Sauvage X and Valiev R Z 2016 J. Mater. Sci. 51(1) 33.
[2] Morozova A, Mishnev R, Belyakov A and Kaibyshev R 2018 Rev. Adv. Mater. Sci. 54(1) 56.
[3] Vinogradov A, Suzuki Y, Ishida T, Kitagawa K and Kopylov V I 2004 Mater. Trans. 45(7) 2187.
[4] Liu P, Kang B X, Cao X G, Huang J L, Yen B, Gu H C 1999 Mater. Sci. Eng. A 265262.
[5] Zhu C, Ma A, Jiang J, Li X, Song D, Yang D and Chen J 2014 J. Alloys Compd. 582 135.
[6] Borodin E N, Morozova A, Bratov V, Belyakov A and Jivkov A P 2019 Mater. Charact. 156 109849.
[7] Humphreys F J, Prangnell P B, Bowen J R, Gholinia A and Harris C 1999 Phil. Trans. Royal Soc. A 357(1756) 1663.
[8] Voce E 1948 J. Inst. Met. 74 537.
[9] Angella G, Zanardi F and Donnini R 2016 *J. Alloys Compd.* **669** 262.

Studying the Impact of Blast Treatment on the Structure and Mechanical Properties of Carbon Steel

G S Nagicheva*, A V Nokhrin, N N Berendeev, N V Melekhin, A V Piskunov, A N Sysoev, M Y Gryaznov

Lobachevsky State University of Nizhny Novgorod, 23 Gagarina ave., Nizhny Novgorod, 603950, Russian Federation

*Corresponding author: Nagicheva@mail.ru

Abstract. The study has covered the effects of blast treatment on the microstructure parameters and mechanical performance of carbon steel. The calculation results of examining the structure and properties of carbon steel in its initial state and after explosive hardening in various sections of cylindrical carbon steel workpieces have been verified with the finite-element analysis using the ANSYS Workbench software environment. Blast treatment has been demonstrated to lead to a perceptible change in the structural state of carbon (cementite) in steel.

1. Introduction

Explosive hardening is currently one of the most promising ways to improve the stress-strain behavior of steel [1]. The use of blast treatment coupled with advanced methods of modelling the blast propagation conditions in steel workpieces of complex geometry can be expected to accomplish the objective of hardening steel workpieces of complex geometry, such as rails. It is important to note that, apart from controlling the phase and structural state of steel and strengthening it, blast treatment helps, at the same time, to control the magnitude of internal stresses that exercise a noticeable impact on the stress-strain properties and service characteristics of rail steels [2].

The purpose of this study is to provide computer simulation and to analyze the impact of blast treatment on steel hardness and wear resistance.

2. Materials and methods

U8A Russian industrial steel (Fe – balance, 0.76%C, 0.27%Si, 0.23%Mn, the Russian analog of W108 steel by AISI) was the object of the study. The initiation site was placed on an attachable wheel. Test modes differed in the supporting plate thickness as well as the content and density of its material.

The finite-element analysis using ANSYS Workbench 19.2 was used to simulate the blast propagation within the steel workpiece. For computer simulation purposes, Steel 4340 behavior was used for the steel; Adiprene behavior, for the supporting plate; and Octol behavior, for the explosive. Detonation of a steel cylinder placed on a concrete support (CONC140MPA) was simulated. The values of the supporting plate density (ρ), its thickness (δ), and shear modulus (G) were varied during the tests. The Johnson-Cook model was used in solving the problem [3].

A Leica DM IRM metallographic microscope was used to study the steel microstructure. HVS-1000 and Rockwell durometers were used to measure microhardness (HV) and Rockwell macrohardness (HRC). A Buehler Auto Met 250 honing machine was used to test abrasive wear resistance. Weight loss over a given time interval was used as a measure of wear rate. A Sartorius CPA 225D analytical balance was used to take mass measurements. The structural condition of carbon in steel was studied through the internal friction method using the inverse torsion pendulum in the continuous heating mode at the rate of 10 °C/min and at a frequency of 9 Hz according to the technique [5]. Research was performed on cylindrical samples with a diameter of 19.8 mm and a height of 11 mm that were cut from the original or blasted workpiece. This cutting-up layout allowed for thorough studies of the impact of nonuniform blast propagation throughout the workpiece. Those samples were primarily selected for research, where the simulation results demonstrated the greatest changes in properties.

3. Experimental results

Analysis of the stress-strain properties distribution along the workpiece demonstrated no noticeable hardening in most of the cylinder. The maximum deformation and flow stress are observed near the explosives area (Figure 1). These values decrease with distance from the explosives area in all directions.

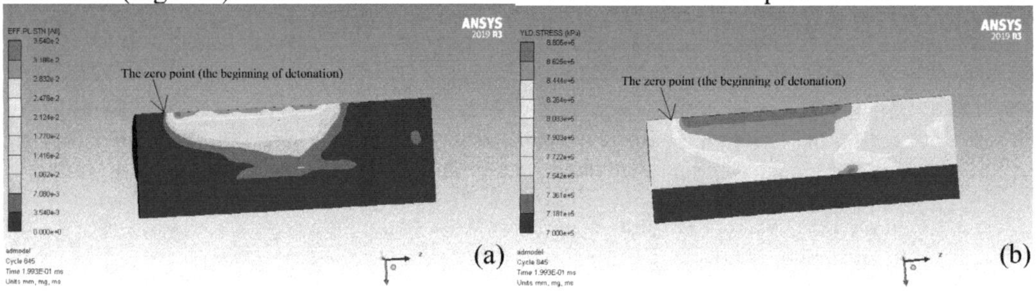

Figure 1. Distribution of plastic deformations (a) and flow stress (b) along the cylindrical workpiece made of U8A steel. Case of loading without a supporting plate

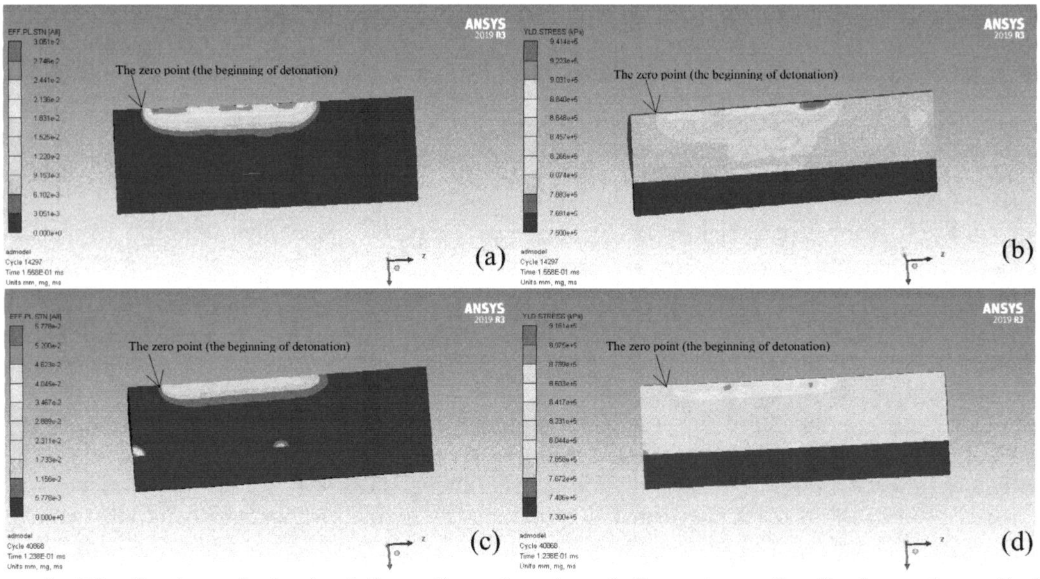

Figure 2. Distribution of plastic deformations (a, c) and flow stress (b, d) along the cylindrical workpiece made of U8A steel. Case of loading with a supporting plate of δ=2 mm, ρ=1.73 g/cm^3, G=3.309 MP (a, b) and a supporting plate of δ=4 mm, ρ=1.32 g/cm^3, G=2.206 MPa (c, d)

The blue-colored rectangular region in Figures 1b, 2b and 2d is a concrete support where a steel cylinder placed.

While modelling blast treatment, the thickness and density of the supporting plate were varied (Figure 2). Analysis of the simulation results demonstrated that the maximum plastic deformation (6.01%) is achieved during explosive hardening of the workpiece when the supporting plate has a thickness of δ=1 mm, density of 1.32 g/cm^3, shear modulus of G=2.206 MPa, while the maximum flow stress (941.4 MPa) is observed when these parameters are at δ=2 mm, ρ=1.73 g/cm^3, G=309 MPa. The optimal ratio of the maximum flow stress (916.1 MPa) and plastic deformation (5.78%) is achieved with the supporting plate of δ=4 mm, ρ=1.32 g/cm^3, G=2.206 MPa. Aggregation of the computer simulation results suggests that an increase in the supporting plate thickness and a decrease in its density can lead to hardening of the steel workpiece. Notably, the flow stress in each of the scenarios increased by about 1.1 times, which indicates a slight hardening of the material in certain areas of the cylinder.

The simulation results indicated that thermal effects are insignificant, and that the local heating of the workpiece does not exceed 100 °C. It should also be noted that heat tints, which could have signaled a noticeable temperature increase, were not observed on the surface of the workpiece.

Studies of the microstructures showed that, following blast treatment, U8A steel has a platelike pearlitic structure (with a varying dispersion of cementite plates) with areas of spheroidized pearlite. It has been established that explosive hardening leads to an increase in areas of microstructure with spheroidized pearlite. This structural feature can negatively affect the stress-strain behavior (lower hardness and wear resistance) [4]. In addition, chaotically directed "sharp" cracks can be spotted in the samples treated in Mode 1 (without a supporting plate). It should be noted that in case of blast treatment with a supporting plate, no cracks were found in the microstructure (see Figure 3).

The average microhardness of steel in its initial state was ~2,100 MPa. Microhardness in areas with a platelike pearlite structure is ~2,500 MPa, while areas of spheroidized structure have lower hardness of 2,100 MPa. Blast treatment does not result in a noticeable change in macrohardness and wear resistance of steel, yet it is accompanied by a slight increase in microhardness in the areas of platelike pearlite (up to ~2,800 MPa) and spheroidized pearlite (up to ~2,300 MPa). Maximum steel hardness was observed in the explosives area, i.e. where the blast wave impact was the greatest.

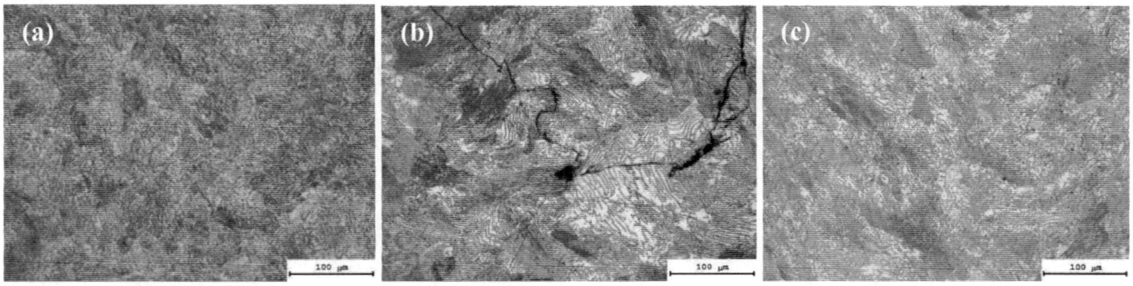

Figure 3. Photos of No.1 samples microstructure in the initial state (a), after treatment without a supporting plate (b) and with a supporting plate (c)

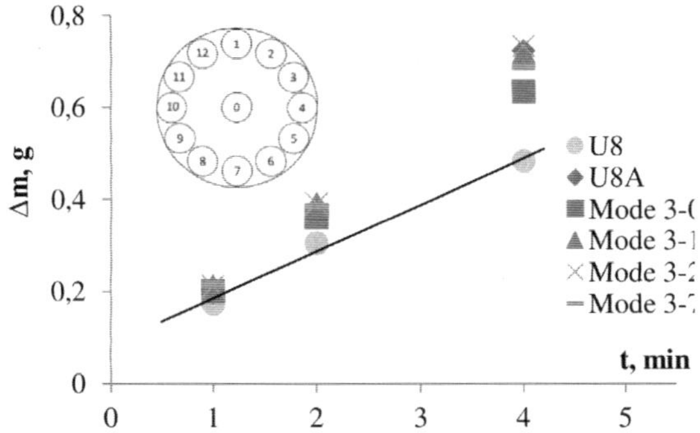

Figure 4. Mass loss dependence of the duration of abrasive wear tests. In Figure 4:

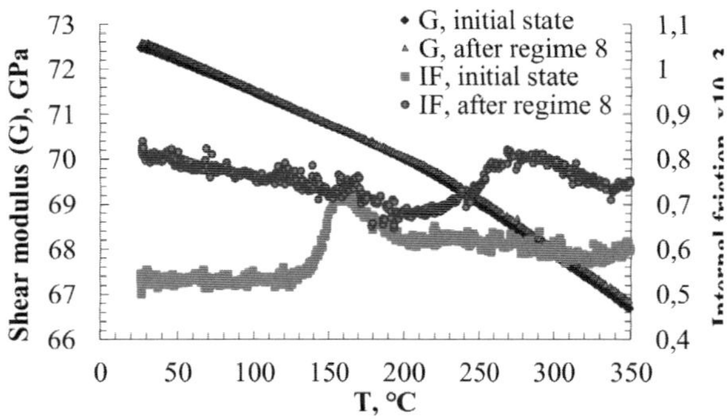

Figure 5. Temperature dependences of the shear modulus and internal friction for U8A steel in its

Mode 3 means the case of loading with a supporting plate of δ=2 mm, ρ=1.9 g/cm3, G=2.758 MPa. 0, 1, and 7 in diagram legend denote sample numbers

initial state and after blast treatment (with a supporting plate)

The distribution of macrohardness was fairly homogeneous in the cross-section of the workpiece for each of the blast treatment modes. The average Rockwell hardness is at 16-18 HRC. The distribution of macrohardness in the cross-section of the workpiece is fairly homogeneous, with a slight decrease by 2-3 HRC in the central part. Wear tests showed that the loss of mass for the U8A steel samples in the initial state and after blast treatment is greater than for samples of U8 normalized steel (see Figure 4). In addition, blast treatment did not result in any marked increase in wear resistance of the steel.

The studies of internal friction have shown that in the initial state there is a peak of internal friction at the heating temperature of ~160 °C, probably of a dislocation nature (Figure 5). Blast treatment results in an increase in internal friction, which is likely to be caused by an increase in the density of lattice dislocations. The internal friction and temperature dependence diagram clearly shows an intense Snoek-Köster peak at ~270 °C. The result indicates that blast treatment increases carbon concentration at the cores of lattice dislocations, which is likely due to the "withdrawal" of carbon atoms from the plates of Fe_3C cementite. This result is qualitatively well in line with the previously described effect of spheroidization and strain-induced dissolution of cementite plates in carbon steel [6].

4. Conclusions

It is shown that the explosive loading process leads to relatively insignificant plastic deformation and an increase in the flow stress of U8A carbon steel. The finite-element analysis demonstrates that deformation is uneven in the longitudinal and cross-section of the workpiece while increasing the supporting plate thickness and reducing its density leads to the hardening of the steel workpiece.

It has been established that blast treatment does not lead to a noticeable change in the parameters of the U8A steel microstructure but causes a slight increase in hardness in the areas of platelike and spheroidized pearlite.

The internal friction method demonstrated that blast treatment increases the concentration of carbon in the cores of lattice dislocations, which is likely due to the "withdrawal" of carbon atoms from the plates of Fe_3C cementite. This leads to their partial spheroidization and reduces the wear resistance of the steel. The process of spheroidization and strain-induced dissolution of cementite plates negatively affects wear resistance of carbon steel and "compensates" for a possible increase in wear resistance and hardness of carbon steel during its deformation hardening through blast treatment.

Acknowledgments
The study was performed with the support of the Ministry of Science and Higher Education of the Russian Federation (project No. 075-03-2020-191/5).

References

[1] Liu F C, Lv B, Zhang F C, Yang S 2011 Enhanced Work Hardening in Hadfield Steel during Explosive Treatment *Mater. Lett.* **65** 2333–36
[2] Wiedorna J, Davesa W, Ossbergerb U, Ossbergerb H, Pletz M 2019 Investigation of deformation mechanisms in manganese steel crossings using FE models *Tribol. Inte.* **138** 424–34
[3] Johnson G R and Cook W H 1985 Fracture characteristics of three metals subjected to various strains, strain rates, temperatures and pressures *Eng. Fract. Mech.* **21** 31–48
[4] Gulyaev A P 1977 *Metal science* (Moscow: Metallurgy) pp 429-430
[5] Sysoev A N, Gryaznov M Y, Chuvil'deev V N 2000 *Determination of the diffusion coefficient of carbon atoms in the α-Fe lattice (Solid State Physics)* (Nizhny Novgorod: Lobachevsky State University Press) pp 3–26
[6] Schastlivtsev V M, Mirzaev D A, Yakovleva I L, Okishev K Y, Tabatchikova T I, Khlebnikova Y V 2006 *Perlit in carbon steels* (Yekaterinburg: Ural Branch of RAS) pp 137-144

Cracking of Ti_2AlNb-based alloy after laser beam welding

D O Panov[1,4*], S V Naumov[1,5*], V S Sokolovsky[1], E I Volokitina[1], N Kashaev[2], V Ventzke[2], R Dinse[2], S Riekehr[2], E A Povolyaeva[1], E B Alekseev[3], N A Nochovnaya[3], S V Zherebtsov[1], G A Salishchev[1]

[1] Laboratory of Bulk Nanostructured Materials, Belgorod State University, 85 Pobeda Str., 308015, Belgorod, Russian Federation
[2] Department of Laser Processing and Structural Assessment, Institute of Materials Research, Materials Mechanics, Helmholtz-Zentrum Geesthacht, Max-Planck-Str. 1, 21502 Geesthacht, Germany
[3] Federal State Unitary Enterprise "All-Russian Scientific Research Institute of Aviation Materials", 17 Radio Str., 105005, Moscow, Russia

*Corresponding author: panov_d@bsu.edu.ru, NaumovStanislav@yandex.ru

Abstract. Crack formation process in the Ti-23Al-23Nb-1.4V-0.8Zr-0.4Mo-0.4Si (at.%) alloy based on orthorhombic titanium aluminide (Ti_2AlNb), especially in the weld zone (WZ), heat-affected zone (HAZ), and base metal (BM), during laser beam welding (LBW) at room temperature was considered. It was determined that the crack spread in the WZ throughout the crystallites of β-phase. In the HAZ, cracks spread throughout the globular β-phase crystals body and along the edges of globular α_2- and O-phases. The cracking stopped in BM due to a mechanism similar to shear ligament toughening. Cracking can be effectively suppressed by increasing the LBW temperature.

1. Introduction

Alloys based on orthorhombic titanium aluminide (Ti_2AlNb) are one of the most promising high-temperature alloys for the aerospace and automotive industries due to their high specific strength characteristics, creep and oxidation resistance, low density, and usage temperatures up to 750 °C [1]. However, the fabrication of structures of these alloys requires the usage of different welding techniques. Meanwhile, welding of Ti_2AlNb-based alloys usually results in the formation of pores or cracks in the welded joint. Cracks not only reduce the welded joint strength [2], but also lead to premature destruction of the welded structure [3] and, therefore, are unacceptable defects.

Crack formation in the Ti_2AlNb-based alloys depends strongly on the welding method and subsequent heat treatment mode [3,4]. To prevent cracking, special approaches such as solder design, alternating nanometric layers, welding using high currents, various structures of metal interlayers (brazing and diffusion welding), preheating are used [5,6]. Solders and interlayers can reduce residual stress and formation of brittle phases, but the intermediate metal affects the mechanical properties of the weld. In electron beam welding [7,8], local heating can be used to, for example, to reduce the residual stresses after welding by 30%. At the same time, they can also increase the grain size in the weld and heat-affected zone [7,8].

Laser beam welding (LBW) can be considered as one of the most promising techniques for the Ti_2AlNb-based alloys due to high energy density, relatively narrow welds, and no need for the inert gas atmosphere for the process. It is known that during laser welding of titanium alloys, weld ductility deteriorates due to gas saturation, and heat-affected zone is subjected to greater atmospheric influence and local stresses due to low thermal conductivity [9]. If local stresses exceed the critical values at the temperatures when the material exhibits limited ductility - cracking occurs [9,10].

Thus, to understand the ways of solving the problem of cracking during LBW, it is necessary to study the process of welding cracks formation. This article is devoted to the investigation of the crack formation process and growth in a welded joint made of an alloy based on Ti_2AlNb after LBW.

2. Experimental

The Ti-23Al-23Nb-1.4V-0.8Zr-0.4Mo-0.4Si (at.%) alloy based on Ti_2AlNb with the initial globular structure was chosen as the program material in the present study. Plates with the following dimensions were used: length 22 mm, width 15 mm, and thickness 2 mm. LBW of two plates was carried out in the lower position (PA ISO 6947) in a butt joint without grooving and filler materials at room temperature (RT) and 400 °C. LBW was performed using an 8.0 kW fiber laser in a controlled atmosphere chamber (Ar 4.6, gas flow rate 40 L / min). The following optimal LBW modes were determined in the experiment - a laser power of 2.5 kW, a welding speed of 3-5 m / min, with a laser defocusing distance of 0.0 mm. Images of the microstructure were obtained on a FEI ESEM Quanta 200 scanning electron microscope using a backscattered electron (BSE) detector at an accelerating voltage of 30 kV. The fine structure was examined on a JEOL JEM-2100 transmission electron microscope (TEM) at an accelerating voltage of 200 kV. Foils for TEM were made from 0.3 mm plates with subsequent thinning to a thickness of 0.1 mm. Next, these foils were subjected to jet electrolytic polishing on Struers Tenupol-5 with an electrolyte (60 ml $HClO_4$, 600 ml methanol, 360 ml butanol) at a voltage of 27 V and temperature of -30 °C. After electrochemical polishing, the foils were washed in distilled water and ethanol and dried.

3. Results and discussion

The initial structure of the examined alloy consisted of equiaxed α_2- and O-phase grains/particles with a mean size of 1.5 μm (Figure 1a). β-phase layers with a thickness of 0.4 ± 0.1 μm were found at the boundaries of the α_2/O grains/particles. TEM observations have revealed a rather high dislocation density inside the O-phase grains/particles. In some cases, dislocations were arranged in subgrain boundaries. In turn, the grains of the α_2-phase were mostly free of dislocations. However, many bending contours inside the grains indicated strong internal stresses (Figure 1b).

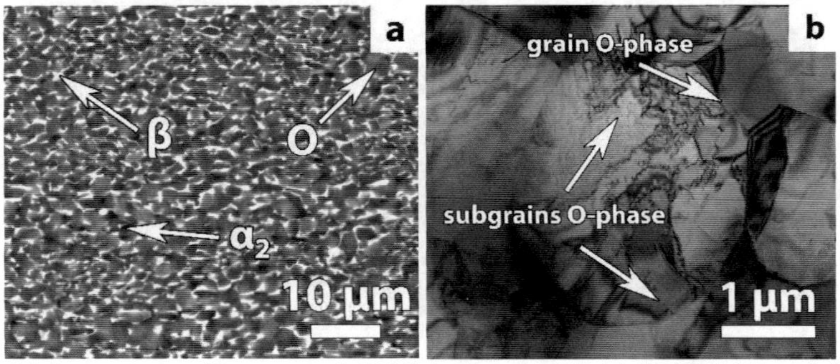

Figure 1. The initial structure of base metal (BM) before LBW (a) SEM-BSE and (b) TEM images.

After the LBW of the studied alloy at RT, a transverse crack appeared, propagating over the entire width of the weld zone (WZ) and the heat-affected zone (HAZ) that ended only in the base metal (BM) (Figure 2). The crack appeared in the center of the joint symmetrically to the weld axis (Figure 2a). Most likely, the crack was formed due to the tensile stresses that were generated because of the bending of the welded plates. The plates were likely bent because the laser heated them from only one side during welding. The crack propagated in step-wise nature; the crack boundaries were sharp. Besides, secondary smaller cracks were found near the main one.

Fracture in the WZ area occurred predominantly throughout crystallites of the β-phase. The crack in this area had a stepwise shape. In the HAZ, the crack propagated along the β-phase globular particles (Figure 2b). In this case, the formation of secondary cracks parallel to the main one was also observed.

In BM, cracking propagates mainly along the edges of the β-phase interlayers and globular crystals of α_2 and O-phase (Figure 2c).

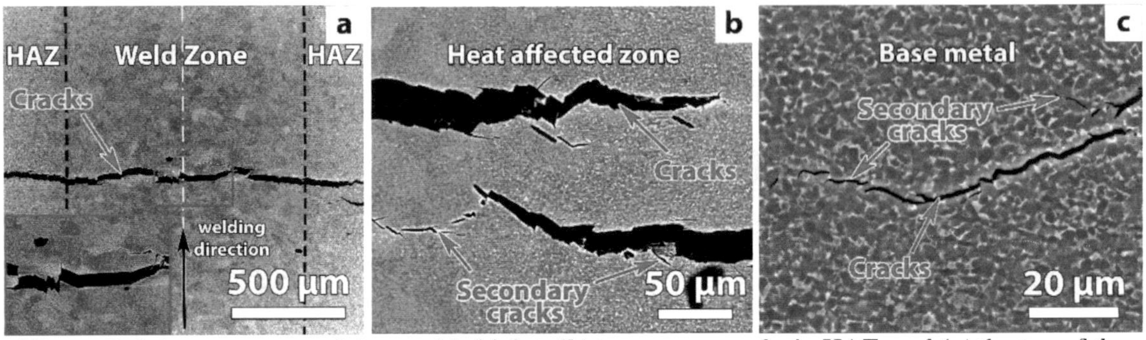

Figure 2. *(a)* transverse crack in a welded joint, *(b)* transverse cracks in HAZ, and *(c)* the top of the crack in base metal after LBW.

Cracking of Ti_2AlNb-based alloys with transverse cracks formation has been associated with high HAZ cooling rates (> 400 K/s) that resulted in the propagation of cracks into the base metal. The cracks deviated to an angle of 45° before blunting and stopping [10]. Crack development in the HAZ caused brittle transgranular cleavage [7], while crack propagation in the BM was prevented by a mechanism similar to shear ligament toughening with a deviation at an angle of 45° [11].

To avoid cracking in the Ti_2AlNb-based alloys after the LBW, it was necessary to reduce the HAZ cooling rate. This can be effectively done by pre-welding heating. For example, photographs of the welds obtained at RT and 400 °C are shown in Figure 3. The presence of cracks perpendicular to the welding direction after RT LBW is evident. In turn, no defects were found after LBW at 400 °C. This example shows that cracking can be effectively suppressed by increasing LBW temperature. However, attention must be paid to the effect of the welding temperature on other aspects, in particular on the structure and mechanical properties of the weld.

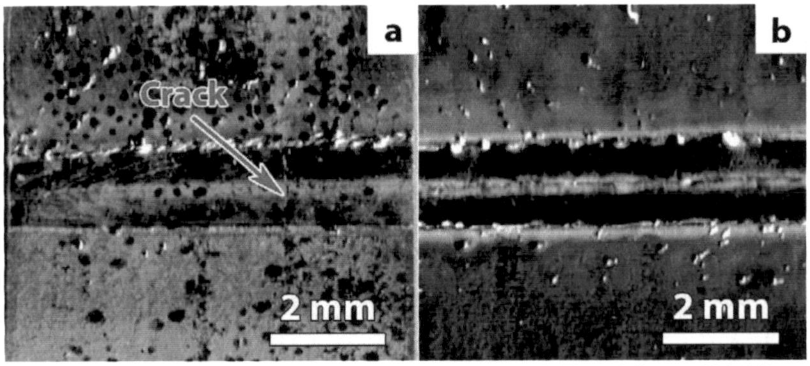

Figure 3. Photographs of joints welded (a) at RT and (b) at 400 °C.

4. Conclusions

The Ti-23.0Al-23.0Nb-1.4V-0.8Zr-0.4Mo-0.4Si (at.%) alloy after laser beam welding at room temperature was examined. Extensive cracking after welding was found. The crack growth in the WZ area occurred predominantly throughout crystallites of the β-phase. The crack had a stepwise profile. In the HAZ, the crack propagated along the β-phase globular particles. The formation of the secondary crack occurred parallel to the main one. In the BM, cracking propagated mainly along the edges of the β-phase interlayers and globular crystals of α_2 and O-phase. The crack growth in the BM stopped due to a mechanism similar to shear ligament toughening. It was also demonstrated that cracking can be effectively suppressed by increasing the welding temperature to 400 °C.

Acknowledgments
The authors gratefully acknowledge the financial support from the Russian Science Foundation Grant no. 19-79-30066.The authors are also grateful to Joint Research Center "Materials and Technologies", Belgorod State University, for the assistance with instrumental analysis.

References

[1] Banerjee D, Gogia A K, Nandi T K, Joshi V A 1988 A new ordered orthorhombic phase in a Ti3AlNb alloy *Acta Metall.* **36** 871-882
[2] Kulkarni A., Dwivedi D K, Vasudevan M 2020 Microstructure and mechanical properties of A-TIG welded AISI 316L SS-Alloy 800 dissimilar metal joint *Mater. Science and Engineering A* **790** 139685
[3] Cai D, Chen J, Mao X, Hao C 2013 Reheat cracking in Ti2AlNb alloy resistance spot weldments *Intermetallics* **38** 63-69
[4] Li Y-J, Wu A-P, Li Q, Zhao Y, Zhu R-C, Wang G-Q 2019 Mechanism of reheat cracking in electron beam welded Ti2AlNb alloys *Transactions of Nonferrous Metals Society of China (English Edition)* **29** 1873-1881
[5] Li H 2020 The causes and control of porosity in titanium alloy welding *Proceedings 3rd International Conf. on Electron Device and Mechanical Engineering ICEDME 2020* **9122191** 490-493
[6] Chen G, Zhang G, Yin Q, Zhang B, Feng J 2020 Research Progress on Welding Crack Control of TiAl Alloy (Review) [TiAl合金焊接裂纹控制研究进展] *Cailiao Daobao/Mater. Reports* **34** 05115-05119
[7] Lobanov L M, Asnis E A, Piskun N V, Vrzhizhevskii É L, Mikhodui O L, Statkevich I I 2020 Structure, Mechanical Properties, and Stressed State of Weld Joints of the Tial(Nb, Cr, Zr) Intermetallic System *Strength of Mater.* **52** 199-204
[8] Zhan Z, Chen Y, Wang S, Huang Y, Mao Y 2019 Prevention of Crack Formation in Electron-Beam Welded Joints of Dissimilar Metal Compounds (TiNi/Ti6Al4V) *Metal Science and Heat Treatment* **61** 373-378
[9] Li H 2020 Analysis and control of laser weld crack *Proceedings 3rd International Conf. on Electron Device and Mechanical Engineering ICEDME* 2020 **9122119** 192-195
[10] Patterson R A, Martin P L, Damkroger B K, Christodoulou L 1990 Titanium aluminide: electron beam weldability *Welding Journal* **1** 39–44
[11] Chan K S 1995 Evidence of Shear Ligament Toughening in TiAl-Base Alloys *Metall. And Mater. Trans. A V* **26A** 1407-1418

IOP Conf. Series: Materials Science and Engineering 1014 (2021) 012036 doi:10.1088/1757-899X/1014/1/012036

Investigation of the impulse impact effect in the process of additive manufacturing of metal products by plasma cladding

S D Neulybin [1*], Yu D Shchitsyn [2], G L Permyakov [2], D S Belinin [2] and S V Varushkin[2]

[1] Institute of Continuous Media Mechanics of the Ural Branch of Russian Academy of Science, Academica Koroleva, 1, Perm, 614013, Russian Federation

[2] Perm National Research Polytechnic University, Komsomolsky av., 29, Perm, 614990, Russian Federation

*Corresponding author: sn-1991@mail.ru

Abstract. Additive technologies (AT) or layered synthesis technologies are one of the most dynamically developing areas of "digital" production. A common problem of additive technologies is ensuring the proper microstructure of the synthesized material and the eliminating defects. The use of filler wire as a working material eliminates the problems associated with the low productivity of existing methods, the high cost of the equipment used, the limited types of materials used, due to the use of powder systems. Products made of stainless chromium-nickel steels are widely used in various industries. The main problem of additive technologies is to ensure the properties of the laminates are not lower than those obtained by traditional methods. Characteristic defects of laminated materials obtained by surfacing are increased porosity, non-metallic inclusions, a decrease in ductility, and for high-alloy steels, a loss of special properties. This predetermined the development of research in the field of additional technological measures to improve the final properties of the product. This paper presents the results of a study of the effect of ultrasonic vibrations on the structure and properties of the deposited steel 12Cr18Ni10Ti. For welding was used the wire 12Cr18Ni10Ti, process was carried out by plasma arc in a protective argon gas. It was found that ultrasonic treatment has an effect on the final grain size, structure formation and hardness, as well as on the geometry of the deposited layer. Studies show that the use of ultrasonic vibrations in the process of surfacing can be applied in the design of equipment for the implementation of processes of additive production.

1. Introduction

One of the development ways in modern engineering is improve durability of manufactured products [1]. Recently, additive technologies have become common, but their widespread introduction into production is discouraged by insufficient information about the effect of layer-by-layer metal surfacing on the properties of finished products and the lack of technological recommendations for producing a defect-free structure and required mechanical properties. This predetermined the need to search for technological methods that provide necessary properties of the metal of products during additive manufacturing [2-3].

Obtaining a fine-grained structure of the surfaced metal allows, as practice shows, to provide high mechanical properties of the welded joint, as well as to increase the technological strength, corrosion

IOP Conf. Series: Materials Science and Engineering 1014 (2021) 012036 doi:10.1088/1757-899X/1014/1/012036

resistance and other special properties, increase the service life of the product as a whole [4]. Application of various impacts on the molten pool during surfacing is one of the possible mechanisms for modifying the deposited metal [5].

The existing specifics of cast metal crystallization during welding and surfacing allow us to establish the harmonic frequency of solidification of the metal. Some researchers believe that the main reason for cyclicity of crystallization is the release of latent heat of melting that inhibits the advance of the solidification front, others consider concentration overcooling in front of the interface boundary, and still others consider the periodic heat supply to the boundary to be the reason. However, there is still no consensus [6].

Currently, there is a number of technological methods enabling to influence crystallization of the metal during welding and surfacing in order to obtain a grinding structure. For these purposes we use: vibration of the welded product, energy input from a pulsed heat source, pulsed filler wire in different spatial positions, sources with modulation of the welding current acting on the liquid metal of the tail section of the pool, as well as, in some cases, the input of separate heat sources influencing the heat-affected zone near the seam axial line, external electromagnetic effects on the melt, both in continuous and in pulsating mode with different frequencies of the magnetic field, etc.

Some authors attributed the conditions for the formation of the weld metal structure to the effect of heat supply on the solidification front. Other researchers believed that vibrations of the molten metal in the welding pool played a significant role in the formation of the structure.

2. Research methodology

For manufacturing critical structures, steel grade 12Cr18Ni10Ti is most often used among high-alloy steels. This can be explained by the fact that austenitic chromium-nickel steels, such as steel 12Cr18Ni10Ti, have high ductility and resilience, high mechanical properties at low temperatures, resistance to oxidation in air and in the atmosphere of fuel combustion products at temperatures below 900 °C and good weldability. However, having a unique complex of physical and mechanical properties, the steel under consideration has several disadvantages: relatively low strength and hardness, susceptibility of intercrystalline corrosion of the welds. The authors in [7] established that ultrasonic action has the most favorable effect at a 30 mm distance from the waveguide.

This paper proposes a method for improving the quality of deposited layers of steel 12Cr18Ni10Ti with layer-by-layer plasma surfacing during additive manufacturing of parts. To study the possibilities of modifying the structure, we used an impulse impact action on the deposited material on a EWM Tetrix 400 Plasma welding machine.

Plasma surfacing was made on 10 mm thick plates of steel 12Cr18Ni10Ti. The plasma arc was performed according to the mode given in Table 1. The impulse impact is applied directly in the arc gap.

Table 1. Plasma surfacing mode

Middle surfacing current, A	Surfacing voltage, V	Plasma argon consumption, l/min	Feed rate, m/min	Action frequency, Hz	Distance from sample to welding torch nozzle section h, mm
195	22	3	2,5	0-15000	10

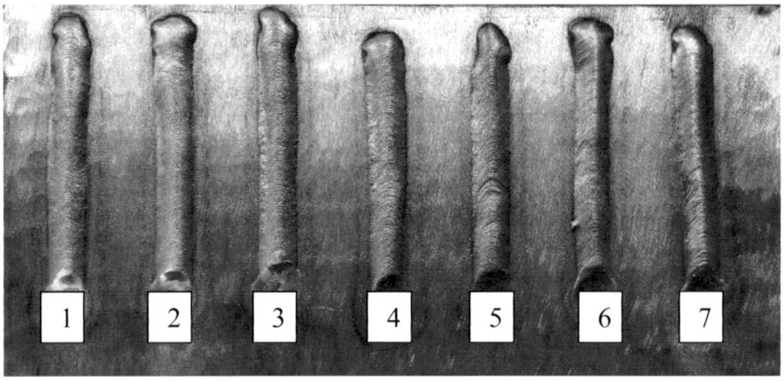

Figure 1. Samples of plasma surfacing. 1 – 15000 Hz; 2 – 10000Hz; 3 – 5000 Hz; 4 – 1000 Hz; 5 – 500Hz; 6 – 50 Hz; 7 – not frequency

3. Results and its consideration

At the first stage, the comparative parameters are the width and height of the deposited bead, relative deviations in width and height. These studies will help to understand the stability of the roll formation over time and to choose the optimal mode for layer-by-layer synthesis. Table 2 shows the results of measurements of the geometric parameters of the rollers depending on the frequency.

Table 2. Results of measurement of parameters of surfacing welds of studied samples

№ sample	Surfacing weld parameters			
	Weld width, mm	Weld height, mm	Width relative deviations, %	Height relative deviations, %
1	8.27-8.84	1.12-1.37	93.55	81.75
2	7.57-9.14	1.14-1.35	82.82	84.44
3	8.08-8.92	1.02-1.41	90.58	72.34
4	8.28-9.68	1.06-1.55	85.53	68.38
5	8.63-8.82	1.11-1.27	97.84	87.40
6	7.78-9.26	1.06-1.55	84.01	68.38
7	8.15-10.17	1.02-1.38	80.13	73.91

4. Conclusions

1. Studies of the effect of impulse action during the surfacing cycle on the constancy of the geometry of the deposited bead have been carried out;
2. It was found that the greatest stability in the formation of a single bead, all other conditions being equal, is provided by modes 1 and 5, which corresponds to frequencies of 15000 Hz and 500 Hz;
3. For the final summing up, more detailed studies of metallography of the seams with the measurement of the micro hardness of characteristic areas and grain size are required.

Presumably, mode 1 should provide a more finely dispersed structure due to high-frequency exposure, due to which a large number of new crystallization centers are formed.

Acknowledgments

The reported study was funded by RFBR, project number **№ 19-38-60043\19,** and The work was carried out with the financial support of the Ministry of Science and Higher Education of the Russian Federation **FSNM-2020-0028.**

References

[1] Shchitsyn Yu D, Trushnikov D N, Akulova S N, Myshkina A V, Neulybin S D and Dushina A Yu 2019 Development of a technology for layer-by-layer growing of a workpiece made of high-alloy steel by plasma surfacing *Metallurg* **2** pp70-77

[2] Sutyrin G V 1975 Investigation of the mechanism of the effect of low-frequency vibration on the crystallization of the weld pool *Automatic welding* **5** pp7-9

[3] Aristov S V Russo V L 1982 Crystallization of the weld metal at low-frequency vibrations of the melt *Welding production* **11** pp42–44

[4] Petrov A V Birman U I 1968 Crystallization of weld metal during pulse-arc welding *Welding production* **6** pp1-3.

[5] Shchitsyn Yu D, Krivonosova E A, Neulybin S D, Olshanskaya T V, Nikulin R G, Fedoseeva E M and Terentyev S A 2019 The use of plasma surfacing for the additive formation of billets from aluminum alloys *Bulletin of the Perm National Research Polytechnic University. Mechanical engineering, materials science* **21** pp63-73

[6] Alov A A Vinogradov V S 1957 Improving the quality of weld metal *Welding production* **7** pp 9–10

[7] Krivonosova E A, Schitsin Yu D, Trushnikov D N, Myshkina S N, Akulova A V, Neulybin S D and Dushina A Yu 2019 Influence of surfacing technologies on structure formation of high-temperature nickel alloys *Metallurgist* **63** pp197-205

IOP Conf. Series: Materials Science and Engineering 1014 (2021) 012037 doi:10.1088/1757-899X/1014/1/012037

Studying Corrosion Resistance of Weld Joints of Ultrafine-Grained Titanium Alloys

A V Nokhrin[1], P V Andreev[1*], C V Likhnitskii[1], M M Vostokov[1], V I Kopylov[1,2], V N Chuvil'deev[1], D A Gudz'[1], M K Chegurov[1] and A A Murashov[1]

[1] Lobachevsky State University of Nizhny Novgorod, 23 Gagarina ave., Nizhny Novgorod, 603950, Russian Federation

[2] Physical and Technical Institute, National Academy of Sciences of Belarus, 10 Kuprevich st., Minsk, 220141, Belarus

[*]Corresponding author: andreev@phys.unn.ru

Abstract. Ultrafine-grained (UFG) samples of PT-3V near-α titanium alloy were subjected to diffusion welding using the Spark Plasma Sintering system. It has been shown that the destruction of welded UFG samples under hot salt corrosion (HSC) is two-staged – first, intercrystalline corrosion is manifested, which then graduates to pitting corrosion. It has been established that the corrosion resistance of weld joints is determined by vanadium concentration at the grain boundaries, the size and volume ratio of β-phase particles and the presence of pores in the weld joints. It has been shown that welded UFG samples possess higher hardness and corrosion resistance than coarse-grained samples.

1. Introduction

Currently, forming ultrafine-grained (UFG) structure in titanium alloys by severe plastic deformation (SPD) represents one of the promising ways to improve their performance. UFG titanium alloys are characterized by increased strength and better creep resistance and fatigue strength [1, 2]. Welding of UFG titanium alloys is the most important challenge of their use. The commonly used argon-arc or electron-beam welding will not maintain the UFG structure. Spark plasma sintering (SPS) [3, 4] is one of the latest techniques for solid state welding of structural materials, including titanium alloys [5, 6]. An important advantage of the SPS system is the ability to implement high heating rates (up to 2500 °C/min), which will maintain the UFG structure with high mechanical properties in titanium alloys.

The purpose of this study is to research the decay behavior of titanium near-α alloy weld joints under hot salt corrosion (HSC). It should be noted that HSC is one of the worst hazards for titanium alloys in nuclear power engineering [7]. This type of corrosion occurs when salt deposits are formed on the surface of heat exchange equipment operating at elevated temperatures (250-270 °C).

2. Methods and materials

PT-3V coarse-grained (CG) near-α titanium alloy (Ti-4.73Al-1.88V) is the object of the study. The UFG structure was formed by Equal Channel Angular Pressing (ECAP). A Dr. Sinter SPS-625 spark plasma sintering was used to weld 7×7×3.5 mm samples. Heating rate (V_h) was varied from 10 to 350 °C/min, pressure (P) – from 50 to 100 MPa, welding time – from 0 to 90 min. Welding was performed in the α-phase field (600 °C), near the (α↔β)-phase boundary (700 °C), at the beginning of the two-phase (α+β)-field (800 °C) and in the β-phase field (1030, 1140 °C). Welding was performed in a vacuum, without molds, which caused tangent stresses, perpendicular to the pressure axis.

A Jeol JSM-6490 scanning electron microscope were used for microstructure studies. A Shimadzu XRD-7000 diffractometer was used for X-ray diffraction analysis (XRD). A Duramin Struers 5 durometer was used for microhardness (H_v) measurements. HSC tests were performed in a 300:1 mixture of NaCl and KBr crystal salts at a temperature of 250 °C over 500 hours. A Leica DM IRM metallographic microscope was used to assess corrosive damage to the surface of samples.

3. Experimental results

Structure studies show incomplete fusions of ~50–70 μm (Figure 1a,b) at the edges of CG samples welded at low pressures, low temperatures or high heating rates. At the top of macrodefects there are large pores, the volume fraction of which decreases farther from the edge of the sample (away from the areas of the tensile stresses). The remaining larger area of the weld joint has micropores, the volume fraction of which depends on the SPS mode (Figure 1 c,d). A low volume fraction of pores is observed following diffusion welding of CG samples at 700-800 °C and at low pressures (50 MPa).

There are virtually no macrodefects in the weld seams of UFG alloys, the volume fraction of macropores and micropores is small and is observed chiefly after welding at low temperatures (600 °C) or at high pressures (70-100 MPa). Varying basic SPS parameters has little effect on the average grain size in CG samples with an α/α'-structure near weld joints where β-phase particles vigorously precipitate along the boundaries of elongated grains of the α'-phase.

The volume fractions of β-phase particles in the welded metal are greater than in the original CG alloy. Increased heating rate, pressure or time of exposure reduce the size of the β-phase particles. It should be noted that the grain size in the welded UFG samples is 3-4 times less than in the CG samples while occasional β-phase particles of 2-3 μm are observed at the grain boundaries. An increase in the SPS temperature and time leads to an increase in the particle size of the β-phase and grain size of the UFG alloy. The heating rates have no marked impact on the size of the β-phase particles while producing smaller UFG grains in the alloy.

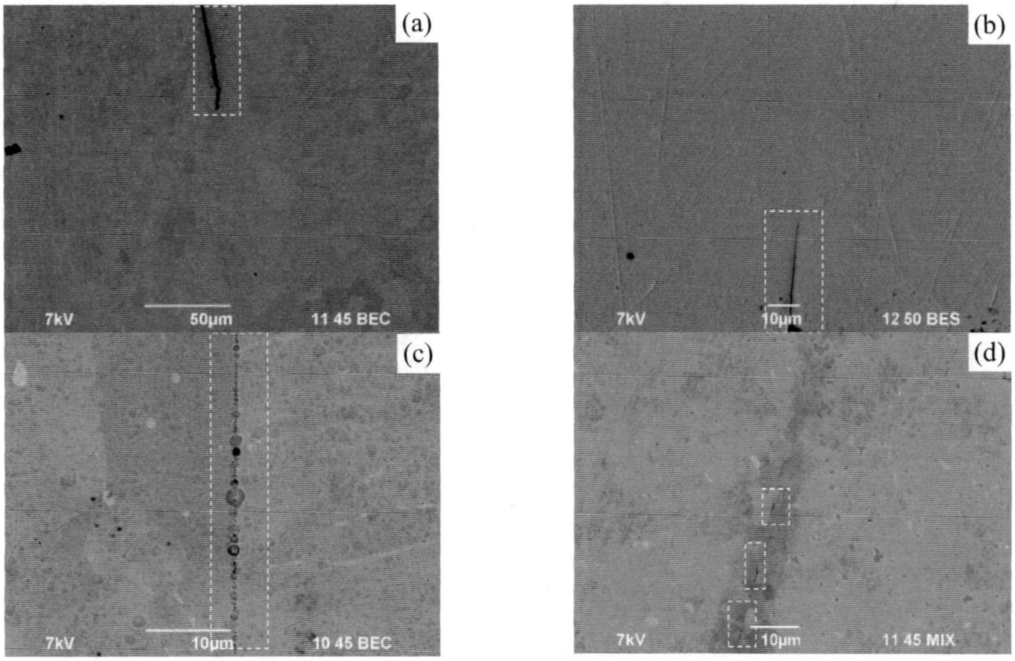

Figure 1. Typical defects in weld joints of CG (a, c) and UFG (b, d) alloys after diffusion welding (50 °C/min, 800 °C, 10 min): (a, b) defective weld, (c, d) micropore

XRD results demonstrate that salt deposits on the surface of the samples after HSC tests are mixtures of NaCl, titanium oxides (TiO_2, TiO) and alumina, as well as vanadium-based constituents ($VO_{1.15}$, $V_2Ti_3O_9$) and Al_3V. Instead of vanadium-based constituents, aluminum-containing constituents ($TiAl_2Cl_8$, $AlTi_3$)and traces of VO_2 were detected in corrosion products of samples welded at temperatures above 1000 °C.

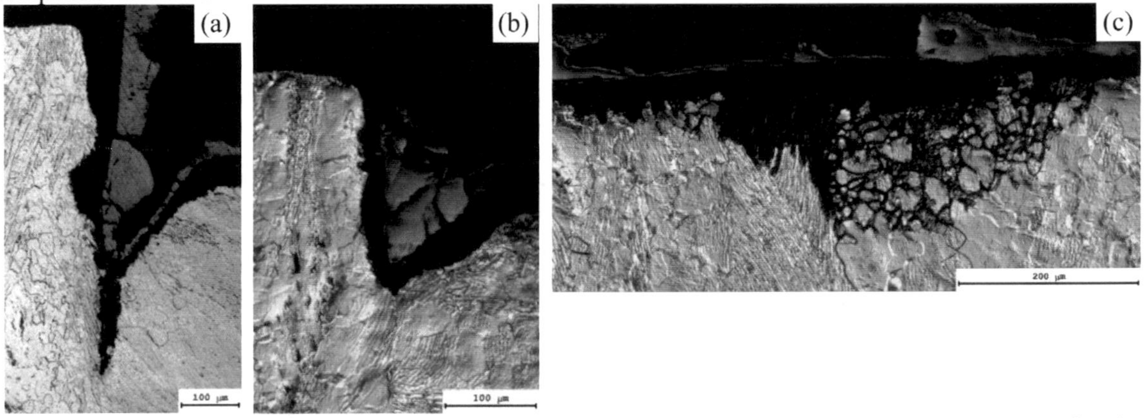

Figure 2. Corrosion defects on the surfaces of CG samples after HSC tests: (a, b) crevice corrosion in Zone I; (c) intercrystalline corrosion in Zone II

The results of the study of weld joints demonstrate that corrosion mechanisms are different in the weld joint and outside it. In the area of CG alloy joints, there is a combination of pitting and crevice corrosion (Figure 2a, b), with crevice corrosion more than 300 μm deep in highly porous seams. Farther away from weld joints, there is a combination of intercystalline corrosion (ICC) and pitting corrosion. The heaviest corrosion is observed after welding in the α'-phase field with an increased volume fraction of β-phase particles, which leads to a heterogeneous distribution of corrosion pits on the surface of the samples. ICC can be observed under large corrosion pits (Figure 2c). Consequently, ICC sets in fine-grained α'-phase fields during the first stage of HSC and pitting corrosion develops next. The pressure rates and time of exposure have the highest impact on corrosion depth in weld joints of the CG samples. It should be noted that the size of corrosion defects in CG samples is smaller than in the PT-3V alloy in its initial state (see [7]).

UFG samples manifested crevice corrosion only in weld joints with higher porosity. Farther away from the welding line, a combination of ICC and pitting corrosion is observed (Figure 3), as in the case of CG samples. An increase in welding time and temperature leads to an increased corrosion depth in UFG alloys weld joints (Table 2). For lower welding temperatures, heating rates or pressure, the corrosion was quite shallow without exceeding the ICC depth in post-ECAP UFG alloys (see [8]). Welding at 700 °C, near the (α + β)-phase boundary at low pressure (50 MPa) and a heating rate of 10 °C/min proves to be optimal – with an average corrosion depth of 110-130 μm and no crevice corrosion in weld seams. It should be emphasized that for these welding modes the average pitting depth in UFG samples was 1.5-2 times less than that in CG samples. Hence, it can be concluded that UFG samples have a higher corrosion resistance compared to CG samples.

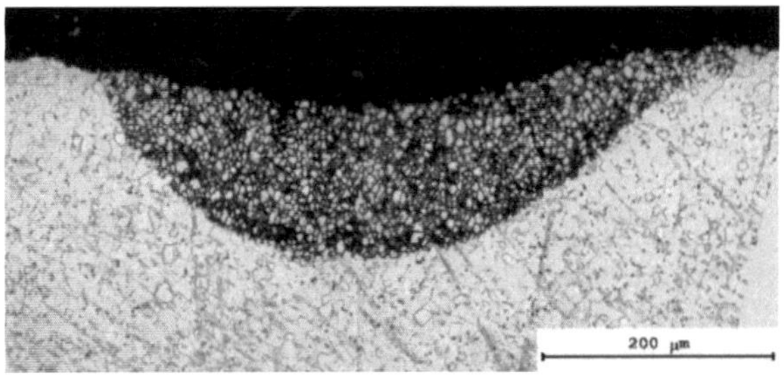

Figure 3. Corrosive defects on the surface of UFG samples after HSC tests

The results of HSC tests on UFG samples demonstrate that increasing welding temperature from 600 to 800 °C leads to an increase in the average depth of corrosion pits from 124±34 μm up to 285±59 μm, with the maximum corrosion depth growing from 188 up to 400-420 μm. Increasing heating rates from 50 to 350 °C/min reduces the average corrosion depth from 304±51 down to 189±36 μm. The applied pressure and time of exposure had low-to-no impact on corrosion depth in the UFG samples.

Hardness studies have suggested that SPS modes have little impact on H_v of seams and the parent metal in the CG samples. Hardness of the CG samples is 2.4-2.6 GPa. Higher hardness of the samples versus PT-3V alloy in its initial state (~2.1–2.2 GPa) may be due to mechanical hardening of the alloys. This finding is indirectly confirmed by seam hardness increasing to 2.7 GPa when welded at a pressure of 100 MPa. The hardness of the seams and parent metal in the UFG samples is higher than that of the CG samples by ~0.4–0.8 GPa. The highest seam hardness (~3.0–3.2 GPa) is observed in UFG samples obtained at low temperatures, high heating rates and fast times of exposure. Increasing heating rates to 350 °C/min can form a high-density seam with UFG structure and high hardness (2.9-3.1 GPa).

4. Conclusions

UFG samples have been shown to have better weldability than coarse-grained alloys, weld joints of which manifested incomplete fusions, causing crevice corrosion. There were virtually no seam macrodefects in the UFG samples while seam porosity was very low, which ensures high resistance of weld joints to crevice corrosion in the UFG samples.

The corrosion resistance of the CG welded samples is higher than that of PT-3V alloy in its initial state. Corrosive defects are distributed across the surface of CG samples unevenly and are primarily concentrated in the field of the fine-grained α'-phase, the grain boundaries of which manifest a higher volume fraction of the β-phase. Destruction of CG samples under HSC is two-staged – first, ICC is manifested, which then graduates to pitting corrosion.

Acknowledgments

The study was performed with the support of the Russian Science Foundation (Grant 19-73-00295).

References

[1] Zherebtsov S, Kudryavtsev E, Kostjuchenko S, Malysheva S, Salishchev G A 2012 Strength and ductility-related properties of ultrafine grained two-phase titanium alloy produced by warm multiaxial forging *Materials Science and Engineering* **A536** 190-196.

[2] Chuvil'deev V N, Kopylov V I, Nokhrin A V, Tryaev P V, Kozlova N A, Tabachkova N Yu, Lopatin Yu G, Ershova A V, Mikhaylov A S, Gryaznov M Yu, Chegurov M K 2017 Study of mechanical properties and corrosive resistance of ultrafine-grained α-titanium alloy Ti-5Al-2V *Journal of Alloys and Compounds* **723** 354-367.

[3] Chuvildeev V N, Panov D V, Boldin M S, Nokhrin A V, Blagoveshensky Yu V, Sakharov N V, Shotin S V, Kotkov D N 2015 Structure and properties of advanced materials obtained by Spark Plasma Sintering *Acta Astronautica* **109** 172-176.

[4] Tokita M 2013 *Spark Plasma Sintering (SPS) Method, Systems, and Applications* (Academic Press) pp 1149-1177.

[5] Chuvil'deev VN, Nokhrin AV, Kopylov VI, Boldin MS, Vostokov MM, Gryaznov MYu, Tabachkova NY, Tryaev P2019Sparkplasma sintering for high-speed diffusion bonding of the ultrafine-grained near-α Ti-5Al-2V alloy with high strength and corrosion resistance for nuclear engineering *Journal of Materials Science*. **54** 14926-14949.

[6] Yang J, Wang GF, Jiao X, Gu Y, Liu Q, Li Y 2018 Current-assisted diffusion bonding of extruded Ti-22Al-25Nb alloy by Spark Plasma Sintering: Interfacial microstructure and mechanical properties *J. Mater. Engin. Performance* **27** 3035-3043.

[7] Chevrot Th 1994 *Pressure effects on the holt-salt stress-corrosion cracking of titanium alloys*. (Cranfield University. School of Industrial and Manufacturing Science) 215 p.

Influence of nitriding parameters in radio-frequency inductive discharge plasma on the structural-phase state and properties of H12MF steel

V Y Novikov[1*], I Y Goncharov[2], M N Yapryntsev[1] and M Y Gazizova[1]

[1] Belgorod State National Research University, Belgorod 308015, Russia.
[2] Belgorod State Technological University named after V.G. Shoukhov, Belgorod 308012, Russia.

*Corresponding author: novikov_v@bsu.edu.ru

Abstract. The influence of technological parameters of plasma nitriding on the structural-phase state and mechanical properties of H12MF steel has been investigated in this work. The possibility of improving the corrosion resistance of the reported material was demonstrated by changing the content of nitride phases (Fe_3N, Fe_4N, CrN) in the surface layer, depending on the nitriding mode.

1. Introduction

The growing manufacturing of ultra-high-strength steels with tensile strength above 1500 MPa requires improvement of the life cycle of tools and dies. Die tool steel H12MF (the equivalent of the AISI D2) has become widespread in industry because of its high mechanical properties, especially due to the high abrasion resistance [1]. However, this steel has a relatively low corrosion resistance [2]. One of the most effective surface modification techniques to improve the main performance characteristics of tool steels is plasma nitriding [3, 4]. In this work, the nitriding process was performed by the high efficient plasma source based on a radio-frequency inductive discharge ("transformer coupled plasma" termed a TCP). This approach allows to create and continuously maintain a high-density gas discharge and makes it possible to independently control the energy and density of the ion flux bombardment to the substrate [5].

2. Materials and methods

For this study, samples with a size of $15 \times 20 \times 3$ mm^3 of H12MF steel were used as the base material. These samples were ground using several grades of sandpaper, and then polished using a 1 μm diamond suspension. The nitriding process was performed with "NIKA 2012" multifunctional vacuum unit. Cleaning was carried out by a RFPG-128 radio-frequency plasma source in the atmosphere of argon, supplied at a rate of 4 l/h. The pressure in the chamber was 0.2 Pa. The power of the source was 900 W, the cleaning time was 10 min, and the sample temperature did not exceed 150 °C, respectively. Nitriding was made in a nitrogen atmosphere, which was supplied at the rate of 4.5 l/h. The pressure in the chamber was 0.2 Pa. Nitriding time was equal to 180 min. The process was performed at a constant temperature of about 450 °C. A pyrometer was used for the control of the temperature. The temperature was maintained by supplying a different potential to the sample, depending on the power of the RFPG-128 source. There were prepared 3 series of samples at various values of the power source and the negative bias potential on the sample: sample 1 - 700W/300V, sample 2 - 800W/250V, sample 3 - 900W/200V.

The microstructure of the samples was studied by a FEI Quanta 200 3D scanning electron microscope (SEM). The phase-structural state was studied on a Rigaku Ultima IV diffractometer in Cu-Kα radiation. The microhardness was measured with a Vickers diamond pyramid using microindentation on an Instron microhardness test device 402MVD with a load of 100 g. The corrosion test was performed by an electrochemical method using a multichannel potentiostat-galvanostat P-20X8 and a standard three-electrode cell. A silver chloride electrode was used as a reference electrode, a carbon electrode acted as an auxiliary. All studies were performed at room temperature in the isostatic solution of 0.9% NaCl. The area of the sample surface was 1.0 cm^2. Studies of changes of a surface morphology before and after tests on corrosion resistance were performed using SEM.

3. Results and discussion

The initial microstructure of steel H12MF is represented by ferrite (α-Fe) and M_7C_3 carbides. The structure of the nitride layer consists of two zones: the compound layer and the diffusion zone (Figure 1). The compound layer is a layer on the surface with a thickness of ~5-10 µm, composed of the base metal and alloying nitride elements. Further, a smooth transition from the modified layer to the base material is observed, that is, the zone of nitrogen diffusion into the depth of the material. The depth of the modified layer is observed in the SEM images for all series of samples in the range of ~60-80 µm. This spread is associated with an irregularity in the thickness of the diffusion layer caused by the irregular distribution of carbides over the volume of the initial material. In elongated colonies of carbides, diffusion of nitrogen is hindered and the thickness of the diffusion zone near their accumulations is much lower.

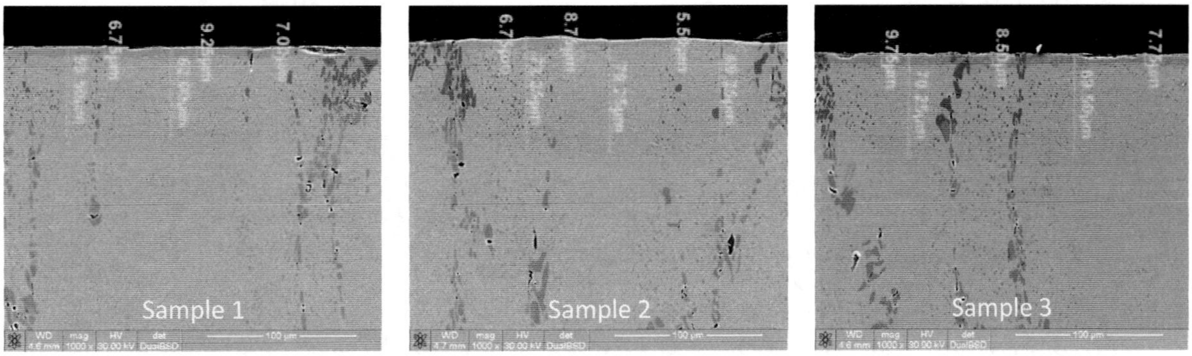

Figure 1. SEM micrographs of the transverse section of nitrided samples.

The depth of nitriding, determined by the microindentation method, was equal to 140 µm (Figure 2) for all modes. The value of hardness was near the surface for samples 1, 2, and 3 is 1054, 938 and 1013 HV, respectively. With increasing depth, the hardness gradually decreases to an average value of 424 HV down to 140 µm.

The phase composition of the surface layer for all samples is represented by nitrides ε-Fe_3N, γ′-Fe_4N and CrN (Figure 3). However, depending on the mode of the process, its ratio changes over a wide range. With an increase in the RFPG power from 700 to 900 W, the content of Fe_3N decreases from 82 to 51%, and the content of Fe_4N and CrN increases from 10 to 33% and from 7 to 15%, respectively (Table 1).

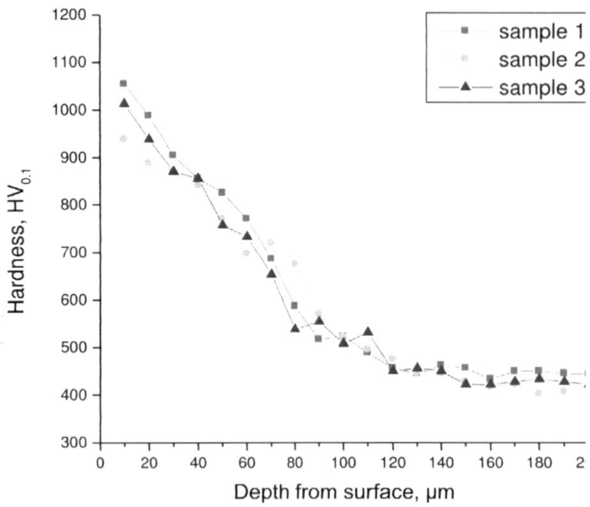

Figure 2. Hardness profiles of nitrided samples.

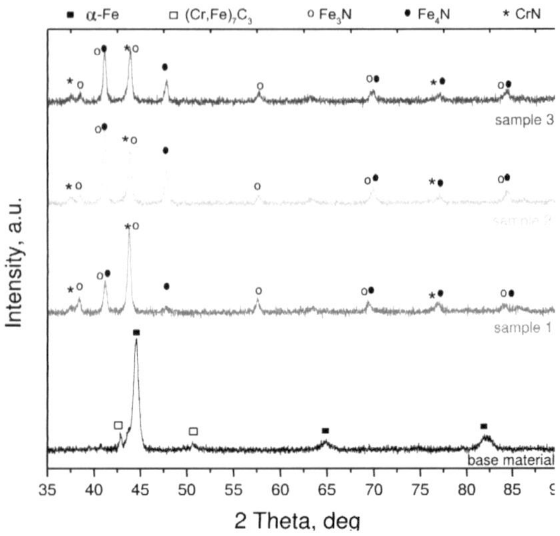

Figure 3. XRD of reported nitrided samples.

Table 1. Phase composition analysis of nitrided samples.

	Nitriding parameters (RFPG power / bias voltage)	Phase (%)		
		Fe_3N	Fe_4N	CrN
sample 1	700 W / 300 V	82.2	10.6	7.2
sample 2	800 W / 250 V	62.2	29.2	8.6
sample 3	900 W / 200 V	51.2	33.2	15.6

The initial H12MF steel dissolution occurs first of all, and pitting is formed on the surface in areas with structural defects (Figure 4b). In the case of samples after nitriding, an improvement in corrosion resistance is achieved by forming a dense layer of nitrides. Analysis of the polarization curves showed that the formation of the nitride layer on the surface of H12MF steel leads to a shift in the value of the corrosion potential E_{cor} to the region of positive values. So, for the state before nitriding, the corrosion potential was -90 mV after nitriding, it varies in the range from 820 mV to 960 mV depending on the process parameters (Figure 4a). This change in potential indicates an improvement in corrosion resistance (Figure 4c). The study of the surface morphology of the samples showed the presence of traces of pitting corrosion, both before and after nitriding.

It should be noted that the samples in the initial state also underwent continuous corrosion, which explains the difference in the values of the current density before and after nitriding. After surface modification, the current density decreases. The greatest potential for corrosion corresponds to the nitriding mode when the power of radio-frequency plasma generator is 900 W. The current density, which characterizes the intensity of dissolution, in this case corresponds to the minimum values, which indicates a higher corrosion resistance. Nitriding at a lower power of 700 W leads to a lower performance corrosion potential and high performance of current density. A decrease in corrosion resistance with an increase in the power of the radio-frequency plasma generator from 700 to 900 W is associated with an increase in the proportion of CrN and γ′-Fe_4N in the modified layer [6].

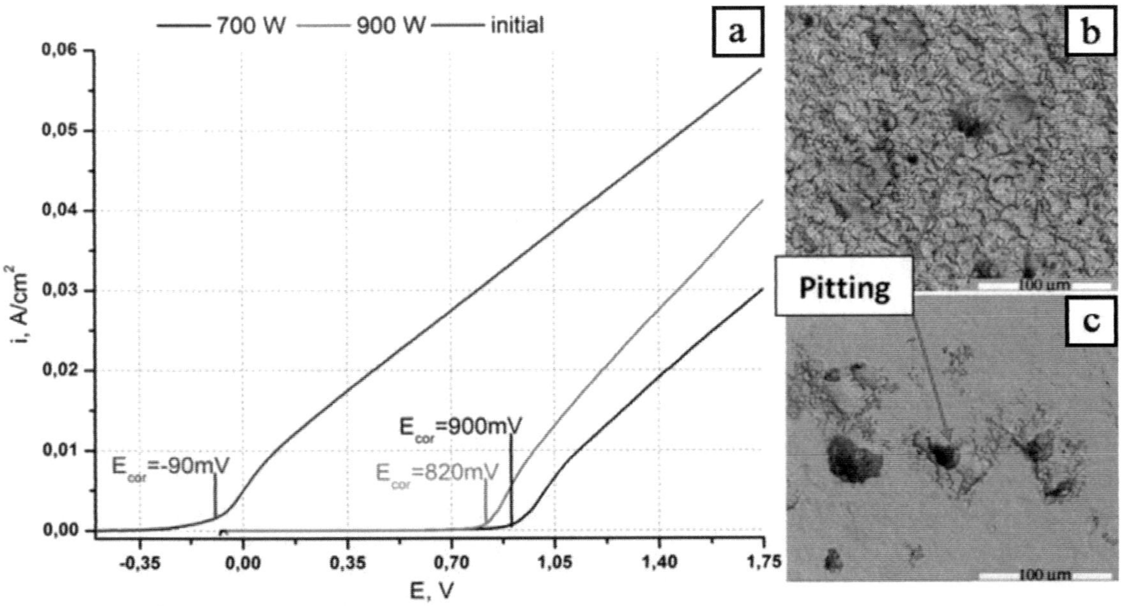

Figure 4. Potentiodynamic polarization curves (a) and surface SEM images of nitrided samples after corrosion test: initial (b); 700W (c).

Thus, with an increase in the power of the radio-frequency plasma source and a simultaneous decrease in the bias potential on the sample at a constant temperature of the process, an increase in the content of CrN and γ′-Fe_4N in the surface layer occurs, which is accompanied by an increase in corrosion resistance while maintaining without significant changes in the thickness of the modified layer, determined mainly by diffusion processes.

References

[1] Ramírez G, Mestra A, Casas B, Valls I, Martínez R, Bueno R, Góez A, Mateo A and Llanes L 2012 *Surf. Coat. Tech.* **206** (13) 3069
[2] Atık E, Yunker U and Merıç C 2003 *Tribology International* **36** (3) 155
[3] Wen D C 2009 *Appl. Surf. Sci.* **256** (3) 797
[4] Díaz-Guillén J C, Naeem M, Hdz-García H M, Acevedo-Davila J L, Díaz-Guillén M R, Khan M A, Iqbal J and Mtz-Enriquez A I 2020 n *Surf. Coat. Tech.* **385** 125420
[5] Berlin E V, Grigoriev V Y and Seidman L A 2018 *Inductive Sources of High-density Plasma and their Technological Applications* (Moscow: Technosphere) p 6
[6] Hubicki R, Richert M and Wiewióra M 2020 *Materials* **13** 2311

Effect of carbon on microstructure and mechanical properties of titanium

M S Ozerov[1*], D N Klimenko[1], L P Rtishcheva[2], V N Kopylov[2], N D Stepanov[1] and S V Zherebtsov[1]
[1] Laboratory of Bulk Nanostructured Materials, Belgorod National Research University, Belgorod 308015, Russia
[2] Central Research Institute of Structural Materials Prometheus, National Research Center Kurchatov Institute, St. Petersburg 191015, Russia

*Corresponding author: ozerov@bsu.edu.ru

Abstract. The microstructure and mechanical properties of cold rolled titanium with 0.0, 0.1 and 0.2 wt.% of carbon were studied after annealing in a temperature range of 500-800 °C. The addition of carbon resulted in slower grain growth kinetics and less intensive softening during annealing of the cold rolled alloys in the temperature range 500-800 °C. Besides, the alloys with carbon are considerably stronger in comparison with pure titanium; cold rolling resulted in an increase in the yield stress from 540 MPa for titanium to 630 and 605 MPa for Ti-0.1C and Ti-0.2C, respectively. However, ductility of the carbon doped titanium was also found to be decreased to ~32% in comparison with 40% for pure titanium.

1. Introduction

Titanium alloys are extremely important structural materials widely used in chemistry, medicine, automobile, shipbuilding and aerospace industry [1, 2]. Titanium based alloys are primarily used in structures that require a combination of high specific strength and corrosion resistance. Alloying of titanium is an effective method for controlling their structure and properties [3-5]. It was found that carbon can increase strength of titanium significantly; however the most effective is solid solution strengthening, while precipitation (carbide) strengthening does not contribute into strength notably [6-9]. Besides, the carbides formation can result in a drop in ductility. Another possible effect of carbon can be associated with the preferential distribution of carbon at grain boundaries thereby decreasing their mobility. This can be used for increasing the stability of a fine-grained structure at elevated temperatures.

In the present work, titanium with 0.0, 0.1 and 0.2 wt.% of carbon was cold rolled and then annealing at temperatures 500-800 °C for one hour. Structure, microhardness and grain growth kinetics were examined as a function of the annealing temperature.

2. Materials and methods

Pure titanium (99.1% purity) and titanium 0.1 and 0.2 wt.% of carbon (Ti-0.1C and Ti-0.2C, respectively) were studied. Since the maximum solubility of carbon in titanium at room temperature is ~0.2 wt.%, this amount of carbon should form solid solution in both cases. The alloys were obtained by vacuum arc melting. Flat samples were rolled at room temperature to a total strain of 1.2 (70% thickness reduction). Then they were annealed at 500, 600, 700 and 800 °C for 60 min.

Samples for metallographic studies was prepared by mechanical grinding with finish polishing using a colloidal suspension of silicon dioxide (SiO_2, the crystallite size of about 0.04 μm). For chemical etching, a mixture of 0.25 ml of HF, 1.75 ml of HCl and 53 ml of water was used. Microstructural studies were carried out using a Quanta 600 field emission gun (FEG) scanning electron microscope (SEM) and a Nova NanoSEM 450 SEM.

Vickers microhardness was measured at room temperature using 200g load for 10s; 10 measurements were made for each specimen.

3. Results and discussion

3.1. Initial microstructure

The as-cast microstructure in all program materials was rather coarse with an average grain size of ~1 mm (Figure 1a-c). XRD showed that pure titanium and Ti-0.1C, Ti-0.2C (Figure 1d) consisted of hcp α-Ti solely without carbide phases.

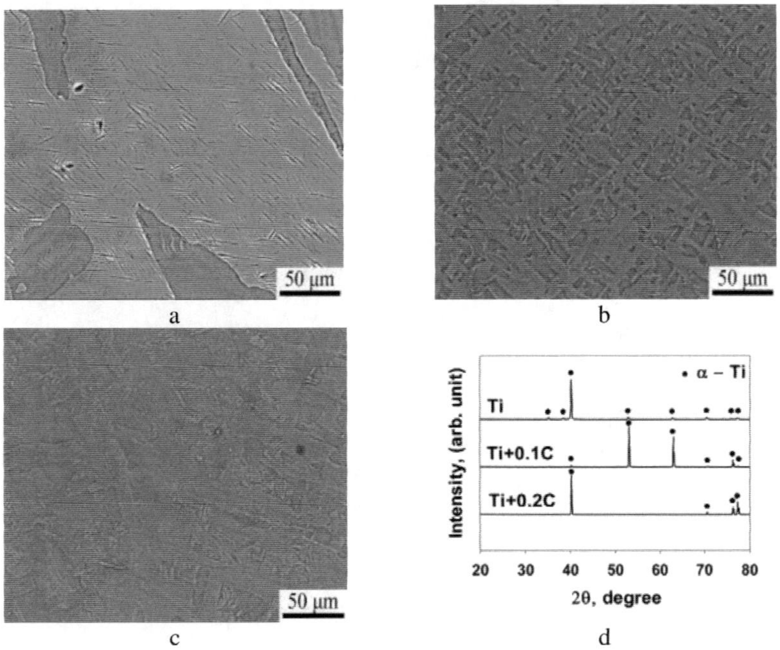

Figure 1. Microstructure images (a – c) and XRD pattern (d) of Ti (a), Ti-0.1C (b) and Ti-0.2C (c) in the initial condition.

3.2. Microstructure evolution after rolling and during subsequent annealing

Annealing of the cold rolled titanium alloys at 500-800 °C resulted in the formation of an equiaxed structure in all studied alloys (Figure 2). However, the addition of carbon to titanium led to a decrease in the kinetics of grain growth during annealing at 700 and 800 °C (Figure 3). The smallest grain size was detected at the carbon content of 0.1 percent. It was found that after rolling at room temperature and annealing, the grain size of the alloy with 0.1wt.% of C was ~ 40 percent smaller than that in pure titanium: 45μm *vs.* 73 μm at 800 °C and 5μm *vs.* 3 μm at 500 °C. Similar result after addition of 0.05 and 0.15% C was earlier reported in [7, 8]. It should be noted that grains in the Ti-0.2C alloy was slightly larger than those in Ti-0.1C. However, this difference was within the measurement error. Thus, alloying of pure titanium with carbon results to inhibition of the grain growth kinetics after cold rolling and annealing in the temperature range 500-800 °C.

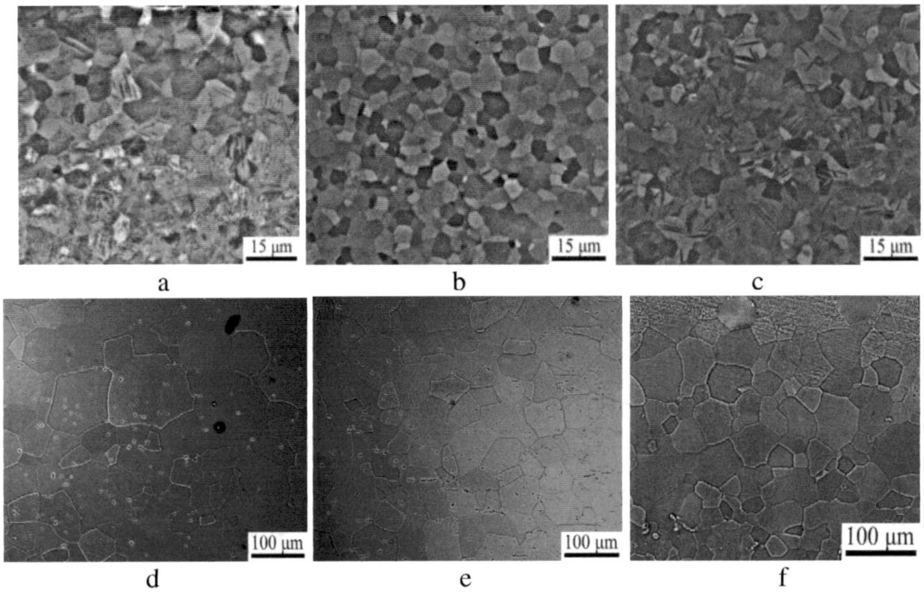

Figure 2. Microstructure images of cold rolled Ti (a, d), Ti-0.1C (b, e) and Ti-0.2C (c, f) after annealing at 500 °C (a – c) and 800 °C (d – f).

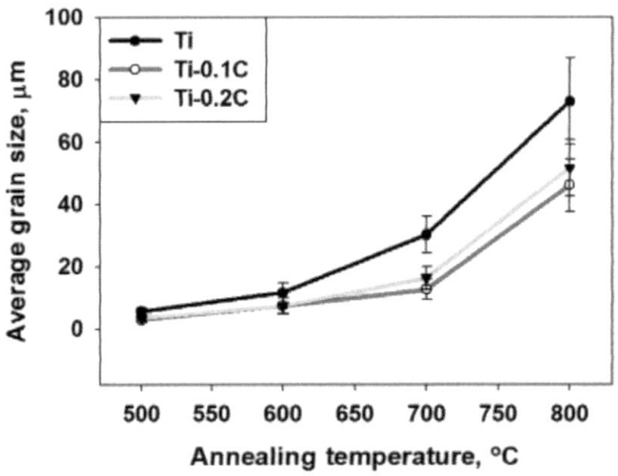

Figure 3. Effect of annealing temperature on the average grain size of cold rolled titanium and Ti-0.1C, Ti-0.2C alloys.

3.3. Mechanical properties of the alloys after cold rolling and subsequent annealing

In the as-cast condition the yield stress of the alloys with carbon was ~55% higher in comparison with that of pure titanium (~300 MPa *vs.* 170 MPa) (Figure 4a). The yield stress values for the Ti-0.1C and Ti-0.2C alloys were almost identical. Rolling at room temperature led to an increase in the yield stress to 540 MPa for titanium and to 630 and 600 MPa for the Ti-0.1C and Ti-0.2C alloys, respectively (Figure 4a). Tensile elongation for titanium in the as-cast was 48%. Ductility of the alloys with 0.1wt.% and 0.2wt.% of carbon was slightly lower: 38% and 40%, respectively (Figure 4b). Cold rolling led to a considerable decrease in the ductility values by 32-40% in all studied alloys.

Microhardness analysis also shows a considerable hardening of all alloys after cold rolling (Figure 4c). The microhardness value of pure titanium increased from 117 HV in the as-cast state to 240 HV after cold rolling. The alloys with 0.1 and 0.2% of C had in the initial state microhardness of 157 and 153 HV, respectively. Cold rolling increased these values to 257 and 252 HV, respectively. The obtained results confirm the strengthening effect of carbon without significant change in ductility [7-9].

Annealing in the temperature range 500-800 °C led to a decrease in the microhardness for all states, however, the decrease in the microhardness of pure titanium was more intensive (microhardness dropped from 240 to 118-89 HV) compared to Ti-0.1C and Ti-0.2C alloys (microhardness values from 257 HV and 252 HV in the initial state decreased to 184-171 HV and 178-166 HV after annealing, respectively).

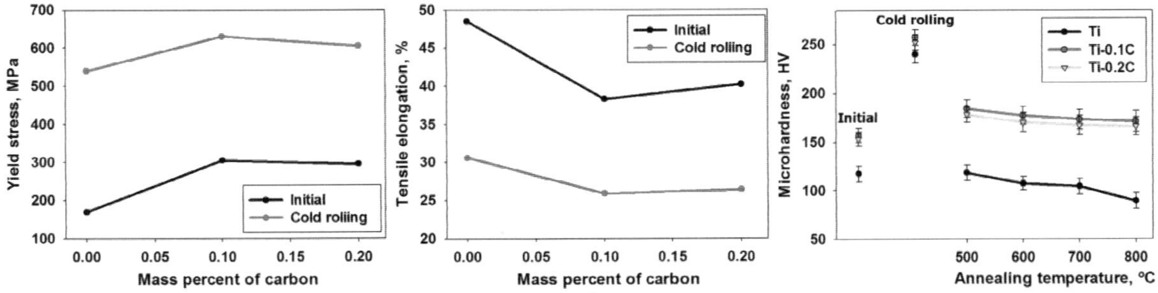

a b c

Figure 4. Effect of carbon content (a, b) and annealing temperature (c) after rolling and subsequent annealing on mechanical properties of titanium and Ti-0.1C, Ti-0.2C alloys.

4. Conclusions

Based on the results of the study, the following conclusions were drawn:

i) Alloying of pure titanium with carbon led to inhibition of the kinetics of grain growth during annealing at 500-800 °C of the cold rolled alloys;

ii) The yield stress of alloys with carbon was ~55 percent higher in comparison with pure titanium (~300 MPa *vs.* 170 MPa); cold rolling led to an increase in the yield stress to 540 MPa of titanium and to 630 and 605 MPa of Ti-0.1C and Ti-0.2C alloys, respectively;

iii) Annealing in the temperature range 500-800 °C resulted in a decrease in the microhardness of all the alloys, however, the decrease in the microhardness of pure titanium was more intensive in comparison with Ti-0.1C and Ti-0.2C.

Acknowledgment

The work was supported by the Russian Science Foundation under Grant № 19-79-30066. The authors are also grateful to Joint Research Center "Materials and Technologies", Belgorod State University, for the assistance with instrumental analysis.

References

[1] Leyens C, Peters M 2003 *Titanium and Titanium Alloys: Fundamentals and Applications* (Weinheim: Wiley-VCH)
[2] Gorynin I V 1999 *Mater. Sci. and Eng. A* **263** 112-116
[3] Zwicker U 1974 *Titanium and Titanium Alloys* (Berlin: Springer)
[4] Ozerov M, Klimova M, Kolesnikov A. Stepanov, N, Zherebtsov S 2016 *Mater. Des.* **112** 17–26
[5] Zherebtsov S, Ozerov M, Klimova M, Moskovskikh D, Stepanov N and Salishchev G 2019 *Metals* **9** 1175
[6] Finlay W L, Snyder J A 1950 *JOM* **42** 277-286
[7] Solonina O P, Ulyakova N M 1974 *Met. Sci. and Heat Treat.* **1** 310-312
[8] Chu M, Jones I P and Wu X 2005 *J. of Mater. Eng. and Perf.* **14** 735-740
[9] Gee E A, Sutton J B, Barth W J 1950 *Ind. & Eng. Chem.* **42** 243-249

Deformability assessment of beta-solidifying TiAl-based alloy with Zr, Cr, and Gd content variability

P V Panin[1*], E B Alekseev[1] and E A Lukina[2]

[1] Titanium Alloys Lab., Federal State Unitary Enterprise "All-Russian Scientific Research Institute of Aviation Materials", 17 Radio Str, 105005, Moscow, Russia
[2] Department of Materials Science and Technologies of Materials Treatment, Moscow Aviation Institute (National Research University), 4 Volokolamskoe Shosse, 125993, Moscow, Russia

*Corresponding author: PaninPaV@yandex.ru

Abstract. An experimental deformability assessment of a new six-component TiAl-based alloy Ti-44.5Al-2V-1Nb-(0.5…1.5)Zr/(1.5…2.5)Cr-(0…0.1Gd) (at.%) with variable Zr, Cr, and Gd content has been accomplished under the conditions of isothermal upsetting in accordance with the uniaxial compression scheme in the temperature range 1000–1300°C and strain rates of 0.001–0.1 s^{-1}. It has been revealed that Cr-containing compositions possess significantly higher technological plasticity compared to Zr-containing compositions. Microadditions of Gd in the amount of 0.1 at.% also promote the reduction of stresses upon deformation. It is shown that with the upsetting temperature increase, the stress values decrease in the entire range of deformation rates. It has been established that the most noticeable decrease in stress levels is observed with an increase in the compression temperature from 1100 to 1200 °C – by 2...2.5 times, while with an increase in the compression temperature from 1000 to 1100 °C – only by 1.1…1.3 times, and from 1200 to 1300 °C – by 1.4...1.5 times.

1. Introduction

Heat-resistant intermetallic alloys based on gamma titanium aluminide (TiAl-based alloys) are promising light-weight materials with service temperatures up to 750–800°C for high-temperature applications, e.g. aero-engine components [1].

It is well-known that the optimal combination of mechanical properties of TiAl-based alloys (low-temperature plasticity, strength, fracture toughness, and creep resistance) can be achieved due to a fine-grained lamellar structure with a grain (or intragranular lamellar colonies) size less than 100 μm. The required grain size is possible to be formed by microalloying with structure modifiers, in particular boron, carbon, and rare-earth elements (REE) [2–4]. But since the size of grains/colonies strongly depends on cooling rates, it is not always possible to provide such a structure in cast products. Another way for grinding coarse-grained ingot structures and casting microporosity elimination is the use of thermomechanical treatment [4,5]. However, the production of defect-free wrought semi-finished products of gamma alloys proves to be a complicated technological task because of natural brittleness and low plasticity thereof. In this work the deformability assessment studies were performed on a new Russian intermetallic β-solidifying TiAl-based alloy [3,6] of variable composition Ti-44.5Al-2V-1Nb-*x*Zr/*y*Cr-(0…0.1)Gd (at.%) (x = 0.5–1.5 at.%; y = 1.5–2.5 at.%) with the REE (Gd) microadditions.

2. Materials and methods

Experimental studies of deformation behavior have been carried out on six various alloy compositions (table 1). The initial triple remelted ingots with a diameter of 70 mm and weight of approx. 2.8 kg each were produced by double vacuum-arc remelting followed by vacuum induction melting and final gravity casting in a steel chill mold (VAR + VAR + VIM). Cylindrical samples measured 10 mm in diameter (d) and 15 in height (h) were cut from the as-cast ingots without any additional treatment.

The isothermal uniaxial compression tests were performed in laboratory air at temperatures of 1000, 1100, 1200, and 1300 °C, and constant strain rates ($\dot{\varepsilon}$) of 10^{-3}, 10^{-2}, and 10^{-1} s^{-1} on a Walter+Bai LFV-100 servohydraulic testing machine equipped with a radial heating furnace.

Table 1. Actual chemical compositions of the studied TiAl-based alloy samples (at.%).

No.	Ti	Al	V	Nb	Cr	Zr	Gd
Alloy 1	Bal.	44.3	1.91	1.02	1.53	-	-
Alloy 2	Bal.	44.4	2.01	1.00	1.45	-	0.11
Alloy 3	Bal.	44.3	1.97	0.99	2.46	-	-
Alloy 4	Bal.	44.3	1.98	1.03	2.52	-	0.10
Alloy 5	Bal.	44.2	1.96	1.02	-	1.50	-
Alloy 6	Bal.	44.3	1.95	0.97	-	1.53	0.09

3. Results and discussion

The analysis of isothermal upsetting results has shown similar appearance of the obtained flow curves (in the "stress–strain" coordinates) for all of the tested compositions (figure 1). It has been found that at all investigated temperatures and strain rates, the highest stress level is observed for the compositions with 1.5 at.% Zr (alloys 5 and 6), while the smallest stress level was detected for the compositions with 2.5 at.% Cr (alloys 3 and 4). The compositions containing a lower amount of Cr (1.5 at.%; alloys 1 and 2) occupy an intermediate position. It is important to note that the doping with Gd in the amount of 0.1 at.% exhibits a positive effect on all of the three groups of alloys and, in general, provides a decrease in stress levels.

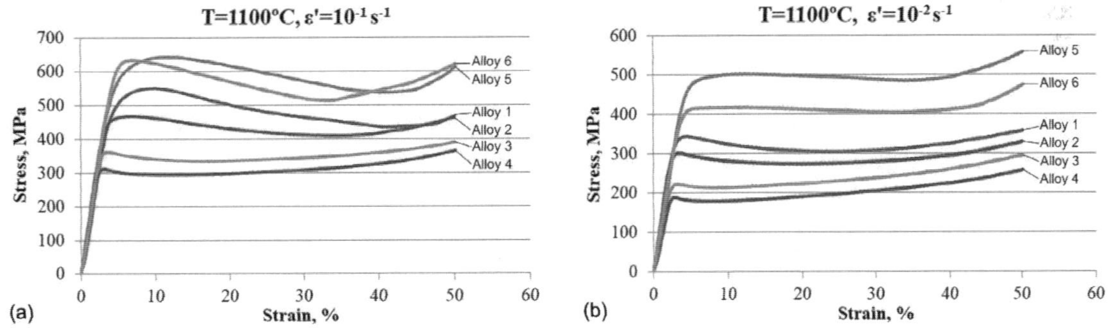

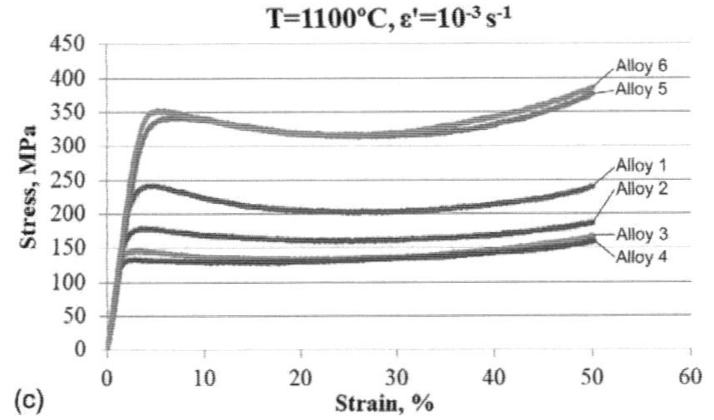

Figure 1. Stress–strain rheological curves of the six investigated alloy compositions (see table 1) for compression tests of cylindrical samples (d = 10 mm, h = 15 mm) at T = 1100°C and strain rates $\dot{\varepsilon} = 10^{-1}$ s^{-1} (a), $\dot{\varepsilon} = 10^{-2}$ s^{-1} (b), and $\dot{\varepsilon} = 10^{-3}$ s^{-1} (c).

A high level of stresses leads to the formation of a large number of cracks along the surface generatrix of the Zr-containing samples (alloys 5 and 6), while almost no cracks were observed on the Cr-containing samples (alloys 3 and 4) (figure 2).

Figure 2. Appearance of cylindrical samples (d = 10 mm, h = 15 mm) upset at T = 1100°C and strain rate $\acute{\varepsilon} = 10^{-2}$ s^{-1}: (a) alloy 1; (b) alloy 2; (c) alloy 3; (d) alloy 4; (e) alloy 5; (f) alloy 6.

The maximum number of deep diagonal cracks is observed at a maximum strain rate of 10^{-1} s^{-1}, which corresponds to the highest stress values. A decrease in the upsetting rate to 10^{-3} s^{-1} leads to a significant reduction or complete absence of cracks on the samples surface. The intermediate strain rate of 10^{-2} s^{-1} ensures the production of practically defect-free upset items from Cr-containing alloys 3 and 4, and alloy 2 with 1.5 at.% Cr + 0.1 at.% Gd (see figure 2). It should be noted that alloy 4 containing 2.5 at.% Cr + 0.1 at.% Gd shows the best technological plasticity and almost complete lack of surface cracks on the samples compressed even at a strain rate of 10^{-1} s^{-1}.

Thus, the strain rates of 10^{-2}...10^{-3} s^{-1} are preferable for deformation processing of the investigated TiAl-based alloy with variable compositions, since almost defect-free workpieces (samples) can be obtained. However, taking into account the duration of the upsetting process (for the tested samples measured d = 10 mm and h = 15 mm the upsetting time is about 50 s at $\acute{\varepsilon} = 10^{-2}$ s^{-1} and about 500 s at $\acute{\varepsilon} = 10^{-3}$ s^{-1}), the most rational strain rate will be $\acute{\varepsilon} = 10^{-2}$ s^{-1} from the point of view of the subsequent scaling up to industrial conditions.

It has been established that with an increase in the upsetting temperature of the samples, the stress values decrease for each strain rate applied. At the same time, it has been found that the most noticeable decrease in stress levels is observed with an increase in the compression temperature from 1100 to 1200 °C – by 2...2.5 times, while with an increase in the compression temperature from 1000 to 1100 °C – only by 1.1…1.3 times, and from 1200 to 1300 °C – by 1.4...1.5 times. It can be assumed that an intense stress level decrease in the temperature range 1100–1200°C takes place due to the high rate of the $\alpha_2(D0_{19}) \rightarrow \alpha(A3)$ phase disordering process at 1200°C and above. At a temperature of 1300°C accelerated oxidation of the samples begins. In addition, at such high temperatures the deformation process becomes more energy intensive which leads to an increase of the processing costs. Thus, for carrying out deformation treatment of the studied TiAl-based alloy, it is rational to choose 1200°C as the initial upsetting temperature.

The limiting degree of isothermal deformation of the TiAl-based alloy samples during upsetting before the first cracks appearance is estimated. Cracks at a deformation degree of 65% were not detected for the compositions containing 1.5–2.5 at.% Cr and 0.1 at.% Gd after upsetting at a temperature of 1200°C and strain rates of 10^{-2} and 10^{-3} s^{-1}. In Zr-containing compositions the first cracks upon upsetting at T = 1200°C and $\dot{\varepsilon} = 10^{-2}$ s^{-1} appear when the degree of deformation reaches 50%.

4. Conclusions

It has been revealed that Cr-containing compositions possess significantly higher technological plasticity under the studied conditions of deformation compared to Zr-containing compositions. It is shown that the microaddition of Gd in the amount of 0.1 at.% results in a positive effect on the manufacturability of all of the studied TiAl-based alloy compositions, reducing the levels of yield ($\sigma_{0.2}$) and flow stresses.

It is notable that with an increase in the upsetting temperature, the stress values decrease in the entire range of strain rates. The most noticeable decrease in stress levels is observed with an increase in the compression temperature from 1100 to 1200 °C – by 2...2.5 times, while with an increase in the compression temperature from 1000 to 1100 °C – only by 1.1…1.3 times, and from 1200 to 1300 °C – by 1.4...1.5 times.

Acknowledgement

This research project was supported by the Russian Science Foundation (grant No. 18-79-10249).

The work was performed as a part of the implementation of the complex scientific direction 7.1 “Intermetallic Alloys Based on Titanium” (“Strategic Directions for the Development of Materials and Technologies of Their Processing for the Period until 2030”) [7].

References

[1] Kim Y-W and Kim S-L 2018 Advances in gammalloy materials–processes–application technology: successes, dilemmas, and future *J. Metals* **70** 553–60

[2] Wang X, Luo R, Liu F, Zhu F, Song S, Chen B, Zhang X, Zhang J and Chen M 2017 Characterization of Gd-rich precipitates in a fully lamellar TiAl alloy *Scr. Mater.* **137** 50–4

[3] Panin P V, Kochetkov A S, Zavodov A V and Lukina E A 2020 Effect of Gd addition on phase composition, structure, and properties of beta-solidifying TiAl-based alloy with Zr and Cr content variability *Intermetallics* **121** 106781

[4] Appel F, Paul J D H and Oehring M 2011 *Gamma Titanium Aluminide Alloys: Science and Technology* (Weinheim: Wiley-VCH Verlag & Co, KGaA) p 745

[5] Sokolovsky V S, Stepanov N D, Zherebtsov S V, Nochovnaya N A, Panin P V, Zhilyakova M A, Popov A A and Salishchev G A 2018 Hot deformation behavior and processing maps of B and Gd containing β-solidified TiAl based alloy *Intermetallics* **94** 138–51

[6] Panin P V, Nochovnaya N A, Lukina E A and Kochetkov A S 2019 Effect of chemical composition variability on phase composition and structure of beta-solidifying TiAl-alloy in as-cast condition *Inorg. Mater.: Appl. Res.* **10** 316–21

[7] Kablov E N 2015 Innovatsionnye razrabotki FGUP “VIAM” GNTs RF po realizatsii “Strategicheskikh napravleniy razvitiya materialov i tekhnologiy ikh pererabotki na period do 2030 goda” [Innovative developments of FSUE “VIAM” SSC of RF on realization of “Strategic Directions for the Development of Materials and Technologies of Their Processing for the Period until 2030”] *Aviacionnye materialy i tehnologii* **1(34)** 3–33 (in Russian)

Precipitation-hardened refractoryTi-Nb-Hf-Al-Ta high-entropy alloys

E Panina[*], N Yurchenko, S Zherebtsov, G Salishchev and N Stepanov

Laboratory of Bulk Nanostructured Materials, Belgorod State University, Belgorod 308015, Russia

[*]Corresponding author: panina_e@bsu.edu.ru

Abstract. This study reports the structure and mechanical properties of new refractory $Ti_{37.5}Nb_{12.5}Hf_{25}Al_{25}$, $Ti_{40}Nb_{30}Hf_{15}Al_{15}$, and $Ti_{40}Nb_{20}Ta_{10}Hf_{15}Al_{15}$ (at.%) high entropy alloys. After annealing at 1200 °C for 24 h, the program alloys had a single-phase B2 structure. Further annealing at 600 °C for 24 h resulted in the formation of Widmanstatten orthorhombic particles (O-phase) in the bcc matrix. The $Ti_{40}Nb_{30}Hf_{15}Al_{15}$ and $Ti_{40}Nb_{20}Ta_{10}Hf_{15}Al_{15}$ alloys annealed at T = 1200 °C showed moderate strength and good ductility (>50%) at 22 and 600 °C; while the $Ti_{37.5}Nb_{12.5}Hf_{25}Al_{25}$ alloy was stronger, but less ductile at both temperatures. Subsequent annealing at T = 600 °C significantly increased the strength of the $Ti_{40}Nb_{30}Hf_{15}Al_{15}$ alloy at 22 and 600 °C, maintaining compressive sufficient ductility at room-temperature .

1. Introduction

Recently introduced refractory high entropy alloys (RHEAs), which demonstrated an outstanding capability to maintain high strength at temperatures up to T = 1600 °C [1], seem promising candidates for next-generation turbines. Despite the attractive high-temperature strength, one of the major drawbacks of most RHEAs is the modest mechanical performance at ambient temperatures.

It is believed that balanced properties can be obtained by developing precipitation-strengthened RHEAs. One of the promising examples of such RHEAs are alloys with the bcc/B2 structure [2, 3]. The superalloy-like microstructure with a bcc solid solution, strengthened by coherent cuboidal B2 nanoparticles, provides a good combination of strength and ductility at temperatures T≤600 °C. However, the development of bcc/B2 RHEAs is a non-trivial task due to weakly established composition-structure relationships and poor fidelity of thermodynamic modeling in the case of the B2 phase [4, 5].

In this work, we presented a series of RHEAs strengthened by nanosized Widmanstatten orthorhombic (O-phase) particles. The O-phase precipitation was found to be an attractive option to achieve a notable strength enhancement both at room and elevated temperatures with only a slight loss in ductility.

2. Materials and methods

Ingots of the $Ti_{37.5}Nb_{12.5}Hf_{25}Al_{25}$, $Ti_{40}Nb_{30}Hf_{15}Al_{15}$, $Ti_{40}Nb_{20}Ta_{10}Hf_{15}Al_{15}$ alloys were produced by vacuum arc melting of pure (≥ 99.9 wt%) elements. The alloys were annealed in quartz tubes at 1200 °C for 24 h. Some samples were further annealed at 600 °C for 24 h. The phase composition and microstructure of the alloys were studied using transmission electron microscopy (TEM) and scanning electron microscopy (SEM). The densities of the $Ti_{37.5}Nb_{12.5}Hf_{25}Al_{25}$, $Ti_{40}Nb_{30}Hf_{15}Al_{15}$, $Ti_{40}Nb_{20}Ta_{10}Hf_{15}Al_{15}$ alloys, determined by the hydrostatic weighing method, were 7.23± 0.03, 7.07 ± 0.03, 7.87± 0.03 g/cm^3, respectively. Isothermal compression was carried out in air at 22 °C or 600 °C using an Instron 300LX test machine.

3. Results

3.1. Microstructure

After annealing at 1200 °C, the $Ti_{37.5}Nb_{12.5}Hf_{25}Al_{25}$,$Ti_{40}Nb_{30}Hf_{15}Al_{15}$, and $Ti_{40}Nb_{20}Ta_{10}Hf_{15}Al_{15}$ alloys had a coarse-grained single-phase microstructure with the average grain size ~300 μm, (Figure 1a-c); no secondary phases were observed.

After further annealing at 600 °C, precipitation of profuse Widmanstatten second phase particles was revealed. The particles located adjust to grain boundaries were coarser than the particleswithin the grain interior. The volume fraction of the second phase was nearly equal in all program alloys and estimated to be ~ 35%.

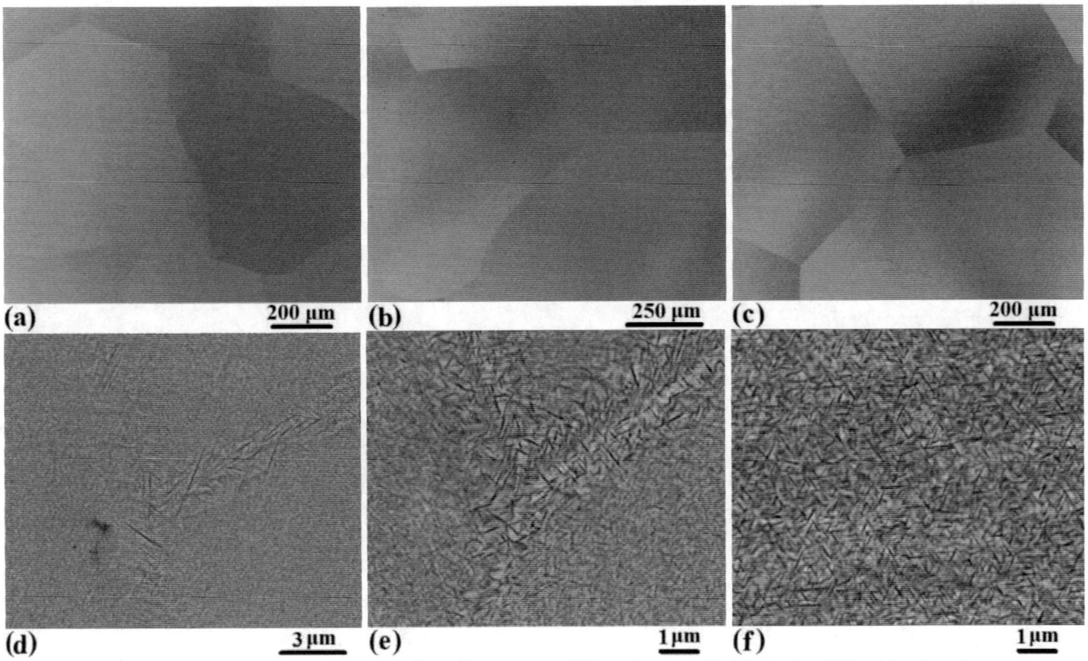

Figure 1. SEM images of the $Ti_{37.5}Nb_{12.5}Hf_{25}Al_{25}$ (a, d), $Ti_{40}Nb_{30}Hf_{15}Al_{15}$(b, e), $Ti_{40}Nb_{20}Ta_{10}Hf_{15}Al_{15}$ (c, f) alloys after annealing at 1200°C, 24 h (a, b, c) and further annealing at 600°C, 24h (d, e, f).

After annealing at 1200 °C, all the program alloys had a single-phase B2 structure as revealed by the selected area electron diffraction (SAED) patterns. TEM analysis revealed numerous anti-phase boundaries in the matrix grains (denoted in figure 2a-c as APBs); no secondary phases were found in all alloys.

The phases found after further annealing at 600 °C, were identified using TEM analysis: the matrix had the B2 ($Ti_{37.5}Nb_{12.5}Hf_{25}Al_{25}$) or bcc ($Ti_{40}Nb_{30}Hf_{15}Al_{15}$ and $Ti_{40}Nb_{20}Ta_{10}Hf_{15}Al_{15}$) structure, whilst the particles were defined as the O-phase.

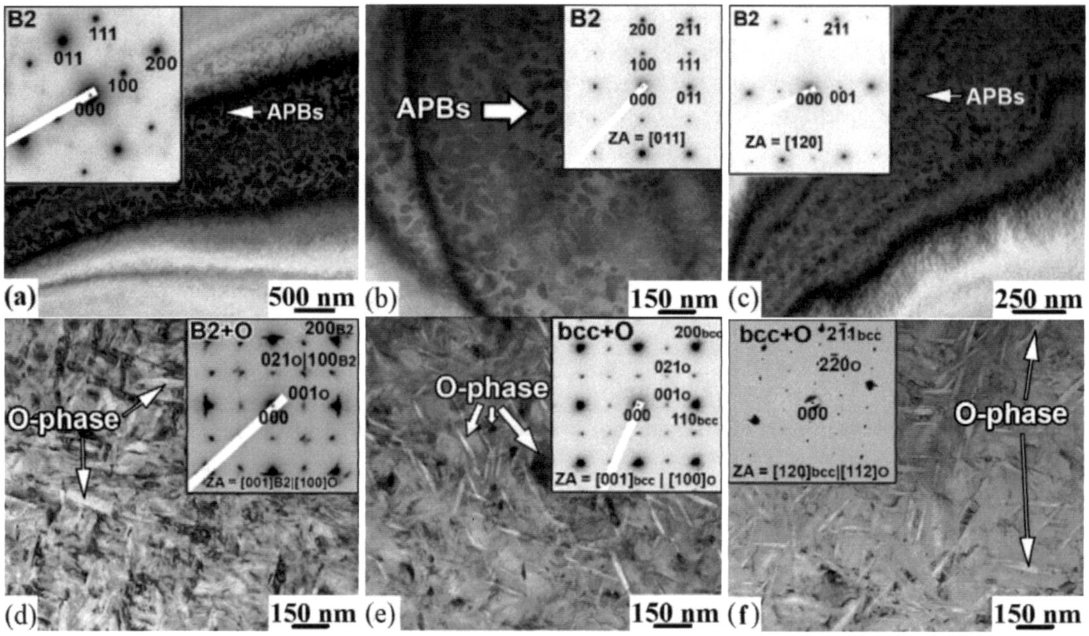

Figure 2. TEM bright-field images of the $Ti_{37.5}Nb_{12.5}Hf_{25}Al_{25}$ (a, d), $Ti_{40}Nb_{30}Hf_{15}Al_{15}$(b, e), $Ti_{40}Nb_{20}Ta_{10}Hf_{15}Al_{15}$ (c, f) alloys after annealing at 1200 °C, 24 h (a, b, c) and further annealing at 600 °C, 24h (d, e, f). Selected area electron diffraction patterns used for phase identification are shown in the inserts.

3.2. Mechanical properties

Stress-strain curves of the program alloys annealed at 1200 °C, i.e. with the single B2 phase structure, are presented in figures 3a, b. At 22 °C, the $Ti_{40}Nb_{30}Hf_{15}Al_{15}$ and $Ti_{40}Nb_{20}Ta_{10}Hf_{15}Al_{15}$ alloys demonstrated moderate strength (830 and 1075 MPa, respectively), high strain hardening capacity and compressive ductility. Meanwhile, the $Ti_{37.5}Nb_{12.5}Hf_{25}Al_{25}$ alloy was the strongest (1645 MPa), albeit fractured after 1% of plastic deformation. At 600 °C, the strength of the $Ti_{40}Nb_{30}Hf_{15}Al_{15}$ and $Ti_{40}Nb_{20}Ta_{10}Hf_{15}Al_{15}$ alloys reduced to 635 and 700 MPa, respectively. The $Ti_{37.5}Nb_{12.5}Hf_{25}Al_{25}$ alloy exhibited a notable decrease in yield strength to 810 MPa with a simultaneous increase in compressive ductility to 12%.

Subsequent annealing at 600 °C (which caused precipitation of the O-phase particles in the bcc/B2 matrix) changed the mechanical properties of the program alloys markedly (Figures 3c, d). The $Ti_{40}Nb_{30}Hf_{15}Al_{15}$ alloy demonstrated a ~50% strength increment both at 22 and 600 °C while maintaining reasonable compressive ductility at 22 °C. A similar annealing effect on room-temperature mechanical properties was also found for the $Ti_{40}Nb_{20}Ta_{10}Hf_{15}Al_{15}$ alloy; however, at 600 °C, the alloy performance degraded inevitably; i.e. both strength and ductility became worse in comparison with annealing at 1200 °C. In the case of the $Ti_{37.5}Nb_{12.5}Hf_{25}Al_{25}$ alloy, annealing at 600 °C increased its high-temperature strength, but drastically diminished compressive ductility. The obtained findings suggest complex relationships between the chemical composition, structure, and mechanical properties of the O-phase strengthened RHEAs since similar microstructural changes (precipitation of the O-phase particles) had a drastically different effect on the mechanical performance of the program alloys. Further studies are required to reliably establish the relationship between the composition and the structural properties of these alloys.

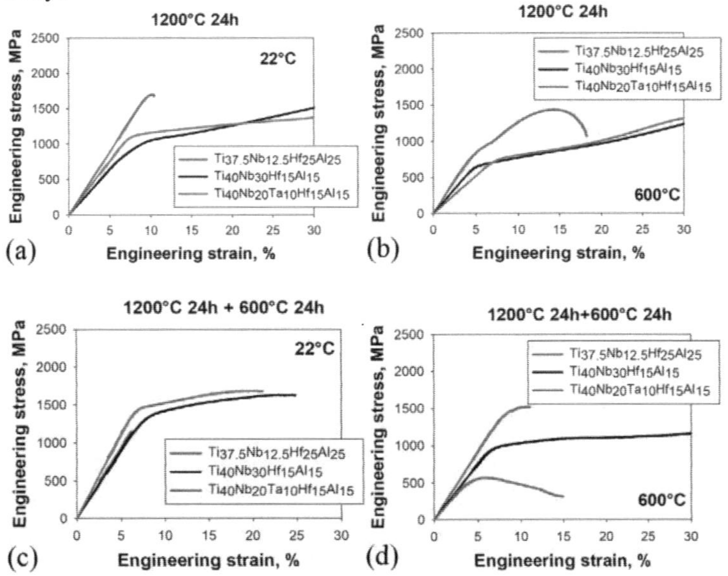

Figure 3. Engineering stress-strain curves of the program alloys after annealing 1200 °C (a, b) and further annealing at 600 °C (c, d) obtained during compression at 22 (a,c) and 600 °C (b,d).

4. Conclusions

Microstructure and mechanical properties of the refractory $Ti_{37.5}Nb_{12.5}Hf_{25}Al_{25}$, $Ti_{40}Nb_{30}Hf_{15}Al_{15}$, and $Ti_{40}Nb_{20}Ta_{10}Hf_{15}Al_{15}$ high entropy alloys were studied. The following conclusions were drawn:

- After annealing at 1200 °C, all the program alloys have a single-phase B2 structure. Further annealing at 600 °C led to the formation of a mixture of the B2 ($Ti_{37,5}Nb_{12,5}Hf_{25}Al_{25}$) or bcc ($Ti_{40}Nb_{30}Hf_{15}Al_{15}$, $Ti_{40}Nb_{20}Ta_{10}Hf_{15}Al_{15}$) matrix and nanosized Widmanstatten O-phase particles.
- In the single-phase state, the $Ti_{40}Nb_{30}Hf_{15}Al_{15}$ and $Ti_{40}Nb_{20}Ta_{10}Hf_{15}Al_{15}$ alloys demonstrated moderate strength and high compressive ductility at 22 and 600 °C. The $Ti_{37.5}Nb_{12.5}Hf_{25}Al_{25}$ alloy showed the highest strength, but very limited ductility. The precipitation of the O-phase enhanced the strength of the $Ti_{40}Nb_{30}Hf_{15}Al_{15}$ alloy at 22 and 600 °C without notable sacrificing in compressive ductility at 22 °C. However, the mechanical properties of the $Ti_{37.5}Nb_{12.5}Hf_{25}Al_{25}$ and $Ti_{40}Nb_{20}Ta_{10}Hf_{15}Al_{15}$ alloys deteriorated after annealing at 600 °C.

Acknowledgments
This study was supported by the Russian Science Foundation, grant № 19-79-30066.

References

[1] Senkov O N, Wilks G B, Scott J M and Miracle D B 2011 *Intermetallics* **19** 698
[2] Soni V, Senkov O N, Gwalani B, Miracle D B and Banerjee R 2018 *Sci. Rep.* **8** 1
[3] Senkov O N, Couzinie J-P,.Rao S I, Soni V and Banerjee R 2020 *Materialia* **9** 100627
[4] Soni V, Senkov O N, Couzinie J-P, Zheng Y, Gwalani B and Banerjee R 2020 *Materialia* **9** 100569
[5] Whitfield T E, Pickering E J, Owen L R, Jones C N, Stone H J and Jones N G 2020 *Materialia* **13** 100858

IOP Conf. Series: Materials Science and Engineering 1014 (2021) 012042 doi:10.1088/1757-899X/1014/1/012042

Methodological approaches to the study of reverse martensitic transformation in metastable austenitic steels

D O Panov[1*], A I Smirnov[2], Y N Simonov[3], N D Stepanov[1], G A Salishchev[1]

[1] Laboratory of Bulk Nanostructured Materials, Belgorod State University, 85 Pobeda Str., 308015, Belgorod, Russian Federation
[2] Department of Materials Science in Engineering, Novosibirsk State Technical University, 20 Prospekt K. Marksa, 630073, Novosibirsk, Russia
[3] Department of Metal Science, Thermal and Laser Processing of Metals, Perm National Research Polytechnic University, 29 Komsomolsky prospekt, 614990, Perm, Russia

*Corresponding author: panov_d@bsu.edu.ru

Abstract. Methodological aspects of the high-resolution dilatometry analysis of the reverse $\alpha \rightarrow \gamma$-transformation in cold-worked metastable austenitic steel Fe-0.07C-18.7Cr-9.2Ni-0.6Ti-1.1Mn-0.4Si (wt.%) were considered. It was shown that the reverse $\alpha \rightarrow \gamma$-transformation occurs in the temperature range from A_S=520 °C to A_F=920 °C. Two consecutive stages corresponding to shear and diffusion transformations were revealed. An abnormal increase in the sample length of the program steel during the reverse $\alpha \rightarrow \gamma$-transformation was found. The recrystallization process also developed during austenitization and was accompanied by a decrease in length of the specimen.

1. Introduction

Dilatometry has established itself as one of the main methods for examination of phase transformations in steels and iron alloys [1–4]. The dilatometry method is based on measuring changes in length of a material under heating/cooling. The measurements provide information about phase transformations and structure changes, i.e. is able to capture all processes that have a volumetric effect. Advances in research equipment make it possible to identify extremely fast phase transformations or record very weak effects with high accuracy [4].

Mathematical analysis of dilatometric data, i.e. using the first derivative, increases the accuracy of determining the phase transformation temperatures critically [5,6]. However, high-resolution dilatometry has revealed that even the derivative often shows a superposition of several peaks from various transformations [6–8]. After proper analysis, the austenitization stages in dual-phase ferrite-perlite steels [9], the bainite-ferrite steels [7], and the martensitic steels [10] were revealed. In these materials, different transformations can occur in the same temperature range but in different micro-volumes. However, this approach has never been used in the case of the reverse $\alpha \rightarrow \gamma$-transformation during heating of cold-deformed metastable austenitic steels. Thus, the purpose of this work is to analyze the reverse $\alpha \rightarrow \gamma$-transformation in cold-worked metastable austenitic steel using high-resolution dilatometry.

2. Experimental

Metastable austenitic steel Fe-0.07C-18.7Cr-9.2Ni-0.6Ti-1.1Mn-0.4Si (wt.%) was used as the program material. Pure nickel was selected as the reference material. Before cold deformation, a rod of the program steel was heated to a temperature of 1050 °C, held for 2 hours and water quenched. Then the rod was deformed at room temperature using a radial forging machine to a true strain of 2.14.

Transmission electron microscopy was performed using a FEI Tecnai 20 G2 TWIN microscope at an accelerating voltage of 200 kV. Samples with a thickness of 300 microns were cut by an electric-discharge machine from the center of the rod in longitudinal and transversal sections. Then they were ground on both sides with abrasive paper with a decreasing grain size. Subsequent electrolytic thinning of disks with a diameter of 3 mm and a thickness of 100 microns was performed using TenuPol-5 in a mixture of 95% acetic acid (CH_3COOH) and 5% perchloric acid ($HClO_4$) cooled to -40 °C. The content of the α-phase was measured using a multifunctional eddy-current tester MVP-2M.

Dilatometry was performed using a Linseis R.I.T.A. L78 quenching dilatometer in a helium (purity of 99.9999%) atmosphere. Cylindrical samples with a diameter of 3 mm and a height of 10 mm were cut from the center of the rod. The long axis of the samples was aligned with deformation direction. The samples were heated to 1000 °C at a rate of 10 °C/s. Analysis of dilatometric curves was performed by obtaining the first derivative of dilatograms (*d(ΔL)/dT)*) with subsequent peaks fitting in the Fityk software [11]. Asymmetric Gaussian curves were used for the approximation of the experimental results. The corresponding R-square values were ≥0.95.

3. Results and discussion

In the initial condition, the program steel had a lamellar dual-phase martensitic-austenitic structure, containing 62.5±0.9 % of deformation-induced martensite. In the longitudinal section, the grains were stretched along the axis of the rod (Figure 1a). Meanwhile, the grains had a fairly equiaxed shape (Figure 1b) with an average diameter of 240±10 nm in the transversal direction. Thus, it seems that the grains had a columnar shape. A more detailed description of the initial structure can be found elsewhere [12,13].

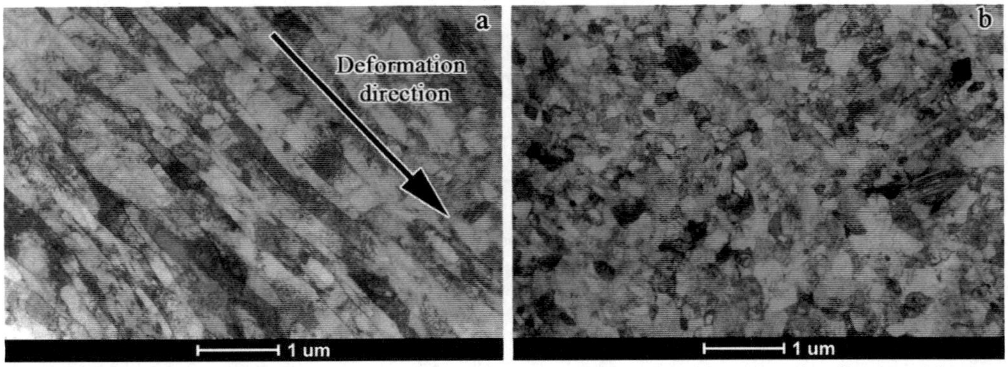

Figure 1. TEM bright-field images of a structure of the program steel in *a* longitudinal and *b* transversal section.

The dilatometric curve (*ΔL*) for the nickel sample was linear without any visible inflections. The first derivative of the dilatometric curve (*d(ΔL)/dT*-curve) also did not demonstrate any evidences of the phase transformation (Figure 2a). At the same time, the *ΔL* curve obtained during heating of the program steel exhibits a series of inflections, associated with the development of the reverse α→γ-transformation (Figure 2b). The temperatures of the start (A_S) and finish of the transformation (A_F) were determined from the *d(ΔL)/dT*. The respective values were A_S = 520 °C and A_F = 920 °C.

The *d(ΔL)/dT*-curve in a A_F - A_S range is obviously the product of a superposition of several peaks from different events (Figure 2c). Fitting the peaks to asymmetric Gaussian curves helped to consider the effect of varying conditions during heating [10]. Note that the beginning and end of the peak on the

d(ΔL)/dT-curve corresponds to the temperatures of the start and finish of the transformation, respectively; the peak area indicates the volume effect of the transformation [14]; and the maximum position corresponds to the highest speed of the process.
The performed analysis has allowed to identify three peaks in the temperature range of the α→γ-transformation (Figure 2c). Peaks #1 and #2 were associated with an increase in the length of the specimen, since they had upward orientation from the baseline. These peaks are likely direct products of the α→γ-transformation. A similar anomalous increase in the sample length during the reverse α→γ transformation was observed in AISI 304 steel [15]. In that case the volumetric expansion was associated with the effect of the gamma fiber texture ({111}<110> and {111}<112>) in the deformation-induce martensite. Furthermore, the α→γ-transformation by shear mechanism occurs at a lower temperature, while the activation of the diffusion mechanism usually requires higher temperatures [16]. Therefore, peak #1 corresponds to the shear transformation, and peak #2 – to the diffusion transformation. Peak #3 was oriented downward from the baseline, which corresponds to a decrease in the sample length. A similar effect was earlier associated with the development of recrystallization during heating of cold-deformed steel [17].

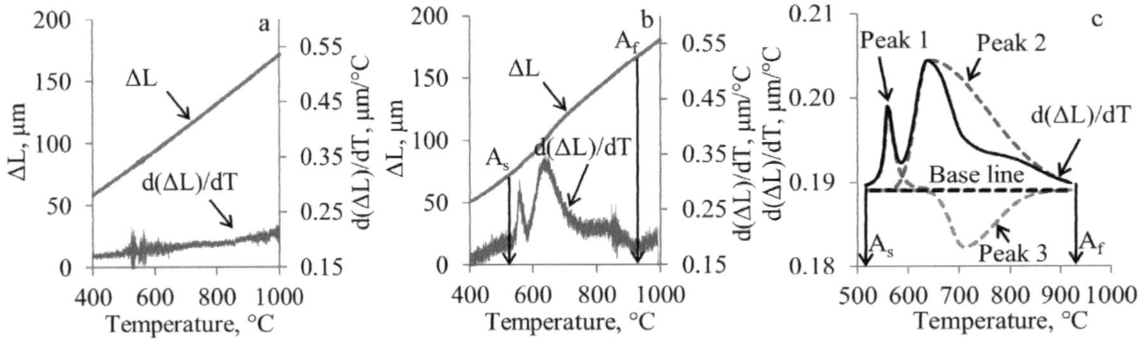

Figure 2. *a* dilatometric curve (*ΔL*) and *d(ΔL)/dT*-curve of the pure nickel sample and b the program steel; *c* the sketch explaining *d(ΔL)/dT*-curve for the program steel.

4. Conclusions
A method for analyzing the reverse α→γ transformation in cold-worked metastable austenitic steel Fe-Fe-0.07C-18.7Cr-9.2Ni-0.6Ti-1.1Mn-0.4Si (wt.%) using dilatometric curves was described. The reverse α→γ-transformation started at A_S=520 °C and finished at A_F=920 °C. Peaks from the reverse α→γ-transformation by the shear and diffusion mechanisms, as well as from the recrystallization process, were identified. The established approach allows us to develop the heat treatment regimes for cold-worked metastable austenitic steels to obtain the required structures and properties.

Acknowledgements
The authors gratefully acknowledge the financial support from the Russian Science Foundation Grant no. 20-79-10094.

References
[1] Bojack A, Zhao L, Morris P F and Sietsma J 2012 In-situ determination of austenite and martensite formation in 13Cr6Ni2Mo supermartensitic stainless steel *Mater. Charact.* **71** 77–86
[2] Wei R, Enomoto M, Hadian R, Zurob H S and Purdy G R 2013 Growth of austenite from as-quenched martensite during intercritical annealing in an Fe-0.1C-3Mn-1.5Si alloy *Acta Mater.* **61** 697–707
[3] Suh D W, Oh C S, Han H N and Kim S J 2007 Dilatometric analysis of austenite decomposition considering the effect of non-isotropic volume change *Acta Mater.* **55** 2659–69

[4] García De Andrés C, Caballero F G, Capdevila C and Álvarez L F 2002 Application of dilatometric analysis to the study of solid-solid phase transformations in steels *Mater. Charact.* **48** 101–11

[5] Li P, Li J, Meng Q, Hu W and Xu D 2013 Effect of heating rate on ferrite recrystallization and austenite formation of cold-roll dual phase steel *J. Alloys Compd.* **578** 320–7

[6] Oliveira F L G, Andrade M S and Cota A B 2007 Kinetics of austenite formation during continuous heating in a low carbon steel *Mater. Charact.* **58** 256–61

[7] Li D, Min Y, Wu X 2010 Calculation of austenite formation kinetics of copper-bearing steel during continuous heating *J. Iron Steel Res. Int.* **1** 62-6.

[8] Son Y Il, Lee Y K and Park K T 2006 Reverse transformation of ferrite and pearlite to austenite in an ultrafine-grained low-carbon steel fabricated by severe plastic deformation *Metall. Mater. Trans. A Phys. Metall. Mater. Sci.* **37** 3161–4

[9] Pawłowski B 2012 Determination of critical points of hypoeutectoid steels *Arch. Metall. Mater.* **57** 957–62

[10] Panov D O and Smirnov A I 2017 Features of austenite formation in low-carbon steel upon heating in the intercritical temperature range *Phys. Met. Metallogr.* **118** 1081–90

[11] Wojdyr M 2010 Fityk: A general-purpose peak fitting program *J. Appl. Crystallogr.* 43 1126–8

[12] Panov D, Pertsev A, Smirnov A and Khotinov V 2019 Metastable Austenitic Steel Structure and Mechanical Properties Evolution in the Process of Cold Radial Forging *Materials* **12** 2058

[13] Panov D O, Smirnov A I and Pertsev A S 2019 Formation of Structure in Metastable Austenitic Steel during Cold Plastic Deformation by the Radial Forging Method *Phys. Met. Metallogr.* **120** 184–90

[14] Cherepin V T 1968 Eksperimental'naia tekhnika v fizicheskom metallovedenii (Kiev: Tekhnika) in Russian

[15] He Y, Zhang J, Wang Y, Wang Y and Wang T 2019 The expansion behavior caused by deformation-induced martensite to austenite transformation in heavily cold-rolled metastable austenitic stainless steel *Mater. Sci. Eng. A* **739** 343–7

[16] Aletdinov A, Mironov S, Korznikova G F, Konkova T, Zaripova R G, Myshlyaev M M and Semiatin S L 2019 Martensite-to-Austenite Reversion and Recrystallization in Cryogenically-Rolled Type 321 Metastable Austenitic Steel *Metall. Mater. Trans. A Phys. Metall. Mater. Sci.* **50** 1346–57

[17] Liu G, Li J, Zhang S, Wang J and Meng Q 2016 Dilatometric study on the recrystallization and austenization behavior of cold-rolled steel with different heating rates *J. Alloys Compd.* **666** 309–16

IOP Conf. Series: Materials Science and Engineering 1014 (2021) 012043 doi:10.1088/1757-899X/1014/1/012043

Structure formation of titanium alloys during vacuum-arc surfacing

G L Permyakov*, T V Olshanskaya, E A Krivonosova, D N Trushnikov

Perm National Research Polytechnic University, 29 Komsomolsky ave., Perm, 614990, Russian Federation

*Corresponding author: gleb.permyakov@yandex.ru

Abstract. The work is based on wire vacuum-arc surfacing as one of the technologies for additive manufacturing. The results of the formation of structure and properties of a high-temperature titanium alloy of the Ti-Al-V-Mo-Zr system during vacuum-arc surfacing are presented. The structure of the surfaced materials consists of dendritic grains that are oriented in the direction opposite to heat removal. Some banding of the structure is observed. There is no porosity in the deposited layers for all surfacing modes. It was determined that a decrease in the surfacing current promotes an increase in the dispersion of the α-phase plates, which is accompanied by an increase in microhardness. With such a thermal cycle of surfacing, conditions are created for the formation of a favorable fine-dispersed martensitic structure with narrow α-colonies of small needles of the α-phase, providing a high microhardness at the level of 400HV.

1. Introduction

The use of additive arc processes makes it possible to synthesize large-sized parts with a minimum allowance for machining and provide mechanical properties of metal at the level of traditional technologies, allows to obtain significant savings in metal and significantly reduce the cost of technological preparation for production [1-7]. The work considers wire-arc surfacing in vacuum as one of the technologies for additive manufacturing.

At the preliminary stage of research, it was revealed that significant disadvantages of titanium alloys in surfacing technologies are unfavourable dendritic structure, anisotropy, banded structure and, as a consequence, a decrease in mechanical properties (for example, the ultimate strength $\sigma_в^{20}$ is below 850 MPa), in particular, the loss of plasticity [8-11].

The purpose of this work is to study the main regularities of the structure formation and properties, as well as the influence of the main surfacing parameters (arc current, filler wire feed rate) on the geometric characteristics of the deposited beads of the heat-resistant titanium alloy (Ti-Al-V-Mo-Zr system) during wire vacuum-arc surfacing.

2. Materials and technology

Surfacing was performed using a 1.2 mm titanium alloy filler wire. The chemical composition of the filler wire is shown in Table 1.

Table 1. Content of elements, wt. %

Al	V	Mo	Zr	Si	Fe	C	N	O	Ti	H
5.5…7.0	0.8…2.5	0.5…2.0	1.5…2.5	< 0.1	< 0.3	< 0.1	< 0.05	< 0.15	85.15…91.4	< 0.015

This alloy is based on the ternary system Ti-Al-V, which is typical for the majority of high-strength titanium alloys that belong to pseudo-α-alloys. In accordance with the principles of heat-resistant alloying of titanium alloys, the multicomponent alloy is alloyed with α-stabilizers (Al), β-stabilizers (V, Mo, Fe, Zr) and neutral hardeners (C, N). This provides an effective combination of dispersion and solid solution hardening mechanisms.

Surfacing was performed with use of equipment designed in Perm National Research Polytechnic University. The process scheme is shown in Figure 1.

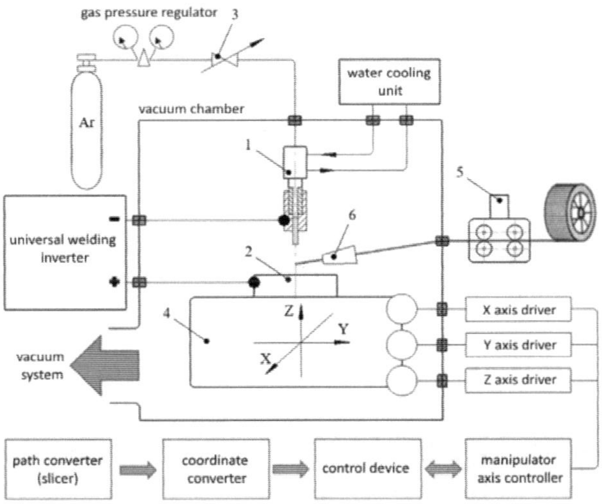

Figure 1. Schematic of vacuum-arc surfacing process: 1 – welding torch; 2 – product (anode); 3 – plasma-forming gas valve; 4 – CNC coordinate table; 5 – wire feed mechanism; 6 – filler wire

The process was implemented using a hollow cathode torch (1) with tungsten WL-15 electrode with 1 mm inner cavity diameter. To ensure the conditions for the formation of an arc discharge with a hollow cathode, a micro flow of argon was supplied through the cavity with use of a plasma-forming gas valve (3).

3. Results

To determine the effect of surfacing mode parameters on the structure and properties of the deposited material cross-sections of the deposited beads were studied. The external view and cross-sections of the deposited beads are shown in Figure 2.

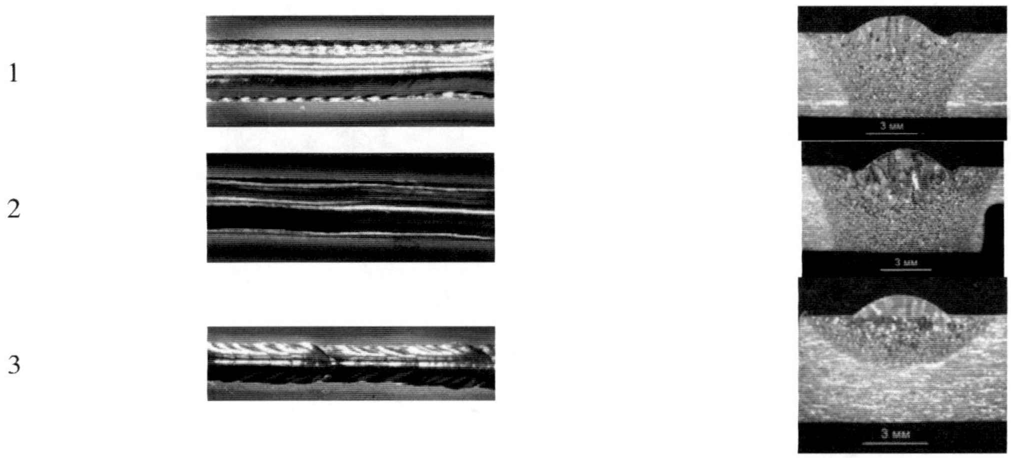

Figure 2. External view (left) and cross-sections (right) of the titanium alloy beads obtained by vacuum-arc surfacing

Table 2 shows the parameters of the surfacing modes and the average values of the deposited layer parameters: height (h), width (b) and the microhardness (HV_{01}).

Table 2. Parameters of vacuum-arc surfacing modes and average values of deposited beads parameters

№	Arc current I, A	Deposition rate V_d, mm/s	Wire feed speed V_f, m/min	Deposited bead height h, mm	Deposited bead width b, mm	Microhardness	
						Interval HV_{01}	Average HV_{01}
1	150	3.5	3.5	0.97	5.70	225…416	340
2	150	3.5	3.2	0.98	5.41	265…457	349
3	120	3.5	2.5	0.91	4.48	372…441	407

The macrostructure of the surfacing materials is dendritic and oriented in the direction opposite to heat removal. Structural banding is observed. There is no porosity in the deposited metal for all processing modes.

Analysis of the microstructure of the deposited material was carried out in different areas of the deposited bead (Fig. 3).

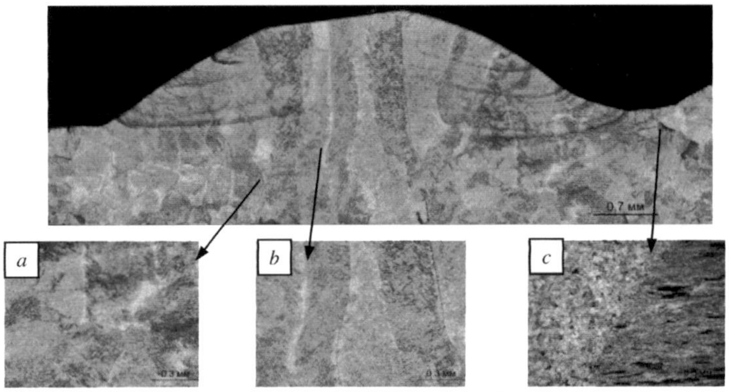

Figure 3. Microstructure of individual sections of the deposited bead: *a*, *b* – central part, *c* – fusion line

The base metal has the typical structure of a titanium alloy in a deformed state: grains elongated in the rolling direction, which are rather long fibers. A change in the structure in the fusion zone is clearly visible: elongated grains of the rolling structure are replaced by an equiaxed, not very coarse grains near the fusion line and larger dendrites in the center of the surfacing. A transcrystalline dendrite adjustment has been marked, which is typical for titanium alloys. The phase composition of the deposited metal has a small amount of the initial β-grains (in fact, it is guessed only by thin interlayers along the grain boundaries), colonies of α-plates of various sizes with intragranular plates of the martensitic α'-phase of different thicknesses. The deposited metal is also characterized by different fineness of the α-colonies plates, and the size of individual needles of 1…100 μm. This difference in the morphology of the structure is explained by the different cooling rates during crystallization and by the stage of solid state transformations.

With low cooling rate, a lamellar structure is formed, represented by colonies of almost parallel α-plates with a microhardness not higher than 360 HV. It was found that a decrease in the current during arc surfacing in vacuum, contributes to the formation of a martensitic structure, an increase in the dispersion of the α-phase plates to fractions of 1 micron, which is accompanied by an increase in microhardness to the level of 400HV. For Ti-Al-V-Mo-Zr system alloys, an increase in the cooling rate promotes the formation of a thinner lamellar structure.

4. Conclusions

The studies have shown that the technology of wire vacuum-arc surfacing prospectively can be effectively used to form defect-free heat-resistant titanium alloys workpieces.

It was found that with an increase in current and wire feed speed, an increase in the height and width of the deposited beads occurred.

Undercuts along the fusion line are observed by upscaling of arc current to 150A and by reducing the wire feed speed. A decrease in the arc current to 120A leads to the disappearance of undercuts, even at a low wire feed speed, and to a decrease in the heat-affected zone size.

A decrease in the arc current, and, consequently, a decrease in the volume of the deposited metal increases the cooling rate, which leads to an increase in the fraction of the martensite phase and an increase in the microhardness above 400 HV.

Acknowledgments
The reported study was supported by the Ministry of Science and Higher Education of the Russian Federation at the base part of the state assignment FSNM-2020-0028), as well as by the Government of Perm Krai research project Nr. S-26/795 of December 21, 2017 and the grant from the Russian Foundation for Basic Research RFBR Nr. 18-08-01016 A.

References

[1] Bintao W, Zengxi P, Donghong D, Dominic C, Huijun L, Jing X, John N 2018 A review of the wire arc additive manufacturing of metals: properties, defects and quality improvement, *Journal of Manufacturing Processes*, vol. 35, pp. 127-139

[2] Tabernero I, Paskual A, Álvarez P, Suárez A 2018 Study on Arc Welding processes for High Deposition Rate Additive Manufacturing, 19th CIRP Conference on Electro Physical and Chemical Machining, *Procedia CIRP*, vol. 68, pp. 358-362

[3] Lockett H, Ding J, Williams S and Martina F 2017 Design for Wire+Arc Additive Manufacture: design rules and build orientation selection, *Journal of Engineering Design*, vol. 28(7-9), pp. 568-598

[4] Oskolkov A A, Matveev E V, Bezukladnikov I I, Trushnikov D N, Krotova E L (2018) Advanced technologies for additive manufacturing of metal product. Bulletin PNRPU. Mechanical engineering, materials science, vol. 20 (3), pp. 90–105 (in Russian)

[5] Krivonosova E A, Schitsin Yu D, Trushnikov D N, Myshkina S N, Akulova A V, Neulybin S D, Dushina A Yu (2019) Influence of surfacing technologies on structure formation of high-temperature nickel alloys, Metallurgist, vol. 63 (1–2), pp. 197-205

[6] Shchitsin Yu D, Krivonosova E A, Neulubin S D, Olshanskaya T V, Nikulin R G, Fedoseeva E M, Terentyev S A (2019) Use of plasma surface for additive formation of aluminum alloys blade Bulletin PNRPU. Mechanical engineering, materials science, vol. 21 (2), pp. 63-72. (in Russian)

[7] Cong B, Qi Z, Qi B, Sun H, Zhao G and Ding J (2017) A Comparative Study of Additively Manufactured Thin Wall and Block Structure with Al-6.3%Cu Alloy Using Cold Metal Transfer Process. *Applied Sciences*, vol. 7(3)

[8] Donoghue J, Antonysamy A A, Martina F, Colegrove P A, Williams S W, Prangnell P B (2016) The effectiveness of combining rolling deformation with Wire–Arc Additive Manufacture on β-grain refinement and texture modification in Ti–6Al–4V, *Materials Characterization*, vol. 114, pp. 103-114

[9] Wu Q, Lu J, Liu C, Fan H, Shi X, Fu J and Ma S 2017 Effect of Molten Pool Size on Microstructure and Tensile Properties of Wire Arc Additive Manufacturing of Ti-6Al-4V Alloy, *Materials*, vol. 10 (7)

[10] McAndrew A R, Rosales M A, Colegrove P A, Hönnige J R, Ho A, Fayolle R, Eyitayo K, Stan I, Sukrongpang P, Crochemore A, Pinter Z 2018 Interpass rolling of Ti-6Al-4V wire+arc additively manufactured features for microstructural refinement, Additive Manufacturing, vol. 21, pp. 340-349

[11] Hirtler M, Jedynak A, Sydow B, Sviridov A, and Bambach M 2018 Investigation of microstructure and hardness of a rib geometry produced by metal forming and wire-arc additive manufacturing, 5th International Conference on New Forming Technology (ICNFT 2018), *MATEC Web of Conferences*, vol. 190

IOP Conf. Series: Materials Science and Engineering 1014 (2021) 012044 doi:10.1088/1757-899X/1014/1/012044

Synthesis and magnetic properties of solid solutions of a diluted magnetic semiconductor $(Zn_{1-x}Fe_x)_3As_2$

V S Zakhvalinskii*, T B Nikulicheva, E A Pilyuk, A A Trubaev, I M Usatyy

[1]Belgorod National Research University, 85 Pobedy St, Belgorod 308015, Russia

*Corresponding author: zakhvalinskii@bsu.edu.ru

Abstract. The magnetic properties of the new semimagnetic semiconductor $(Zn_{1-x}Fe_x)_3As_2$ with $x < 0.04$ have been investigated over a wide temperature range. Single crystals of ZFA were grown using the modified Bridgeman technique. The results of X-ray analyses and the magnetometry data show that the border of solid solutions in this system lay near x = 0.015. Further increasing of x leads to the growing of the second phase, which was identified as Fe_2As. A notable feature of our magnetometry results is the spin-glass-type freezing of magnetic moments in the temperature range of 10 - 50 K. The value of the freezing temperature decreases when the magnetic field increases and in 2 kOe magnetic field the effect is no longer observable. It is likely that the observed freezing phenomena are due to the cluster nature of magnetic properties of ZFA. The thermoremanent magnetization measurements give evidence for two types of clusters in the magnetic system of ZFA.

1. Introduction

Compounds of the type $(A_{1-x}B_x)_3As_2$ with A = Zn, Cd and B = Mn form a class of II_3V_2 diluted magnetic semiconductors (DMS) [1]. In DMSs, a part of cations is replaced by a transition metal or rare earth elements with partially filled d- and f-shells, respectively. The transition metal or rare earth ions are randomly distributed in the host lattice. As a result, two types of interactions take place in the DMS crystal: (i) the sp-d (sp-f) exchange interaction between the band carriers and localized magnetic moments of these ions and (ii) the d-d (f-f) interaction between the ions themselves. One of the types of practical application of a DMS is its use as an injector for the spin polarized current of charge carriers in heterostructures [2]. Solid solution $(Zn_{1-x}Fe_x)_3As_2$ ($x \leq 0.015$) (ZFA) is a new diluted magnetic semiconductor of the A_2B_5 group. ZFA is isomorphic to pure Zn_3As which has the width of the band gap $E_g \approx 1$ eV and tetragonal crystal structure (space group *I*41*cd*) [3].

2. Experimental results and discussion

$(Zn_{1-x}Fe_x)_3As_2$ (ZFA) was synthesized from stoichiometric melts of Zn_3As_2 and Fe in the range of $0.005 < x < 0.04$ by using the modified Bridgeman method (slow cooling in the presence of a temperature gradient). They are isomorphic to pure Zn_3As_2 (x = 0.0), which at room temperature has a tetragonal structure with the space group $I4_1cd$ (the so called α-phase of Cd_3As_2) [4]. The crystal were growing in carbon-coated quartz ampoules with a conical bottom, inserted in sealed quartz ampoules under a vacuum of 10^4 Torr. Large monocrystalline ingots were obtained by using a cooling speed of 2°C/h and a temperature gradient of l-2°C/cm. Powder X-ray investigations showed that the obtained crystals were homogenous up to x = 0.015. However, the diffractograms of the samples with x > 0.02

revealed growth of the second phase, identified as Fe_2As. The dc magnetic measurements were made a with SQUID magnetomiter in the range 4-120 K using magnetic fields up to 2 kOe. A notable feature of our magnetometry results is a spin-glass-type freezing of magnetic moments in the temperature range 10-50 K.

The investigations of the magnetic properties of ZFA were performed using the samples with $0.005 < x < 0.04$. Figure 1 present the temperature dependencies of the magnetic susceptibility χ of ZFA with x = 0.01 in a number of magnetic fields. It is seen from curves 1-3 that the magnetic susceptibility of the specimen, cooled in zero magnetic field χ_{ZFC} starts to grow in the field when increasing the temperature, exhibit a broad maximum and then decreases. Cooling the specimen in the field (FC case) leads to growing the χ_{FC} with the following saturation at temperatures just below the χ_{ZFC} maxima. The appearance of a maximum in the ZFC susceptibility curves suggests that freezing of magnetic moments in the sample takes place. The difference between χ_{ZFC} (T) and χ_{FC} (T) curves becomes smaller and finally disappears together with the shifting to lower temperatures and smearing out of the χ_{ZFC} maxima, which defines the freezing temperature, T_f, when increasing of the magnetic field from 50 Oe to 2 kOe. Such influence of the magnetic field on the freezing phenomenon is the most characteristic feature of spin glasses [5].

The same behaviour, in general, with a change in temperature and applied magnetic field is exhibited by ZFA with x = 0.005, as can be seen in Figure 2. Half as much decreasing of the concentration of magnetic impurity decreased about one order of magnitude the value of χ and slightly shifted the position of χ_{ZFC} (T) maxima to the lower temperatures in comparison with x = 0.01 sample. The saturation of χ_{ZFC} (T) takes place at lower temperatures, than in the previous case, and, moreover, it starts to increase after the saturation when cooling at 500 Oe and 2 kOe. A further increase in the of Fe concentration up to x = 0.015 leads to an increase in χ to about 1.5 times in comparison with x = 0.01 together with the appearance of a high-temperature "shoulder" near the maximum on the χ_{ZFC} (T) curves in magnetic fields up to 50 Oe (see Figure 3). The behaviour of χ_{FC} (T) appeared to be new at T < 12 K when cooling at 5 Oe in comparison with the previous samples: in this region of temperatures χ_{FC} (T) slightly decreases instead of saturation (x = 0.01 case) or instead of growth (case x = 0.005). A feature on the corresponding ZFC curve (Figure 3) at the same temperatures should be noted.

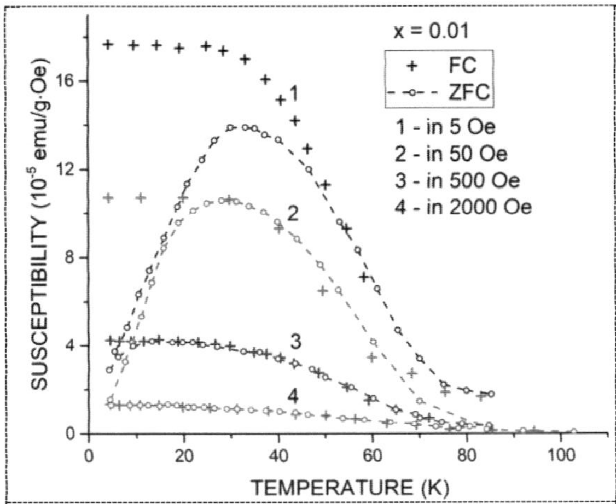

Figure 1. Temperature dependence of the magnetic susceptibility of a ZFA (x = 0.01) single crystal.

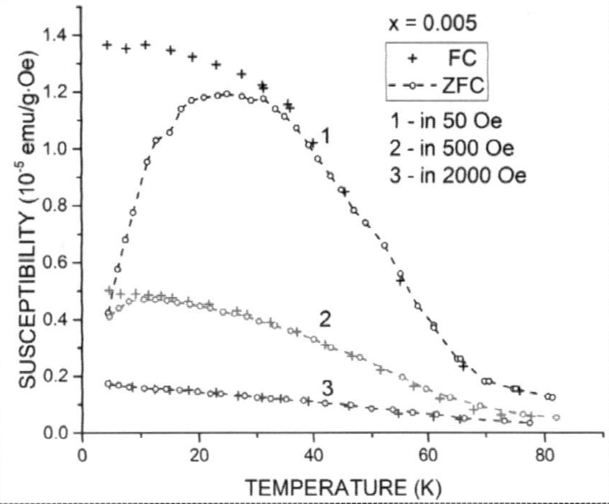

Figure 2. Temperature dependence of the magnetic susceptibility of a ZFA (x = 0.005) single crystal.

The same feature is seen at 5 Oe magnetic field in Figure 4, where the results for the sample with, x = 0.02, are presented. This sample exhibits a decrease in χ and a broadening of the χ_{ZFC} maximum in comparison with the other samples. Full saturation of the susceptibility when cooling was achieved only at 50 Oe applied field. In higher fields growing of χ during cooling in the field was observed. This fact becomes understandable, taking into account the data for specimen with x = 0.04, which are presented in Figure 5. For this composition no maxima exists on the χ (T) dependencies and only the difference between ZFC and FC curves appears below the certain temperature in small magnetic fields. The χ_{ZFC} dependencies are practically similar for all magnetic fields applied and show the existence of the strong antiferromagnetic interactions in this material, as it can be seen in Figure 6, where the temperature dependencies of the inverted ZFC susceptibility of this sample are presented. The estimation of the Curie temperature gave the magnetic field dependent negative values of -94 K and 89 K for 5 and 50 Oe correspondingly (where the freezing phenomena were observed) and - 90 K for higher fields, where the ZFC and FC temperature dependencies are similar.

The prominent feature of spin glasses is a certain dependence of T_f on the magnetic field and the concentration of magnetic impurity [4].

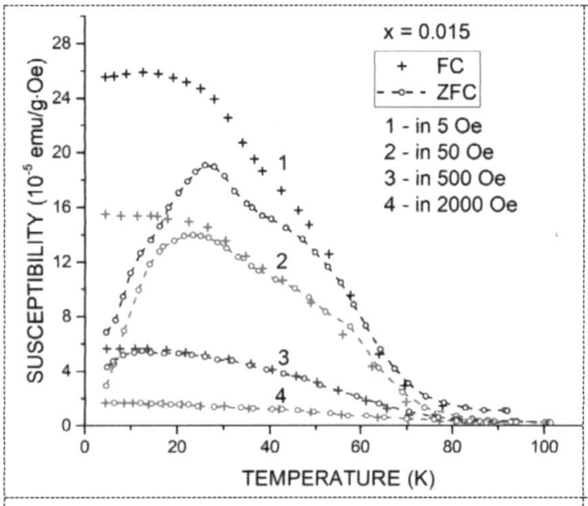

Figure 3. Temperature dependence of the magnetic susceptibility of a ZFA (x = 0.015) single crystal.

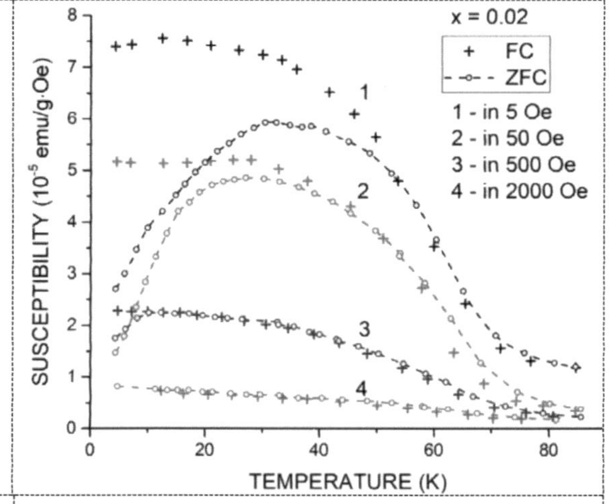

Figure 4. Temperature dependence of the magnetic susceptibility of a ZFA (x = 0.005) single crystal.

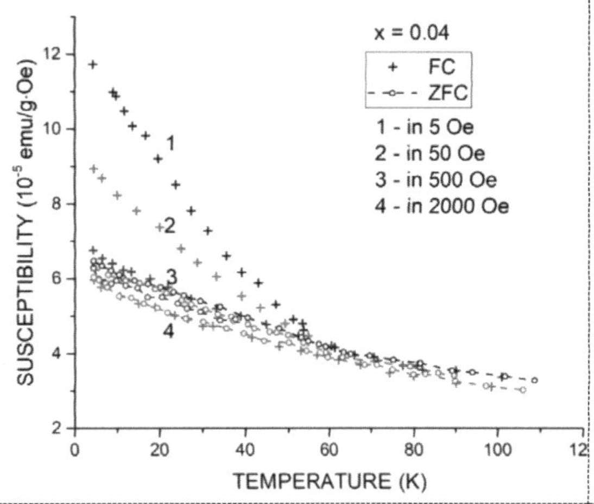

Figure 5. Temperature dependence of the magnetic susceptibility of a ZFA (x = 0.04) single crystal.

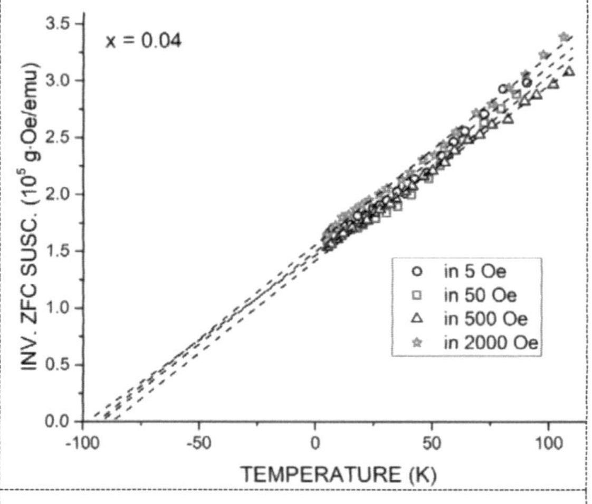

Figure 6. Temperature dependence of the magnetic inverted ZFC susceptibility (x = 0.04) single crystal.

We found that the value of the saturated M_{FC} increases as x change from 0.005 to 0.015 in all applied magnetic fields, but then at x > 0.02 starts to decrease. Moreover, the magnetic field dependence of M_{FC} undergoes a change in it's behaviour at x = 0.04: $M_{FC}(H)$ becomes linear up to 2 kOe instead of an approximately logarithmic dependence in the other samples. This change in the behaviour correlates well with the temperature dependencies of χ in ZFA, showing a considerable influence of antiferromagnetism on the magnetic properties of the specimen with x = 0.04. Magnetic systems exhibiting the freezing phenomena in a field should possess magnetic remanent properties after the field is removed of a field. Two types of remanency are distinguished: thermoremanent magnetization (TRM), which is observed in zero field after cooling the sample in the field, and isothermal remanent magnetization (IRM), which is the effect of field a cycling. All of TRM at 4.2 K just after removing of the field exhibit saturating behaviour, and the full value of TRM was achieved after 500 Oe already.

The value of TRM practically does not changes when increasing the field from 500 Oe to 2 kOe, what means that at 500 Oe all magnetic moments in the system are already orientated and frozen. Another type of remanency isothermal remanent magnetic moment of the specimens with x = 0.005 and x = 0.01, was also studied versus the magnetic field at 30 K as well as the magnetization in the field. The magnetic field dependencies of the specific magnetization M and IRM for x = 0.005 and x = 0.01 at 30 K (near the value of T_f for these specimens) were investigated. The saturation character of M vs. H and IRM vs. H is obvious for x = 0.01, and for x = 0.005 the saturation was achieved in lower fields. The values of the field after which M (IRM) are practically constant are about 500 Oe (1000 Oe) for x = 0.005 and 2.5 kOe (500 Oe) for x = 0.01, correspondingly.

3. Conclusion

One of the most characteristic properties of ZFA is that its magnetic behaviour depends strongly on the history of the sample (see Figure 1-5). This fact suggests the existence of a spin-glass-type freezing of magnetic moments in this material. The same can be concluded from the saturation of IRM vs. H and the small hysteresis. Also the increasing of the ZFC magnetization with temperature and the existence of a maximum in the temperature dependence of the ZFC susceptibility are spin-glass properties. Increasing of the applied magnetic field shifts this maximum to lower temperatures.

The two very different slopes of TRM vs. T (one of which depends on the magnetic field previously applied) in our data indicate the existence of two types of magnetic clusters in ZFA. The same is suggested by the wideness of the ZFC susceptibility maxima in ZFA and even "splitting" in the sample with x = 0.015. Nonstatistical distribution of iron in ZFA cannot be ruled out. In this case the probability of the formation of clusters with strong internal interactions and relatively big magnetic moments grows as well as the existence of the second phase. The changing of the concentration of iron in ZFA allow to observe the transition from the system, containing clusters, mixed with the lost spins (x = 0.005) through the intermediate state (x = 0.01) to the clusters of several types with complex internal interactions (x = 0.015) and even to the appearance of the second phase (x > 0.015).

References

[1] Gaj J A and Kossut J 2010 *Introduction to the Physics of Diluted Magnetic Semiconductors*. (Springer-Verlag Berlin Heidelberg) p. 469

[2] Gould C, Schmidt G, Richter G, R.,Fiederling, Grabs P, Molenkamp LW 2002 *Applied Surface Science* **190**, 395-402

[3] Pietraszko A and Łukaszewicz K 1973 Thermal expansion and phase transitions of Cd_3As_2 and Zn_3As_2 *Physica Status Solidi* (a) **18(2)** 723–30

[4] Stutz E Z, Friedl M, Burgess T, Tan H H, Caroff P, Jagadish C and Fontcuberta i Morral A 2019 *Physica Status Solidi (RRL)–Rapid Research Letters* **13** 1900084

[5] Kumar R, Iyer K K, Paulose P L and Sampathkumaran E V 2019 *Journal of Applied Physics* **126** 123906

IOP Conf. Series: Materials Science and Engineering 1014 (2021) 012045 doi:10.1088/1757-899X/1014/1/012045

The effect additives of magnesium, titanium and zirconium oxides additives on the densification kinetics and structure of alumina during spark plasma sintering

A A Popov*, M S Boldin, A V Nokhrin and V N Chuvil'deev

Lobachevsky State University of Nizhny Novgorod, Nizhny Novgorod, 603950 Russia

*Corresponding author: popov@nifti.unn.ru

Abstract. The results of an experimental study of the effect of metal oxide additives on the sintering kinetics of systems based on alumina by the method of spark plasma sintering (SPS) are presented. The influence of additives at the stage of intensive compaction of powder pressing and at the stage of intensive growth of ceramic grains is analyzed. A qualitative explanation of the mechanism of the effect of magnesium oxide additives on the stabilization of the grain structure of aluminum oxide is given.

1. Introduction

Ceramics based on aluminum oxide have high physical and mechanical properties, low cost and availability. This makes alumina the most promising ceramic material for a wide range of structural applications [1]. An increase in the physical and mechanical properties of this ceramics can be expected when a homogeneous fine-grained structure is formed [2]. SPS technology is able to provide materials with small grain size [1].

Research results [1, 3, 4] show that composite ceramics based on aluminum oxide are suitable for creating modern wear-resistant metalworking tools, making armor protection elements, angular contact bearings, etc.

One of the tools for influencing the sintering kinetics, grain size, and physical and mechanical properties of ceramics is the introduction of a small amount of different additives, for example: MgO, TiO_2, Y_2O_3, SiC, ZrO_2 and others. The aim of this work is to study the effect of such additions on the sintering kinetics, the evolution of the grain structure of alumina ceramics during spark plasma sintering.

2. Experimental and materials

Ceramic samples made of pure α-Al_2O_3 (Taimei Chemicals Co., Ltd, d = 200 nm) and composites based on it with MgO additives (Alfa Aesar - A Johnson Matthey Company, d = 100 nm), TiO_2 (Institute of Electrophysics, UB RAS, d = 100 nm) and ZrO_2 (Pangea Int., Ltd, d = 100 nm).

Spark plasma sintering of powders was carried out on the installation Dr. Sinter SPS-625 "SPS SYNTEX, INC." Heating rate $V = 10 \div 700$ °C/min, mechanical stress during sintering $P = 6 \div 70$ MPa. Sintering was carried out in a vacuum of 5 Pa. The temperature of the sintering process was monitored with an optical pyrometer focused on the surface of a graphite mold with an inner diameter of 12 mm.

The maximum temperature - sintering temperature (T) lay in the range 1150 ÷ 1300°C. Isothermal holding at T was $\tau = 0 \div 30$ min. The density of the sintered samples was measured by hydrostatic weighing in distilled water using a Sartorius CPA analytical balance. The average grain size in sintered ceramics was calculated using the Good Grains software package from micrographs obtained with a Jeol JSM-6490 scanning electron microscope.

3. Results and discussion

The dilatometer measures the change in linear dimensions. During sintering, it integrally measures the shrinkage of the powder compact and the thermal expansion of the materials of the molds and press plungers. The shrinkage graph of the powder was obtained by subtracting from the overall shrinkage graph the thermal expansion graph obtained by repeating the sintering regime with an empty mold. Conversion of shrinkage to compaction value ρ/ρ_{th} was carried out on the basis of the procedure described in [5].

3.1 Analysis of the effect of additives at the stage of intensive compaction.

Let us calculate the activation energy of grain boundary diffusion at the second stage. According to [5], the value of the activation energy of grain-boundary diffusion Q_b can be calculated from the experimental dependence $\rho/\rho_{theor}(T)$, presented in double logarithmic coordinates on the reciprocal temperature T_m/T:

$$\frac{Q_b}{kT_m} = \frac{\partial \ln\ln\left(\left(V_p^{\max}/V_{\text{theor}}\right)\cdot\left(\frac{\rho(T)/\rho_{\text{theor}}}{1-(\rho(T)/\rho_{\text{theor}})}\right)\right)}{\partial(T_m/T)} \quad (1)$$

where Q_b - activation energy of grain boundary diffusion, $V_p^{\max}$ - maximum pore volume, V_{theor} - theoretical pore volume, $\rho(T)$ - density versus temperature, ρ_{theor} -theoretical density, T_m - melting temperature, k - Boltzmann constant.

Table 1 shows the calculations of the activation energy of sintering at the II stage. For the systems Al_2O_3 + 0.5% MgO, Al_2O_3 + 0.5% vol. TiO_2 and Al_2O_3 + 0.5% ZrO_2, the activation energy is $Q_b = 30 \pm 2\ kT_m$ and coincides with the activation energy of sintering of pure Al_2O_3 in the entire range of the studied heating rates (from 10 °C/min to 700 °C/min). Thus, it can be concluded that particles of MgO, TiO_2, and ZrO_2 do not affect the kinetics of sintering of aluminum oxide at the stage of intense shrinkage. It should be noted that the found value of the activation energy for sintering is in agreement with the experimental values of the activation energy for grain-boundary diffusion of oxygen in aluminum oxide ($Q_b \sim 16$–$40\ kT_m$) given in the review [6].

Table 1. Activation energy

Systems:	stage of intensive compaction Q_b, kT_m 10 °C/min	100 °C/min	700 °C/min	*the stage of isothermal holding* Q_b, kT_m
Al_2O_3	30±1.5	30±1.5	30±1.5	29±1.5
Al_2O_3+0,5%vol.MgO	31±1.5	31±1.5	31±1.5	30±1.5
Al_2O_3+0,5%vol.TiO_2,	29±1.5	29±1.5	29±1.5	30±1.5
Al_2O_3+0,5%vol.ZrO_2,	31±1.5	31±1.5	31±1.5	24,5±1.5

Thus, the analysis of the influence of modifier additives on the stage of intense shrinkage showed that the introduced additives do not affect the activation energy of grain boundary diffusion. The activation energy of grain boundary diffusion is $Q_b \sim 30\ kT_m$

In other words, the change in the sintering kinetics of aluminum oxide caused by the segregation of Mg^{2+} and Ti^{4+} ions at the grain boundaries of the sintered ceramics, widely described in the literature [7-10], is not observed.

3.2 Analysis of the effect of additives on grain growth.

Let us consider the effect of the introduced additives at the stage of grain growth. In this case, let us consider the evolution of the grain structure for three systems: Al_2O_3, Al_2O_3 + 0.5%vol.MgO, Al_2O_3 + 0.5%vol.TiO_2, Al_2O_3 + 0.5%vol.ZrO_2, under conditions of isothermal holding. Let us estimate the activation energy of sintering of the systems under study.

As shown in [11, 12], recrystallization processes are carried out by the diffusion of single atoms and their groups. Thus, the mobility of the boundary is determined by a parameter that has the physical meaning of the diffusion coefficient.

Let us analyze the effect of the addition of zirconium oxide and magnesium oxide on the activation energy of grain growth in aluminum oxide under SPS conditions. The expression [11, 12] can be used to describe the kinetics of grain growth during sintering of aluminum oxide:

$$d^n - d_0^n = \frac{2\gamma_b b\delta t}{kT} D_0 exp\left(-\frac{Q_b}{kT_m}\right) \quad (2)$$

where d - grain size, d_0 - initial grain size, n - exponent, t - sintering time, k - Boltzmann coefficient, γ_b - surface energy, b - Burgers vector, δ - grain boundary width.

It can be seen from table 1 that the activation energy of grain-boundary diffusion at the third stage, the stage of grain growth, does not change with the introduction of magnesium and zirconium modifier particles and is $Q_b = 30 \pm 1.5$ kT_m. Adding to aluminum oxide 0.5% vol. titanium oxide leads to a decrease in the activation energy of grain growth to $Q_b = 24.5$ kT_m.

As shown above, the additives do not change the activation energy of grain boundary diffusion during grain growth. However, the introduced additives of magnesium oxide and zirconium slow down the growth of grains (2.75 μm and 3 μm, respectively) in relation to pure aluminium oxide. So, the samples which sintered at 1300°C with 30 min holding time have the average grain size 16 μm for pure alumina oxide, and 2.75 and 3 μm for alumina with magnesium and zirconium additions, respectively.

In this case, the most probable stabilization mechanism seems to be the Zener fixation mechanism. When used as additives MgO and ZrO_2 powders with the same initial particle size (d ~ 100 nm), the stabilized grain size of ceramics Al_2O_3 + 0.5%vol.MgO and Al_2O_3 + 0.5%vol.ZrO_2 must match. However, this contradicts the experimental data obtained, since at high temperatures and long holding times, the average grain size of ceramics Al_2O_3 + 0.5%vol.MgO is approximately two times smaller than the average grain size of the ceramic Al_2O_3 + 0.5%vol.ZrO_2.

It should be noted that the structure of the Al_2O_3 + 5%vol.ZrO_2 is characterized by a comparable average grain size with ceramics Al_2O_3 + 0.5%vol.MgO. This means that the rate of grain growth in these composites was the same:

$$d = \alpha \cdot \left(\frac{d_{MgO}}{f_{MgO}}\right) = \alpha \cdot \left(\frac{d_{ZrO2}}{f_{ZrO2}}\right) \quad (3)$$

where d_{MgO} is the diameter of MgO particles, $f_{MgO} = 0.005$ is the volume fraction of MgO particles, d_{ZrO2} = 50–100 nm is the diameter of ZrO_2 particles, $f_{ZrO2} = 0.05$ is the volume fraction of ZrO_2 particles.

Substituting into expression (5) the volume fractions of the additives used and the average particle size of ZrO_2, we obtain $d_{MgO} = 5 \div 10$ nm.

It should be noted here that in reality we are not talking about MgO particles, but about dispersed particles of spinel MgAl2O4. Since the reaction of its formation during the interaction of MgO and Al_2O_3 lies below the temperature of the onset of sintering [13].

4. Conclusions

The influence of MgO, TiO_2 and ZrO_2 particles on the sintering kinetics and grain growth of Al_2O_3 has been studied. It is shown that MgO, TiO_2 and ZrO_2 have no effect on the kinetics of the stage of intensive compaction of Al_2O_3 at heating rates V = 10 °C/min, 100 °C/min, 700 °C/min. The activation energy of grain boundary diffusion at stage II is $Q_b = 30kT_m$. It is shown that MgO and ZrO_2 do not change the diffusion activation energy at the stage of Al_2O_3 grain growth. The activation energy of grain boundary diffusion at stage III is $Q_b = 30 \pm 1.5$ kT_m. Adding to aluminum oxide 0.5% vol. titanium oxide leads to a decrease in the activation energy of grain growth to $Q_b = 24.5$ kT_m.

It is shown that the grain structure of composites containing 5% ZrO_2 and 0.5% MgO is equivalent. The size of the MgO particles in the sintered sample was estimated. The size of the magnesium oxide particles in the sintered ceramics is 5 ÷ 10 nm. A qualitative explanation of the mechanism of the effect of the addition of MgO on the stabilization of the grain structure of Al_2O_3 is given. The small size of dispersed particles and the uniformity of their distribution determines the effective action of the Zener force even with the introduction of a small amount of additive.

Acknowledgements

This study was performed with the support of the Russian Science Foundation (Grant No. 20-73-10113).

References

[1] Shackelford J F 2008 *Ceramic and Glass Materials* (Davis: Springer) p 201.
[2] Kiyoshi I 2001 *Functionally graded materials in the 21stcentury* (Kluwer: Kluwer Academic Publishers) p 242
[3] Krell A 2007 *Hierarchy of Key Influences on the Ballistic Strength of Opaque and Transparent Armor* vol 28, ed Prokurat L Salem J and Zhu D (Hoboken: Journal of the American Ceramic Society) pp 9201– 9207
[4] Young M 2004 Development of Al_2O_3-SiC composite tool for machining application vol 30 (Amsterdam: Elsevier) pp 2081– 2086.
[5] Chuvil'deev V N 2015 *B comparative study of hot pressing and high-speed electropulse plasma sintering of $Al_2O_3/ZrO_2/Ti(C,N)$ powders* vol 60 (Moscow: Russian Journal of Inorganic Chemistry. pp 987 - 993
[6] Rahaman N 2003 *Ceramic Processing and Sintering Second edition* (Boca Rator: CRC Press) p 876
[7] Shen Z 2002 *Spark Plasma Sintering of Alumina* vol 85 (Hoboken: Journal of the American Ceramic Society) pp 1921–1927
[8] Konstantin L 1999 *Silica and Magnesia Dopant Distributions in Alumina by High-Resolution Scanning Secondary Ion Mass Spectrometry* vol 82 (Hoboken: Journal of the American Ceramic Society) pp 1001–1008
[9] Greskovich C 2001 *Solubility of Magnesia in Polycrystalline Alumina at High Temperatures* vol 84 (Hoboken: Journal of the American Ceramic Society) pp 420–425
[10] Kebbede A 2000 *Anisotropic Grain Growth in a-Al2O3 with SiO2 and TiO2 Additions* vol 83 (Hoboken: Journal of the American Ceramic Society) pp 2845–2851
[11] Chuvil'deev V N 2004 *Non-equilibrium grain boundaries in metals. Theory and applications* (Moscow: Fizmatlit) p 304.
[12] Geguzin Ya 1984 *Sintering physics* (Moscow: Science) p 312
[13] Sako E 2012 *Fundamentals and applications on in situ spinel formation mechanisms in Al_2O_3 MgO refractory castables* (Amsterdam: Elsevier) pp 2243–2251

IOP Conf. Series: Materials Science and Engineering 1014 (2021) 012046 doi:10.1088/1757-899X/1014/1/012046

Influence of titanium and chrome on the structure and heat resistance of layered composites from a multicomponent Nb-alloy and intermetallic compounds of aluminium

D V Prokhorov* and V P Korzhov

Institute of Solid-State Physics RAS, Chernogolovka, Moscow Region, Russia

*Corresponding author: prohorov@issp.ac.ru

Abstract. Investigations of the structure and measurements of the strength characteristics of layered composites of niobium alloys prepared by diffusion welding under pressure have been carried out. It was found that titanium has a strong effect on the formation of microstructure and strength of composites. Higher values of short-term strength and fracture toughness are demonstrated by layered composites with a lower titanium content.

1. Introduction

In several structural materials of modern science, a special place is occupied by heat-resistant alloys, among which alloys based on niobium and its intermetallic compounds are promising. They can possess high mechanical properties up to 1400–1500 °C. Alloys based on the Nb–Al system stand out.

One of the most promising methods for producing heat-resistant alloys based on the Nb–Al system are technologies for creating composites with a layered structure, developed by foreign [1–5] and Russian scientists [6, 7]. The widespread methods are high-speed magnetron sputtering [1–6] and combined methods for obtaining layered structures [3, 5]. But in all layered composites, layers consisting of strong intermetallic compounds alternate with layers of a relatively plastic Nb solid solution. In addition to the above-mentioned methods of obtaining (Nb–Al) alloys, there is a method that combines diffusion welding and rolling of already welded packages.

For heat-resistant composites, it is important to combine high strength at elevated temperatures with satisfactory crack resistance, which characterizes their viscous-plastic state under normal conditions. On the example of alloyed steels, it has long been known that toughness and plasticity are acquired by them with the addition of vanadium. For this purpose, similar studies have been undertaken in this work. This should be facilitated by the Nb–V phase diagram, which is characterized by continuous mutual solubility of the components in liquid and solid states.

2. Materials and experimental technique

The studied composites made of Nb alloys foils of two compositions, differing in the contents of Ti, Cr, Zr, Mo and Si, but to a greater extent in the contents of titanium (Table 1). The fact is that for smelting of both alloys, we used charges of the same composition. The compositions became different after the alloys were smelted by different methods. The first initial niobium alloy (Nb)1 (see Table 1) was obtained by melting in suspension (MiS) with pouring the melt into massive cylindrical Cu-crystallizers and subsequent 2-fold electron-beam remelting (EBR) in a stream water-cooled Cu-crystallizer. The second alloy (Nb)2 was obtained using only the MiS-method with pouring the melt into a flat Cu-mold.

It turned out that after MiS the ingots contained 85.44 wt.%Nb, and after MiS+EBR – 87.70 wt.%Nb. An increase in the concentration of niobium in the alloy was due to the almost complete evaporation of titanium during electron-beam remelting, complete evaporation of chromium and partial evaporation of silicon, as, like aluminum, an intermetallic-forming element with metals. After smelting, the resulting ingots 6.5–8.0 mm thick were rolled on a vacuum rolling mill at 1000–1300 °C with 20% reduction to a thickness of 2 mm. Further rolling of strips to 0.1 mm was carried out already at room temperature.

Table 1. Compositions of Nb alloys according to X-ray spectral analysis.

Alloy	Method	at. or wt.%	Nb	Ti	Cr	Zr	Mo	Hf	Si
(Nb)1	MiS+EBR	at.%	89.08	0.27	0.01	0.73	5.40	2.76	1.03
		wt.%	87.70	0.14	0.01	0.71	5.49	5.22	0.31
(Nb)2	EBR	at. %	82.11	7.72	0.79	0.19	3.26	2.53	3.00
		wt.%	85.44	4.14	0.46	0.20	3.50	5.06	0.94

From pieces of rolled strips of Nb-alloys with a thickness of 100 μm and strips of alloy Al–2 wt. % Si 25 μm thick, multilayer packages (Nb)1/AlSi and (Nb)2/AlSi 2 mm thick were assembled.

The formation of the layered structure of the composite was carried out in the process of diffusion welding under pressure (Nb)/AlSi-packages. Welding was carried out in an installation, which has a water-cooled body and a high-strength graphite heater, designed for temperatures up to 1700 °C.

To obtain the creep characteristics, a series of experiments was carried out in which the deformation of the sample was measured under the action of discretely increasing loads. The sample was held for 10–15 h at each load. The calculation was carried out according to the method proposed in [8, 9].

Microstructure studies and local X-ray spectral analysis were performed on Tescan VEGA-II XMU and CamScan MV230 digital scanning electron microscopes. Tests of specimens for short-term, with a loading rate of 0.5 mm / min, and long-term strength were carried out according to the 3-point bending scheme in a vacuum chamber combined with an Instron testing machine in an atmosphere of high purity argon.

3. Experimental results

3.1 Microstructure of composites

The melted and rolled alloys had a homogeneous structure of solid solutions based on niobium without any other structural components.

The layers inheriting the alloy foils, after diffusion bonding, are identified as an Nb-solid solution of mainly aluminum and silicon in niobium. Its layers contained light inclusions of regular shape 1–2 μm in size (shown by arrows, see Figure 1a), identified as HfO_2, and a large number of pores of various configurations formed in place of brittle inclusions, the composition of which remained unknown.

The microstructure of the (Nb)2/AlSi composite prepared from alloy 2 (Figure 1b) significantly differed from the structure of the composite with the (Nb)1/AlSi alloy. At this stage of solid-phase interaction, the composite contains two structural components that do not have a sharp boundary between them. These are "layers" of the Nb solid solution (points of analysis 3, 4, 8–12, and 15) and 2-phase "layers" (points of analysis 1, 2, 5–7, 13, and 14), consisting of intermetallic compound $(Nb,Me)_3Al$ in the form of a gray matrix and light threadlike or cellular formations from an Nb-solid solution.

Summing up the results of studying the microstructure of the (Nb)2/AlSi composite, we can say that it consists of two alternating 2-phase structural components (structural formation) – $(Nb,Me)_3Al$ + Nb-s. s. and Nb-s. s. + $(Nb,Me)_3Al$. One is the $(Nb,Me)_3Al$ aluminide matrix, which contains elongated thin sections of the Nb-solid solution inclusions. Another structural component is large grains consisting of an Nb-solid solution and $(Nb,Me)_3Al$, with boundaries also from an intermetallic compound $(Nb,Me)_3Al$.

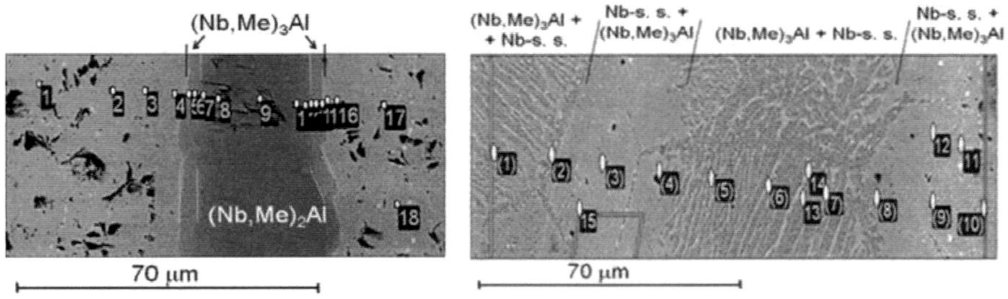

a b

Figure 1. Microstructures of layered composites (Nb)1/AlSi (a) and (Nb)2/AlSi (b): 1–18 and 1–14 – points of analysis.

The third structural object, as in the (Nb)1/AlSi composite, includes finely dispersed HfO_2 inclusions, which in this composite are mainly localized in the $(Nb,Me)_3Al$ compound.

3.2 Mechanical test results

The dependences of the short-term strength of composites on the test temperature in the range from room temperature to 1300 °C are shown in Figure 2. The changes in strength of the (Nb)1/Al and (Nb)2/Al composites differ significantly. The ultimate strength σ_B of the 1st composite at room temperature was 930 MPa, retained a high value equal to 850 MPa at 1100 °C and only at 1300 °C fell to 420 MPa. Composite (Nb)2/AlSi already at room temperature had an average strength of only 370 MPa and practically the same – 375 MPa – at 1300 °C.

Crack resistance K* of composites with alloys 1 and 2 at room temperature turned out to be equal to 12.5–13.0 and 8.5–9.0 MPa·m$^{1/2}$, respectively. The K* values differ so much that the fracture patterns are clearly different.

The results of creep tests of composites at 1300 °C are shown in Figure 3. For the (Nb)1/AlSi composite, the 100-hour strength or stress value causing 1% deformation per 100 hours is 8 MPa with an exponent n = 0.42. The composite 2 had a lower 100-hour strength of 7 MPa with a higher n value of 0.77.

4. Discussion of the results

The analysis of the results obtained is reduced to the effect of Ti and Cr (and, perhaps, Si) on the structure and strength characteristics of composites based on niobium alloy. The alloy, from the foils of which the packages for (Nb)1/AlSi composite were assembled, contained ~0.3% Ti, ~0% Cr, and 1 at. % Si. The (Nb)2/AlSi composite was made from Nb alloy foils with 7.7% Ti, ~0.8% Cr, and 3 at. % Si (see Table 1). It followed from microstructural studies that the presence of these elements accelerated the reactive diffusion between the layers of the Nb alloy and the Al–Si alloy. This resulted in a relatively low volumetric content in the (Nb)2/AlSi composite of the structural component with the Nb-solid solution. According to the results of processing the color image of the surface area of a thin section in the composite instead of Nb-alloy tapes after diffusion welding, about 30 volume percent of the structural component [Nb-solid solution + $(Nb,Me)_3Al$], partially containing viscous plastic (Nb)-solid solution. This amount was extremely insufficient to resist crack propagation in the composite. Hence – the brittle nature of the destruction of the (Nb)2/AlSi composite, its low crack resistance and underestimated, in relation to (Nb)1/AlSi, strength values. The composite became so brittle that it lost its ability to resist bending deformation not only at room temperature, but also at temperatures up to 1300 °C. Therefore, we observed the "independence" of the ultimate strength of the composite from the test temperature, except for an increase in the average value of σ_B by 5 MPa at 1300 °C in relation to the ultimate strength at room temperature.

The layered composite (Nb)1/AlSi, containing ~50 volume percent of the Nb-solid solution, in contrast to (Nb)2/AlSi, was destroyed by the mechanism of the formation of longitudinal cracks, which propagated along the intermetallic layers as less crack-resistant structural components.

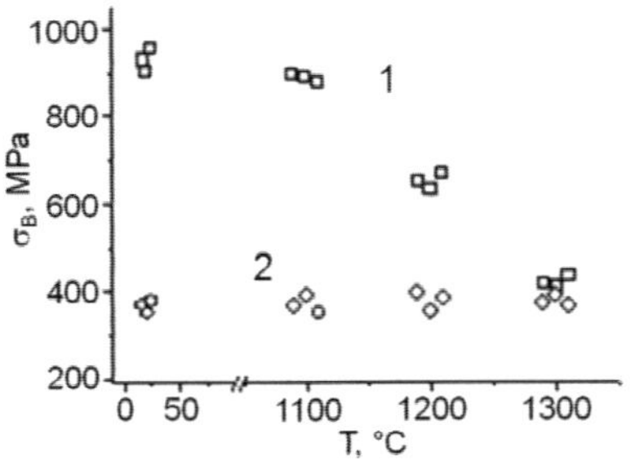

Figure 2. Dependences of short-term bending strength σ_B on the test temperature for composites (Nb)1/AlSi (1) and (Nb)2/AlSi (2).

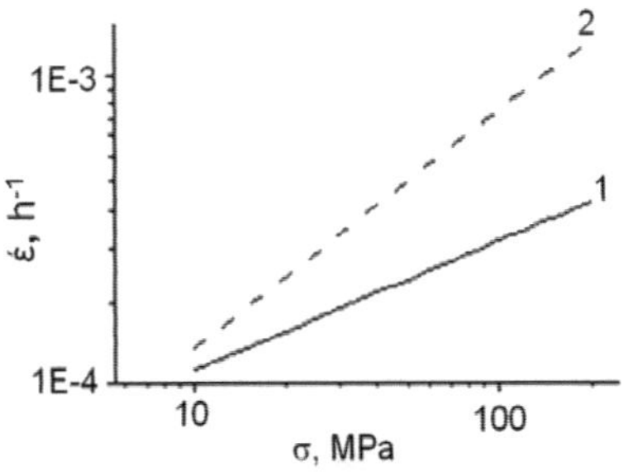

Figure 3. Dependences of the creep strain rate on stress for the (Nb)1/AlSi (1) and (Nb)2/AlSi (2) composite at a temperature of 1300 °C.

The mechanical properties of layered composites during short-term tests and during creep tests turned out to be noticeably lower than those of highly alloyed eutectic alloys of the Nb–Si system under similar conditions. Crack resistance of layered composites and melted Nb–Si alloys had, in general, the same values [10]. However, if in cast Nb-alloys the results are maximally achieved, then in composites they were obtained for the first time and there are opportunities to significantly improve them.

5. Conclusions

1. It was found that titanium and chromium accelerated the reactive diffusion of aluminum in a solid solution based on niobium. The layered structure of the (Nb)2/AlSi composite with a Nb alloy containing slightly more than 4% Ti and ~0.5 wt. %Cr had two structural components separated by diffuse boundaries. One of them consisted of the $(Nb,Me)_3Al$ intermetallic compound with extended inclusions of an Nb-solid solution. Another structural component was from coarse grains of the Nb-solid solution, which contained the compound $(Nb,Me)_3Al$ in the form of inclusions and along the grain boundaries.

2. The nature of the change in strength depending on temperature in composites was significantly different. The ultimate strength of the composite with a low titanium content at room temperature was 930, at 1100 °C – 850, and only at 1300 °C it fell to 420 MPa. Composite with 4.14 wt. % Ti already at room temperature had a strength of only 370 and practically the same – 375 MPa – at 1300 °C.

References

[1] Cao H et al 1994 *Mater. Sci. and Eng. A* **185** 87
[2] Li N et al 2009 *J. Appl. Phys.* **105** 123522-1
[3] Barmak K et al 1990 *J. Appl. Phys.* **67** 7313
[4] Lusadamo G et al 1999 *Mater. Let.* **39** 268
[5] Odette G et al 1992 *Acta Metal. et Mater.* **40** 2381
[6] Korzhov V P Kiiko V M 2015 *Deform. and destr. of mater.* **6** 6
[7] Korzhov V P et al 2013 *High Pres. Phys. and Eng.* **23** 99
[8] Mileiko S T 2002 *Comp. Sci. and Tech.* **62** 195
[9] Mileiko S T Kiiko V M 2004 *Mech. of Comp. Mater.* **40** 523
[10] Karpov M I 2018 *Metal Sci. and Heat Treat.* **60** 7

Effect of solid-solution treatment on the microstructure and mechanical properties of high-nitrogen FeMnCrNiCo high entropy alloys

K A Reunova*, **E G Astafurova, E V Melnikov, S V Astafurov, M Yu Panchenko and V A Moskvina**

Institute of Strength Physics and Materials Science SB RAS, 2/4 Akademicheskii prospekt, Tomsk, 634055, Russia

*Corresponding author: reunova.ksenya@mail.ru

Abstract. The effect of solid-solution treatment on the microstructure and tensile properties of high-nitrogen and nitrogen-free high entropy alloys (FeMnCrNiCo and FeMnCrNiCoN, N=0.36 mass.% and N=0.44 mass.%) has been investigated. It has been established that all materials under study possess an fcc crystal lattice, the lattice parameter of which increases with increasing nitrogen content. In comparison with the cast state, which is characterized by a coarse-grained dendritic microstructure, the alloys after solid-solution treatment show a more homogeneous microstructure. Solid-solution treatment does not significantly affect the mechanical characteristics of the alloys and their fracture micromechanisms.

1. Introduction

High entropy alloys (HEAs) form a new class of the multicomponent materials, which are rapidly gaining the attention of contemporary scientists. The study of HEAs began with the appearance of works introduced by Cantor et. al. [1] and Yeh et. al. [2]. HEAs typically consist of five or more components with atomic concentrations ranging from 5 to 35 at. % [3, 4]. Equiatomic alloys tend to form single-phase structures with face-centered cubic (fcc), body-centered cubic (bcc) and hexagonal close-packed (hcp) crystal lattices, and they possess some remarkable properties: corrosion resistance, high strength, ductility, and fracture toughness at room and cryogenic temperatures [1, 5]. Nevertheless, HEAs have low strength properties, which limit wide industrial applications of these materials [6, 7]. Nitrogen alloying can be used to improve strength properties of HEAs, like it was previously used for steels, because nitrogen is a strong austenite stabilizer, solid-solution strengthener and is useful for enhancement of the pitting corrosion resistance of steels [8]. Currently, there are few works devoted to the study of nitrogen-bearing high-entropy or medium entropy alloys [9-11]. All of them show a positive effect of nitrogen-alloying on the strength properties of the HEAs even in the cast state [10-11].

In this study, we evaluated the influence of solid-solution treatment on the dendritic microstructure, tensile properties, and fracture micromechanisms in high-nitrogen and nitrogen-free high entropy alloys (FeMnCrNiCo and FeMnCrNiCoN, N=0.36 mass.% and N=0.44 mass. %).

2. Materials and methods

High entropy alloys with the chemical compositions $Fe_{19.8}Mn_{19.5}Cr_{18.6}Ni_{21.0}Co_{21.1}$ (HEA, mass. %), $Fe_{19.9}Mn_{19.7}Cr_{18.7}Ni_{21.0}Co_{20.3}N_{0.36}$ (HEA-0.36%N) and $Fe_{20.5}Mn_{20.2}Cr_{19.0}Ni_{21.4}Co_{18.5}N_{0.44}$ (HEA-0.44%N) were obtained by vacuum induction melting in pure Ar atmosphere. Rectangular dumbbell-shaped specimens with the dimensions of the working part 12×2.6×1.4 mm (length×height×thickness) were cut by electrical discharge machining. Part of the specimens was examined in the cast state, the other part was subjected to conventional solid-solution treatment (SST) at a temperature of 1200 °C for 1 hour, followed by water-quenching. All specimens were mechanically grinded and electrochemically polished for final surface purification. The microstructure characterization was carried out using a light microscope (LM, Altami MET 1C). The X-ray diffraction (XRD) structural and phase analyses were done using a DRON 7 diffractometer in the θ-2θ Bragg-Brentano geometry with Co-K_α radiation. Microhardness measurements were performed using a Vickers hardness tester Duramin 5 (Struers) at a load of 1.96 N for 10 s. Tensile tests to failure were carried out at room temperature and an initial strain rate of 5×10^{-4} s^{-1} using an electromechanical machine LFM-125 (Walter + Bai AG). The fracture surfaces were examined with a scanning electron microscope (SEM, Leo Evo, Zeiss).

3. Results and discussion

Typical LM micrographs of the alloys are presented in Figure 1. All investigated alloys have an inhomogeneous dendritic microstructure in the cast state (Figure 1 a, b, c), which is typical for HEAs [11]. Nitrogen alloying promotes partial homogenization of the cast microstructure, but this effect is insufficient. Areas of dendrites are enriched in Fe, Cr and Co, while interdendritic interlayers are characterized by increased contents of Mn and Ni atoms [11]. Conventional SST has a noticeable homogenizing effect but does not provide a fully homogenized microstructure without dendrites (Figure 1 d, e, f). The X-ray measurements testify that all investigated alloys possess a single-phase structure with an fcc crystal lattice. According to XRD analysis, no other phases have not been found both for cast and SSTed specimens. It was shown that nitrogen alloying with the concentrations of 0.36 mass.%N and 0.44 mass.%N is accompanied with an increase in the lattice parameter (*a*) relative to nitrogen-free HEA (Table 1). This behavior indicates the formation of a solid solution of nitrogen in the austenite phase. The comparison of the a-values for cast and SSTed specimens confirms an additional increase in nitrogen concentration in the HEA-0.44%N alloy after SST (Table 1). The latter testifies to the fact that in the cast HEA-0.44%N alloy nitrogen can form a small fraction of nitrides, which dissolve in SST.

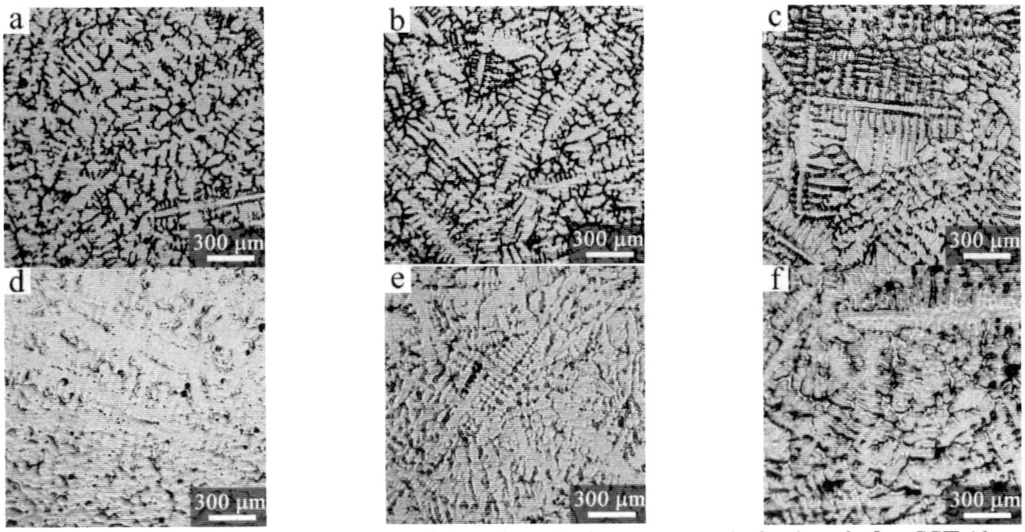

Figure 1. LM images of the microstructure of the alloys in the cast state (a, b, c) and after SST (d, e, f): (a, d) – HEA; (b, e) – HEA-0.36%N; (c, f) – HEA-0.44%N.

Representative tensile stress-strain diagrams for the investigated alloys in the cast state and after SST are shown in Figure 2. Alloying with nitrogen significantly increases both the yield strength (YS) and the ultimate tensile stress (UTS) of the alloys and does not lead to a decrease in the elongation to failure δ (table 1). For cast alloys, the δ-values for nitrogen-bearing specimens possess even higher values relative to a nitrogen-free alloy (Figure 2, table 1). After SST, the elongation of HEA noticeably increased relative to the cast specimens due to partial dissolution of dendrites, and all alloys possessed similar plastic properties. The homogenization of the microstructure during SST assists an insufficient decrease in the strength properties of the alloys (YS, UTS, microhardness), however, no significant changes are observed between these properties of the cast and SSTed specimens.

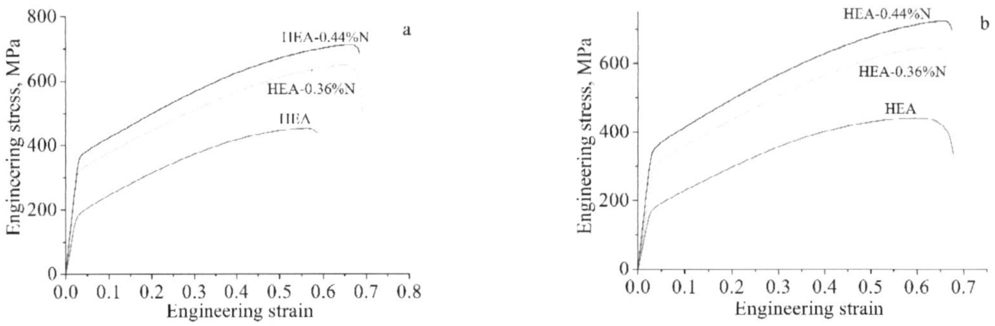

Figure 2. Tensile diagrams for cast (a) and SSTed (b) HEA, HEA-0.36%N, HEA-0.44%N.

Table 1. Mechanical properties and the lattice parameter of the austenitic phase (*a*) in the cast state and after solid-solution treatment of the alloys. YS – the yield strength, UTS – ultimate tensile stress, δ – elongation to failure, ε_u – a uniform elongation, Hμ – microhardness.

Notation	YS (MPa)	UTS (MPa)	δ (%)	ε_u (%)	Hμ (MPa)	*a* (Å)
			Cast Specimens			
HEA	175	710	53	37	1785	3.593
HEA-0.36%N	315	1075	65	41	2484	3.608
HEA-0.44%N	370	1188	63	42	2603	3.609
			Solid-Solution Treated Specimens			
HEA	170	714	63	38	1496	3.595
HEA-0.36%N	300	1063	60	39	2438	3.607
HEA-0.44%N	355	1195	64	41	2365	3.612

Figure 3 shows typical SEM images of the lateral and fracture surfaces of HEA and HEA-0.36%N specimens after tensile tests in the cast state and after SST (the characteristics of the HEA-0.44%N specimens are similar to those of the HEA-0.36%N specimens). The lateral surfaces of all the presented alloys are characterized by a pronounced deformation relief, which testifies considerable plastic deformation prior to the failure of the specimens. Slip traces and slip bands in several slip systems are observed in grain bodies. Microshear bands are clear visible near the fracture zone, and microcracks are obviously formed inside of these bands (Figure 3). No grain boundary cracks were observed on the lateral surfaces of both nitrogen-free and nitrogen-alloyed alloys. Comparison of the deformation relief for N-free and N-bearing alloys shows that nitrogen-alloying stimulates planar slip (as it was previously confirmed in [10]) and inhibits wavy glide.

In the cast state and after SST, the initial HEAs are characterized by ductile transgranular fracture, numerous dimples are observed on the fracture surface (Figure 3 a, c). The mechanism of fracture of nitrogen-containing alloys remains ductile both in the cast state and after SST, however, the micromechanism of fracture is slightly transformed. The dominant fracture micromechanisms of the HEA-0.36%N alloy is dimple fracture, but along with dimples, some plain facets are visible (Figure 3 b, d). This is peculiar for alloys with a planar dislocation structure.

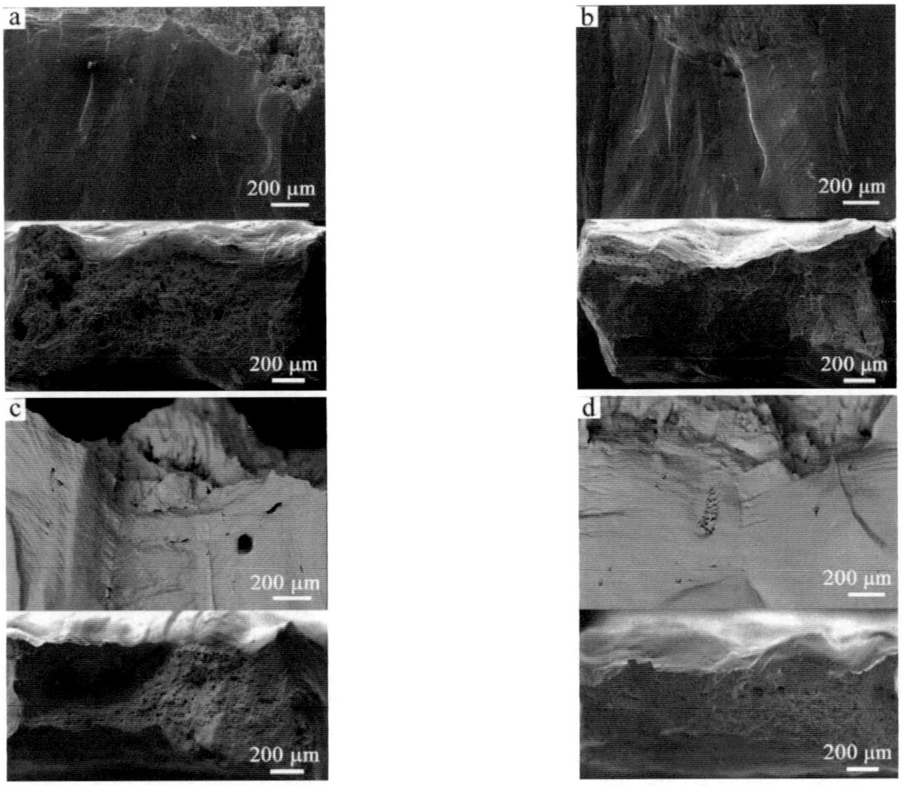

Figure 3. Typical SEM images of the lateral and fracture surfaces of specimens tensile tested in the cast state (a, b) and after SST (c, d): (a, c) – HEA; (b, d) – HEA-0.36%N.

4. Conclusion

The microstructure and mechanical properties of high entropy alloys FeMnCrNiCo and FeMnCrNiCoN (0.36 and 0.44 mass.% of N) have been investigated. Nitrogen-alloying is accompanied by solid-solution hardening of the alloys due to the dissolution of nitrogen in the fcc crystal lattice without detectable nitride formation. Solid-solution treatment promotes a more homogeneous microstructure relative to the cast specimens, but does not contribute to substantial variation in the mechanical properties of the alloys.

Acknowledgements

The research was supported by the Russian Science Foundation (project No. 20-19-00261).

References

[1] Cantor B, Chang I, Knight P and Vincent A 2004 *Mater. Sci. Eng. A* **375-377** 213–218
[2] Yeh J, Chen S, Lin S, Gan J, Chin T, Shun T, Tsau C and Chang 2004 *Adv. Eng. Mater.* **6** 299–303
[3] Miracle D and Senkov O 2017 *Acta Mater.* **122** 448–511
[4] Zhang Y, Zuo T, Tang Z, Cao M, Dahmen K, Liaw P and Lu Z 2014 *Prog. Mater. Sci.* **61** 1–93
[5] Senkov O, Scott J, Senkova S, Miracle D and Woodward C 2011 *J. Alloy. Compd.* 509 6043–6048
[6] Otto F, Dloughy A, Somsen Ch, Bei H, Eggeler G and George E 2013 *Acta Mater.* **61** 5743–5755
[7] Gali A and George E 2013 *Intermetallics* **39** 74–78
[8] Simmons J 1996 *Mater. Sci. Eng. A* **207** 159–169
[9] Moravcik I, Hadrada H, Li L, Dlouhy I, Raabe D, Li Z 2020 *Scripta Mater.* **178** 391–397
[10] Klimova M, Shaysultanov D, Semenyuk A, Zherebtsov S, Salishchev G and Stepanov N 2020 *J. Alloy. Compd.* **849** 156633
[11] Astafurova E, Reunova K, Melnikov E, Panchenko M, Astafurov S, Maier G and Moskvina V 2020 *Mater. Lett.* **276** 128183

IOP Conf. Series: Materials Science and Engineering 1014 (2021) 012048 doi:10.1088/1757-899X/1014/1/012048

Investigation of the formation and distribution of elements in the production of Al-Ti-V-Zr-Nb alloys by the aluminothermic method

A S Russkih[*], S A Krasikov, E M Zhilina and S V Zhidovinova

Institute of Metallurgy of the Ural Branch of the Russian Academy of Sciences, 101, Amundsen street, Ekaterinburg, Russia 620016

[*]Corresponding author: russkih_a_s@mail.ru

Abstract. The results of testing the method for synthesis of a high-entropy AlNbTiVZr alloy using metallothermic reduction of group IV and V metals from their oxides are presented. At the first stage, a joint aluminothermic recovery of metals in a resistance furnace was performed. Then, to achieve uniformity and remove harmful impurities, the metal was melted in a vacuum furnace with an inert atmosphere. The alloy obtained after the second remelting was subjected to x-ray diffraction and electron microprobe (EMA) analysis. The results of the analysis revealed the presence of a solid solution of Zr (Nb, V) with TiAl inclusions, which confirms the prospects of using metallothermic synthesis to produce a high-entropy alloy with a single-phase structure.

1. Introduction

High-entropy alloys with an equiatomic distribution of elements are very promising materials for use in various fields of technology, and in the last decade they have attracted the attention of researchers. The first mention of a high-entropy alloy was made in 2004 [2]. The idea behind these alloys was their high mixing entropy, which suggested a departure from traditional alloying methods, when one or two main base elements are present in the alloy. The new approach provided a transition to the concept of multicomponent alloys, where all elements are present in equiatomic or close to equiatomic ratio to each other. The development of high-entropy refractory alloys is aimed at creating so-called heat-resistant and heat-resistant compositions.

Work on finding rational ways to obtain high-entropy alloys has been carried out for more than fifteen years, all over the world. However, there is still no clear opinion on what methods are most appropriate to produce them. The most well-known methods for producing high-entropy alloys include vacuum arc melting [3], mechanical alloying [4], and laser surfacing [5]. These methods are based on mixing and fusing components with a chemical purity of 99.99 %, which implies the high cost of obtaining these alloys in industrial production. To improve the economic performance of the technology for obtaining alloys, it is possible to use the stage of aluminothermic reduction of metals from relatively inexpensive oxides. In this paper, this approach was tested in the synthesis of a high-entropy AlNbTiVZr alloy in a resistance furnace.

The aim of this work was to study the phase and chemical composition of the AlNbTiVZr alloy sample obtained by aluminothermic reduction of Nb, Ti, V, and Zr oxides under controlled temperature conditions in a resistance furnace. Samples of melting products synthesized in corundum crucibles were examined.

2. Materials and methods of research

Synthesis of the alnbtivzr alloy was performed in a resistance furnace followed by metal remelting in a vacuum furnace with an inert atmosphere. The charge for aluminothermic reduction of niobium, titanium, vanadium and zirconium oxides to metals is calculated using the reaction stoichiometry, taking into account the reducing capacity of each of the elements.

X-ray phase analysis of the samples was performed on an x-ray diffractometer "XRD 7000 diffractometer (Shimadzu)". The samples were taken in filtered monochromated CuKα radiation. Diffractograms were decoded using JCPDS and ASTM databases.

Electron microprobe analysis (EMA) was performed on polished samples using a Carl Zeiss EVO 40 scanning electron microscope and Inca X-Act EMA prefix.

3. The results of research and their discussion

The phase and elemental composition of metal samples obtained by aluminothermic reduction of Nb2O5, TiO2, V2O5 and ZrO2 oxides in corundum crucibles was studied using the example of an alloy after remelting in an inert atmosphere of Ar with the chemical composition, at.%, 53.2 Al, 9.14 Nb, 14.53 Ti, 16.32 V, 6.79 Zr and mass%, 0.12 O, 0.009 N. To achieve the equiatomic composition, it is necessary to adjust the charge mixture and improve the experimental method previously tested in [6, 7].

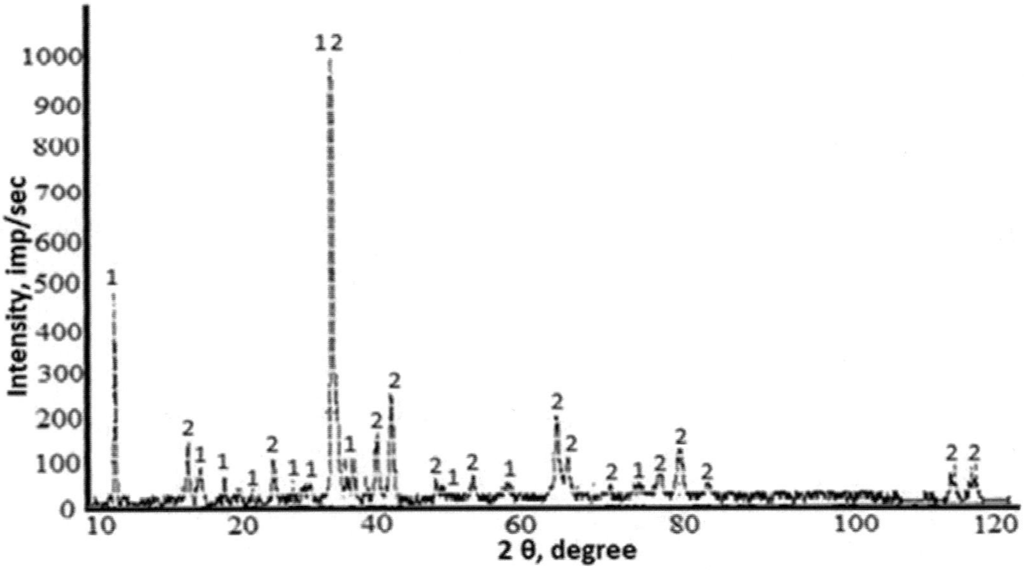

Figure 1. X-Ray analysis results: 1 - Zr(Nb, V); 2 - TiAl)

According to the results of x-ray phase analysis presented in figure 1, the main identified phases in the sample were Zr (Nb,V) and TiAl. According to the EMA results, the microstructure of the alloy (figures 2 and 3, tables 1 and 2) was the base of a solid solution evenly distributed in it, with non-metallic inclusions. The image of the Scanning electron microscope (SEM) shown in figure 2 had an middle position relative to the sample under study. The base of the plume was in the form of a uniform structure with needle-shaped dendrites. The chemical composition of the spectra is shown in table 1. The solid solution on the cross-section of the sample has an area of about 30 -50 microns located under the spectra 1, 2, 7, 9. Under the spectra 3, 6, 8 there are dendrites that have a needle-like structure with a length of

2 - 25 and a width of 1 - 5 microns and have carbon in their composition.The gray areas under the spectrum 4, 5 have a size of 5 - 10 microns, carbon and silicon are present in the composition.

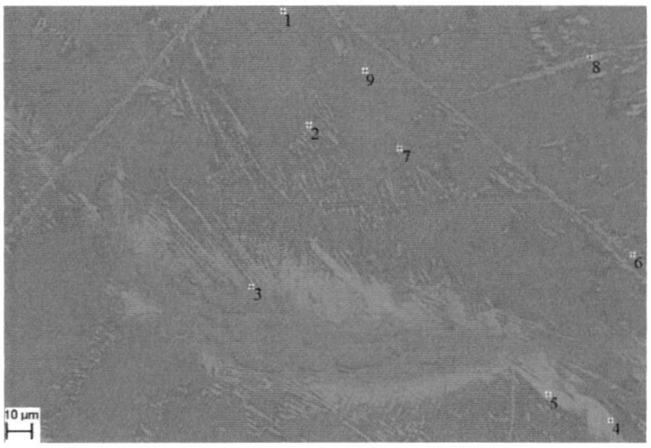

Figure 2. SEM image (middle of the test sample): 1, 2, 7, 9 - solid solution; 4, 5 - a mixture of solid solution of carbides and silicides; 3, 6, 8 - carbides of solid solution.

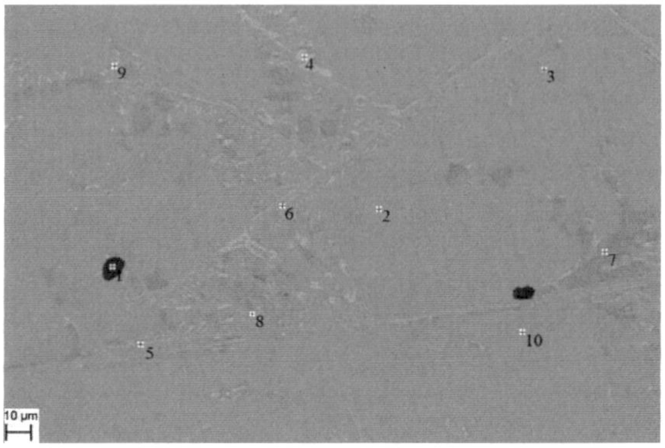

Figure 3. SEM image (edge of the test sample):1 - aluminum oxide; 2, 3, 5, 8, 9, 10 - solid solution; 4 - a mixture of solid solution of carbides and silicides; 6, 7 - carbides of solid solution.

Table 1. Chemical spectral analysis for figure 2 (spectrum-chemical composition in at.%).

№	Al	Nb	Ti	V	Zr	P	Si	C
1	61.67	11.03	11.87	12.90		2.53		
2	51.19	10.04	11.93	21.96		2.57	2.30	
3	27.94	11.02	23.86	13.27			2.34	21.57
4	14.68	12.57	13.04	18.45		2.89	13.32	25.05
5	17.53	10.78	13.00	16.64		2.59	10.36	29.11
6	24.03	7.73	18.33	9.86			1.30	38.76
7	52.71	8.21	11.72	22.45		2.72	2.19	
8	19.91	8.19	18.43	11.49			2.45	39.53
9	61.43	10.87	11.85	12.98		2.87		

Figure 3 shows a SEM image taken at the edge of the sample under study. The image shows a uniform structure with non-pronounced dendrites and the presence of non-metallic inclusions. The chemical composition of the spectra is shown in table 2. The solid solution has the same area as in Figure 1, about 30 - 50 microns, located under the spectra 2, 3, 5, 8, 9 and 10. The dark region under spectrum 1 has oxygen and aluminum in its composition, which indicates the presence of Al_2O_3. The light gray regions under spectrum 4 and the dark gray regions under spectra 6, 7 are represented by a mixture of carbon, silicon and carbon, respectively. Drawing an analogy with figure 1, we can say that it is possible to form carbides in the metal. Carbides can be formed due to CO/CO_2 gases in the resistance furnace. It is possible to exclude the formation of carbides when replacing heating elements made of carbon, for example, molybdenum or tungsten.

Table 2. Chemical spectral analysis for figure 3 (spectrum-chemical composition in at.%).

№	Al	Nb	Ti	V	Zr	P	Si	C	O
1	32.81				0.47			8.53	58.19
2	58.63	9.89	11.46	12.78	7.24				
3	60.78	11.17	12.14	12.53		3.39			
4	21.8	11.83	13.57	19.01		3.31	11.97	18.51	
5	56.86	9.93	13.26	15.82		2.94	1.19		
6	23.97	9.74	23.72	9.51		2.43	1.04	29.59	
7	35.97	6.3	8.73	37.88				11.11	
8	49.11	9.09	11.7	19.7	6.53		1.94		1.93
9	50.95	8.75	11.78	19.34	6.81		2.37		
10	58.81	9.9	11.61	12.06	7.61				

When comparing the test sample in the middle and at the edge, using a SEM, differences in the structure and the presence of non-metallic inclusions in the form of aluminum oxide were observed. To eliminate oxides in the final alloy, it is necessary to perform additional remelting in a vacuum-arc furnace to ensure uniformity and separation of the metal and oxide phases.

Conclusions

The results of the study show that the formation of a high-entropy alloy is possible with the combined aluminothermic reduction of Nb, Ti, V and Zr. In the synthesized alloy, the formation of a solid solution based on Zr (Nb, V) with the presence of intermetallic inclusions of AlTi was achieved. Also, electron microprobe analysis showed that the alloy contains non-metallic inclusions in the form of carbides and oxides, which may not favorably affect the final product.

Acknowledgments

The work was carried out according to the state assignment for IMET UB RAS using equipment of the Collaborative usage centre "Ural-M".

References

[2] Gao M C, Yeh J W, Liaw P K, and Zhang Y 2016 *High-Entropy Alloys* (Germany: Springer International Publishing Switzerland) pp 5-6
[3] Senkov O N, Scott J M, Senkova S V, Meisenkothen F, Miracle D B and Woodward C F 2012 *J. Materials Science* **47** Iss. 9 4062–74
[4] Murty B S and Ranganathan S 1998 *J. International Materials Reviews* **43** Iss. 3 101–41
[5] Qiu X W, Zhang Y P, He L and Liu C G 2013 *J. Alloys and Compounds* **549** 195–9
[6] Allibert M, Gaye H, Geiseler J and Janke D 1995 *Slag Atlas 2nd Edition* (Germany: Verlag Stahleisen GmbH) pp 616–617
[7] Russkih A S, Agafonov S N and Ponomarenko A A 2019 *J. Bytlerovskie Soobscheniya* **60** Iss. 11 11–15
[8] Agafonov S N, Ponomarenko A A and Russkikh A S 2019 *J. Russian Metallurgy* **2019** Iss. 2 173–5

Study of phase formation and properties of high-entropy carbide $HfTaTiNbZrC_5$ obtained by selfpropagating high-temperature synthesis

A S Sedegov*, A R Bobojanov, S Vorotilo, K V Kuskov and D O Moscovskikh

Scientific and research center of Functional Nano Ceramic, NUST MISIS, Leninskii prospect 4, Moscow, 119049 Russia

*Corresponding author: a.sedegoff@gmail.com

Abstract. This work is devoted to the study of the combustion synthesis mechanism in complex multi-component systems using the quenching of the combustion front in a copper block. Bulk specimens were produced by reactive spark plasma sintering (R-SPS) of mechanically activated mixtures as well as by SPS of high-entropy ceramics obtained by combustion synthesis. The resistance of high-entropy carbides towards oxidation was studied in relation to the number of metallic constituents.

1. Introduction

In the last decades there is a surge of interest towards materials with enhanced structural and functional properties. To achieve this goal, the complexity of the composition of materials is increased. Complex-alloyed high-end steels and heat resistant alloys are prime examples of this trend. An increase in the number of alloying elements usually is associated with an increase in the number of phases in the alloy, which complicates the design of the material. This is the case for alloys based on a single main constituent. Recently, a new direction in materials science has emerged – the design of high-entropy materials with near-equimolar ratio of constituents [1-4]. The diversification of this field gave rise to high-entropy carbide and boride ceramics [5-6]. Due to the high refractoriness, high-entropy ceramics are proposed as candidates for ultra-high-temperature applications. This work focused on the production of high-entropy ceramics $HfTaTiNbZrC_5$ and $HfTaTiNbMoC_5$ via combustion synthesis (also known as SHS), which allows for a single-stage low-cost synthesis [7,8] The goal of this work was to investigate the mechanism of phase formation within the combustion front of the related systems and to compare the properties of ceramics produced by SHS and reactive spark plasma sintering.

2. Materials and Methods

The methodology for the synthesis of high-entropy ceramic powders by combustion is described in [7]. Sintering was carried out on an SPS Labox 650 (Sinterland, Japan) spark plasma sintering unit at temperatures of 1800 °C and 2000 °C, the pressure applied to the punches was 50 MPa. The density of samples was measured by hydrostatic weighing. The structure of HEC powders and bulk materials was analyzed using SEM on a JSM 7600F (JEOL, Japan) with a microanalysis system EDX (Oxford Instruments). X-ray phase analysis was carried out on a DRON-4-07 (Burevestnik, Russia).

3. Results and Discussion

The mechanisms of phase and structure formation in the Hf-Ta-Ti-Nb-Zr-C system during SHS were studied by quenching the combustion fronts in a copper block, which was carried out in 3 stages: 1) preparation of reactive mixture; 2) pressing the reactive mixture into green triangles (relative density ~ 60%); 3) ignition of combustion with subsequent quenching. The quenched combustion front was then polished along the axis of the combustion front propagation and studied by SEM and EDS (Figure 1). Figure 1a provides a general view of the quenched combustion front; one can see both the initial component and the combustion products. As the wave propagates through the reactive mixture, the components melt and then the main product phase begin to crystallize (Figure 1b). The after-combustion zone features ring structures (Figure 1c), within which crystals of high-entropy carbide lie. In the secondary structure formation zone, the ring structures experience recrystallization (Figure 1d) into submicron (150-400 nm) grains of high-entropy carbide (Figure 1e).

High-entropy carbide with the composition $Hf_{0.2}Ta_{0.2}Ti_{0.2}Nb_{0.2}Zr_{0.2}C$ (HEC_{Zr}), produced by SHS, was then consolidated by spark plasma sintering (SPS). The sintering of HEC specimens was carried out at a pressure of 50 MPa and at temperatures of 1500 - 2200 °C to ascertain the influence of the sintering temperature on the relative density and mechanical properties of materials. The relative density increased from 90.5 to 94.8%.

In addition, the reactive SPS method was used, which allows one to combine the combustion reaction and conslodiation of the ceramics (Figure 2). Due to the higher vacuum in the SPS apparatus as compared to SHS reactor, the method is expected to yield higher purity products with lower residual oxide content, refined microstructure and higher mechanical properties. The structural refinement is related to the suppression of the reacrystallization processes characteristic for SHS.

The XRD analysis of the combustion products (Figure 3a) demonstrated the formation of single-phase high-entropy carbide with an FCC lattice. The consolidation of combustion products at 1800 and 2000 °C (Figure 3 b,c) did not alter the single-phase FCC solid solution, suggesting that the phase is stable at high temperatures. The R-SPS method yielded an identical solid solution phase (**Figure 3**d)

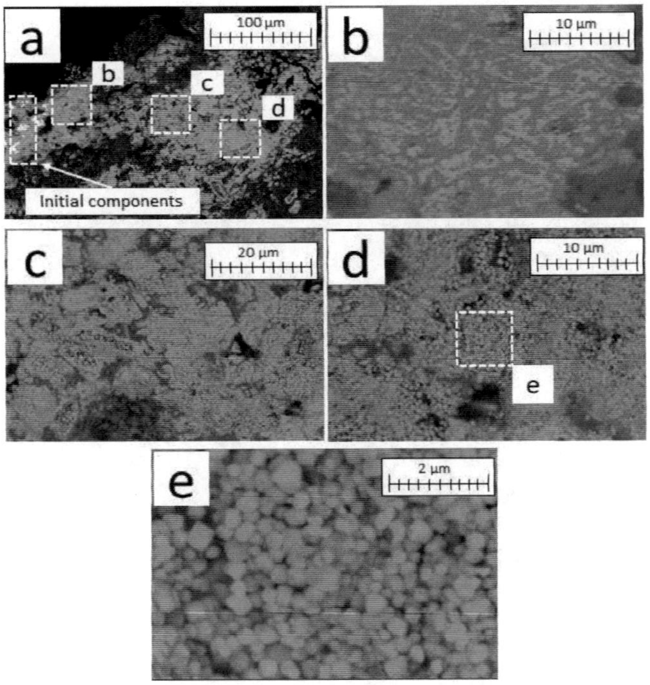

Figure1. SEM images of the quenched combustion front in the Hf + Ta + Ti + Nb + Zr + C system: general view of the quenched front (a), combustion zone (b), afterburning zone (c), zone of secondary phase formation (d, e)

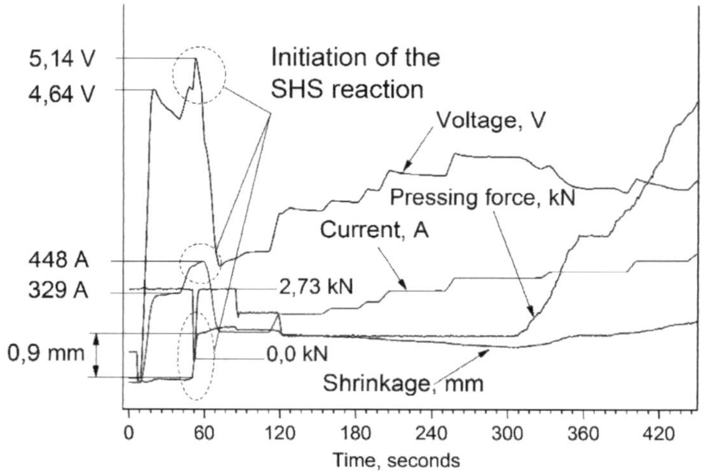

Figure 2. Sintering monitoring curves: voltage-current-pressing force-shrinkage versus time

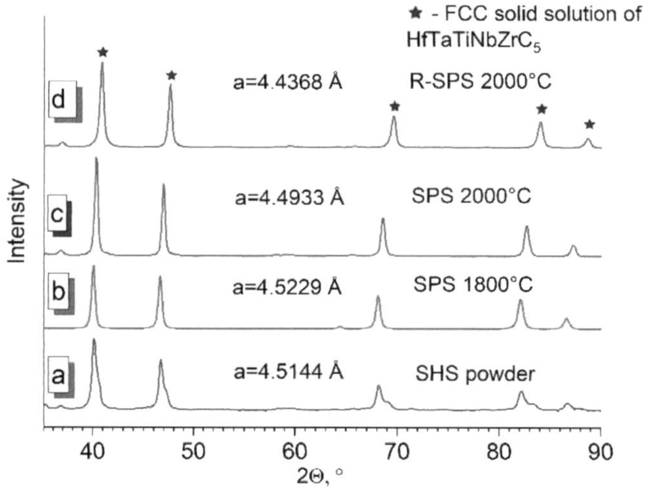

Figure 3. XRD analysis of combustion products and sintered materials of high-entropy ceramics

The combustion kinetics did not show any pronounced dependencies on the relative density of green pellets. Figure **4** provides the morphology of the SHS products HEC_{Zr}, with the 0.1-5 μm crystals form 20-50 μm agglomerates.

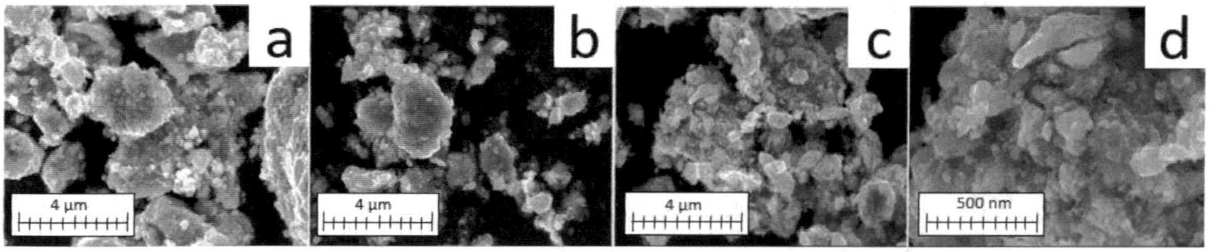

Figure 4. Morphology of HEC_{Zr} powder particles after SHS compressed at different pressures (a - 8 kg·f, b - 80 kg·f, c - d - 160 kg·f)

The structure of the consolidated SHS ceramics is shown in Figure **5**. They feature rounded grains, which become increasingly more coarse at higher temperature (0.5-4 µm at 1800 °C and 1-10 µm at 2000 °C). The density of specimens sintered at 1800 and 2000 °C was 7.75 g/cm^3 и 8.12 g/cm^3. However, the hardness of HEC_{Zr}, sintered at 1800 °C (21.5 GPa) was higher than at 2000 °C (19.1 GPa), presumably due to the grain coarsening. The R – SPS method yielded fine grains with a narrow size distribution (1-3 °C) due to the application of pressure and higher cooling rate realized in the R-SPS method; the specimens demonstrated a density of 8.34 g/cm^3 and a hardness of 26 GPa.

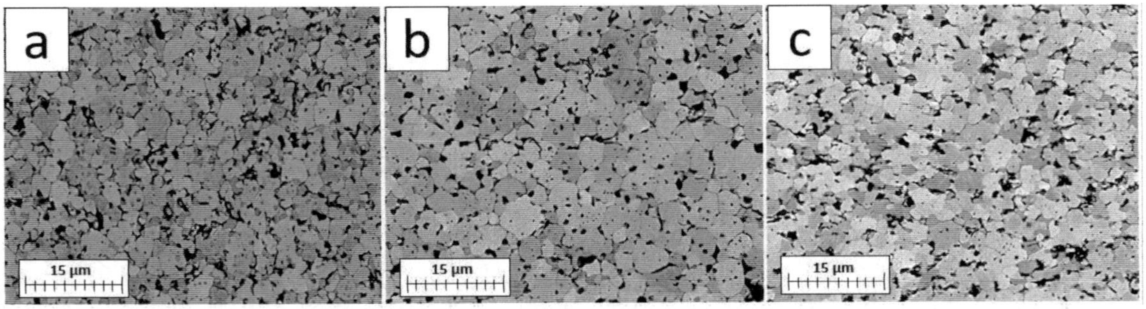

Figure 5. Microstructures of sintered HEC_{Zr} at 1800 °C (a), 2000 °C (b) and reaction sintering at 2000 °C (c)

Acknowledgments
This work was conducted with the financial support of the Russian Science Foundation (grant no. 18-79- 10215).

References

[1] Huang K H and Yeh J W 1996 A study on multicomponent alloy systems containing equal-mole elements M.S. thesis (Hsinchu: National Tsing Hua University)
[2] Yeh J W, Chen S-K, Lin S J, Gan J Y, Chin T S, Shun T T, et al.,Nanostructured high-entropy alloys with multiple principal elements: novel alloy design concepts and outcomes, Adv. Eng. Mater. 6 (2004) 299–303,
[3] Yeh J W, Lin S J, Breakthrough applications of high-entropy materials, J. Mater. Res. 33 (2018) 3129–3137
[4] Miracle D B, Senkov O N, A critical review of high entropy alloys and related concepts, Acta Materialia, 122, 2017,448-511
[5] Gild J, Zhang Y, Harrington T, Jiang S, Hu T, Quinn M C, et al., High-entropy metal diborides: a new class of high-entropy materials and a new type of ultrahigh temperature ceramics, Sci. Rep. 6 (2016) 37946,
[6] Castle E, Csanádi T, Grasso S, Dusza J, Reece M, Processing and properties of high-entropy ultra-high temperature carbides, Sci. Rep. 8 (2018) 8609
[7] Sedegov A et al 2019 IOP Conf. Ser.: Mater. Sci. Eng. 558 012043
[8] Moskovskikh D O, et al., High-entropy (HfTaTiNbZr)C and (HfTaTiNbMo)C carbides fabricated through reactive high-energy ball milling and spark plasma sintering, Ceramics International 46 (2020) 19008–1901

IOP Conf. Series: Materials Science and Engineering 1014 (2021) 012050 doi:10.1088/1757-899X/1014/1/012050

Effect of carbon content on cryogenic mechanical properties of CoCrFeMnNi high entropy alloy

A Semenyuk*, M Klimova, D Shaysultanov, R Chernichenko, S Zherebtsov and N Stepanov

Belgorod State National Research University, Pobeda, 85, Belgorod, 308015, Russia

*Corresponding author: semenyuk@bsu.edu.ru

Abstract. The effect of the carbon content (0-2 at.%) on the structure and mechanical properties at room and cryogenic temperatures of CoCrFeNiMn-based high entropy alloys with reduced Cr concentration was studied. The as-cast alloys were cold rolled to a thickness reduction of 80% and annealed at 800°C for 1 hour. As a result, a fully recrystallized microstructure with a grain size of 6.4 μm was produced in the carbon-free alloy. The recrystallized grain size was much smaller (1.5 μm in the alloy with 2.0 at.% of C) due to the pinning effect of the precipitated $M_{23}C_6$ carbides. The yield strength of the alloys increased with an increase in the carbon concentration from 313 MPa to 636 MPa, while the elongation to fracture slightly decreased from 56% to 43%, respectively, in the alloys with 0 and 2 at.% of C. A decrease in the test temperature to 77K resulted in a significant increase in both the strength and ductility of the alloys. The alloys had high values of impact toughness of 140 J/cm^2 and 85 J/cm^2, respectively, in the alloys with 0 and 2 at.% of C. A decrease in the testing temperature did not have a noticeable effect on the impact toughness.

1. Introduction

High-entropy alloys (HEAs) represent a new class of metallic materials with promising properties. The first mentions of HEAs date back to 2004 [1]; in several works, they are referred to as multicomponent equiatomic alloys. According to the initial definition of HEAs, these are the alloys that consist of at least 5 elements, and the amount of each element should be 5-35 at.%.

Alloys based on 3d transition metals are among the most studied HEAs families. They usually crystallize with a face-centered cubic (FCC) structure and have attractive mechanical properties. Alloys of the CoCrFeMnNi system exhibit high ductility and impact toughness at room temperature [3]. Their mechanical properties are even better at cryogenic temperatures, which makes them attractive for cryogenic applications. However, CoCrFeMnNi-based alloys have relatively low strength [4]. One of the most effective ways to increase strength is alloying with interstitial elements, in particular, carbon [5-7]. Carbon-doped HEAs can benefit both from interstitial solid solution hardening due to dissolved carbon atoms and from precipitation hardening due to the formation of carbides [8]. The equilibrium solubility of carbon in the CoCrFeMnNi solid solution is low due to the presence of a strong carbide-forming element, Cr. A decrease in the molar concentration of chromium can potentially increase the equilibrium solubility of carbon. Meanwhile, carbide particles can also affect the fcc grain size. It was shown in several works [9-10] that secondary phases in CoCrFeNi (Mn, Al, C) alloys can effectively restrict grain growth and promote the formation of a fine-grained microstructure. The fine grain size can also be beneficial for strength. However, the effect of carbon on the cryogenic mechanical properties of CoCrFeMnNi alloys, especially on the impact toughness, at cryogenic temperature is poorly studied. Therefore, in this work, we studied the tensile mechanical properties and impact toughness at room and cryogenic temperatures of CoCrFeMnNi high-entropy alloys with a different carbon content (x = 0, 0.5 and 2.0 at.%) after thermomechanical processing.

2. Materials and methods

CoCrFeMnNi-based high entropy alloys with a reduced concentration of Cr and different carbon concentration (x = 0, 0.5, and 2.0 at.%; further denoted as C0, C0.5, and C2.0 alloys respectively) were produced by vacuum induction melting. The measured chemical composition of alloys is presented in Table 1. The as-cast samples were cold rolled at room temperature to a thickness reduction of 80%. The samples then were annealed in a muffle furnace in an air atmosphere at temperatures of 800°C for 1 hour, followed by cooling in air.

Table 1. Measured chemical composition of the alloys depending on the carbon concentration.

	Concentration of elements, at.%					
	Co	Cr	Fe	Mn	Ni	C
C0	23.29	6.22	23.90	23.09	23.46	0.03
C0.5	23.17	6.42	23.97	23.67	22.24	0.53
C2.0	23.42	6.23	22.41	22.02	23.82	2.11

The microstructure of the alloy was studied using scanning (SEM) and transmission (TEM) electron microscopy in the RD-ND plane (perpendicular to the transversal direction). SEM studies were carried out a using FEI Quanta 600 FEG microscope equipped with a backscattered electron (BSE) detector. TEM investigations were conducted using a JEOL JEM-2100 microscope with an accelerating voltage of 200 kV equipped with an EDS detector. Selected area electron diffraction (SAEDs) patterns were used for the phase identification and results of EDS – for chemical analysis. Tensile tests were carried out on an Instron 5882 universal electronic tensile testing machine with a strain rate of 1×10^{-3} s^{-1} at room (293K) and cryogenic (77K) temperatures using samples with a dimension of 6×3×1.5 mm^3. The KCV impact toughness was determined on an Instron IMP460 pendulum impact machine using samples with a dimension of $2 \times 8 \times 55$ mm^3 and with a V-notch in the center at the same temperatures.

3. Results

The microstructure of the program alloys after cold working and annealing at 800°C is shown in Figure 1. A fully recrystallized microstructure with many annealing twins was observed. The C0 alloy (Figure 1a) had a single-phase structure with a grain size of 6.4 μm. The addition of carbon promoted the precipitation of the Cr-rich $M_{23}C_6$ carbides. The precipitation of the carbides resulted in a decrease of the recrystallized fcc grain size. For instance, in the C0.5 alloy (Figure 1b) with 1.5% of carbides the grain size was 5.2 μm, while in the C2.0 alloy (Figure 1c) the volume fraction of carbides increased to 7.4% and the grain size decreased by 1.6 μm. The size of the carbide particles increased from 89 nm to 117 nm in the C0.5 and C2.0 alloys, respectively. Carbides were found both along the grain boundaries and in grain interiors.

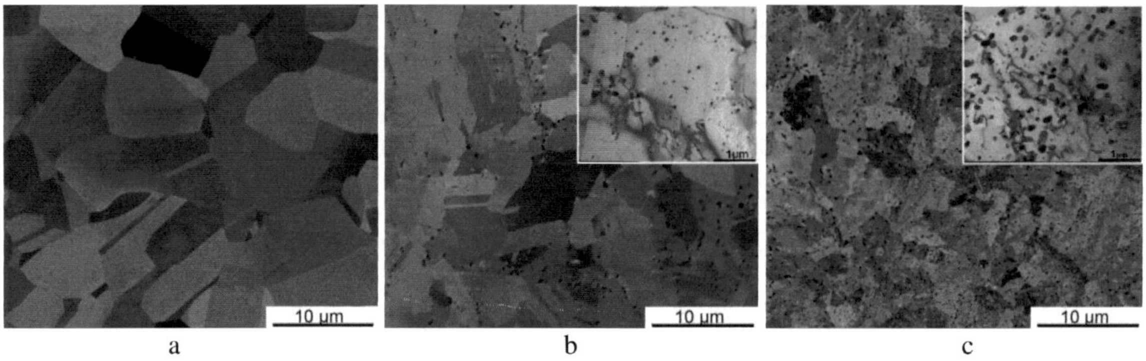

Figure 1. SEM-BSE images of the microstructure of the CoCrFeNiMn alloys with different carbon contents (x): C0 (a); C0.5 (b); C2.0 (c) after annealing at 800°C.

The tensile stress-strain curves of the alloys obtained at room (293 K) and cryogenic (77 K) temperatures are shown in Figure 2. At 293 K the yield strength increased with an increase in the carbon content from 313 MPa for the C0 alloy to 636 MPa for the C2.0 alloy (Table 2). The ductility decreased as the carbon concentration increased from 56% for the C0 alloy to 43% for the C2.0 alloy. Decreasing the testing temperature to 77 K led to a simultaneous increase in both strength and ductility of the alloys. The yield point slightly increased from 460 MPa in the C0 alloy to 480 MPa in the C0.5 alloy and then raised sharply to 786 MPa in C2.0 the alloy.

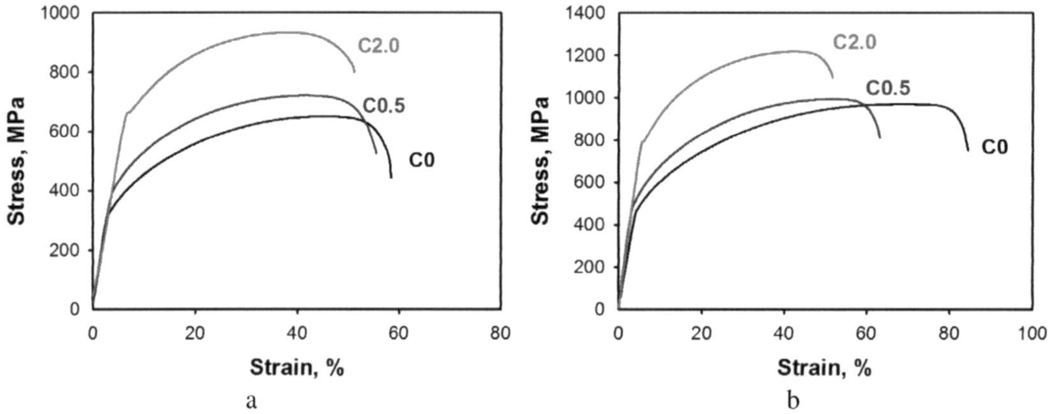

Figure 2. Tensile stress-strain curves of the CoCrFeMnNi alloys with different carbon content after cold rolling and annealing at 800°C obtained at 293 K (a) and 77 K (b).

Table 2. Mechanical properties of the CoCrFeMnNi alloys with different carbon contents (x) of after cold rolling and annealing at 800°C obtained at room and cryogenic temperatures

Temperature	293 K					77 K				
Alloy	YS, MPa	US, MPa	TE, %	UE, %	KCV, J/cm²	YS, MPa	US, MPa	TE, %	UE, %	KCV, J/cm²
C0	313	651	56	41	142	460	970	76	57	140
C0.5	381	721	52	36	119	480	994	62	48	130
C2.0	636	933	43	29	85	786	1218	52	39	86

Impact toughness testing has revealed the following trends: (i) the impact toughness decreased with increasing carbon content. For instance, at room temperature, the respective values of KCV for the C0 and C2.0 alloys were 140 J/cm^2 and 85 J/cm^2. Besides, only the C2.0 alloy completely disintegrated into two parts, which indicates a low resistance to crack propagation. (ii) The testing temperature had a rather weak effect on the impact toughness, for example, the KCV values obtained at 293K and 77K for the C0.5 alloy were 119 J/cm^2 and 130 J/cm^2, respectively. Figure 3 shows the fracture surfaces of the program alloys after impact tests at 293 K (a, b, and c) and 77 k (d, e, and f). The C0 alloy showed ductile fracture appearance with coarse dimples both at room and cryogenic temperatures. The C0.5 and C2.0 alloys also showed a ductile fracture, but with smaller dimples. At 77 K in the C2.0 alloy, brittle fracture takes place near the notch (not shown).

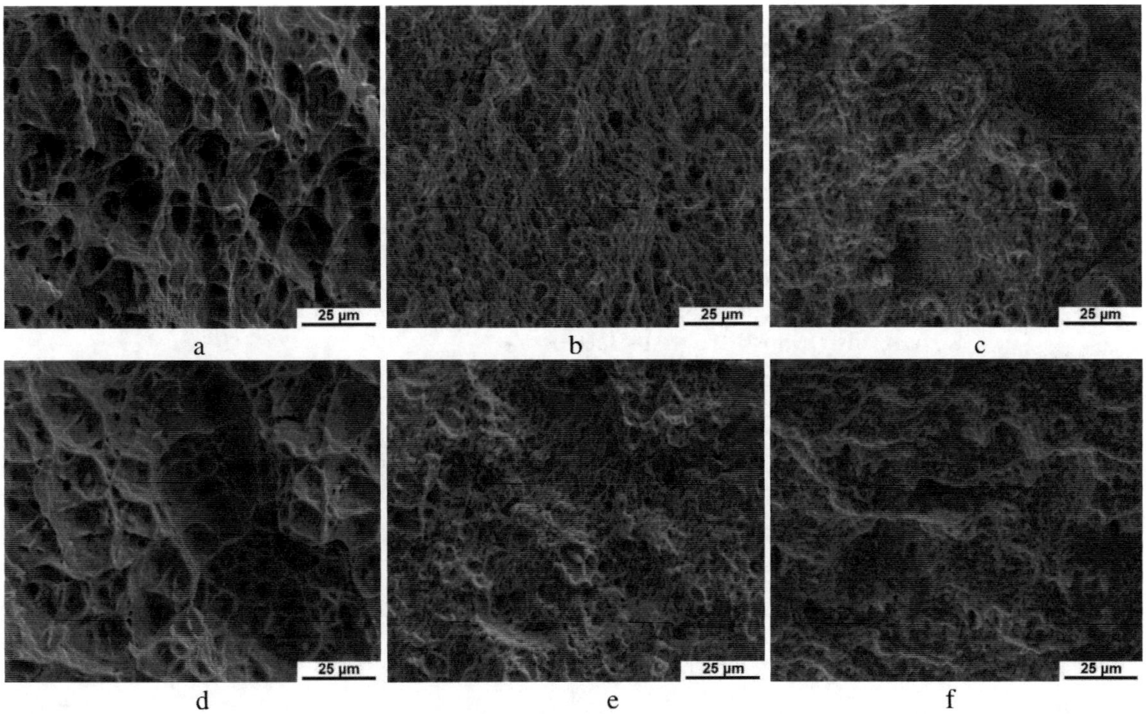

Figure 3. Fractography of the CoCrFeMnNi alloys with different carbon content after impact tests at 293K (a-c) and 77K (d-e): C0 (a, d); C0.5 (b, e); C2.0 (c, f).

Acknowledgments
The authors gratefully acknowledge the financial support from the Russian Science Foundation Grant No. 18-19-00003. The authors are grateful to the personnel of the Joint Research Center, «Technology and Materials», Belgorod State University, for their assistance with the instrumental analysis.

Reference list

[1] Yeh J, Chen S, Lin S, Gan J, Chin T, Shun T, Tsau C and Chang S *Adv. Eng. Mater.* **6** 2004 pp 299-303
[2] Cantor B, Chang I, Knight P and Vincent A *Mater. Sci. Eng. A.* **375–377** 2004 pp 213-8
[3] Otto F, Yang Y, Bei H and George E *A. Mater.* **61** 2013 pp 5743-55
[4] Chen J, Yao Z, Wang X, Lu Y, Wang X, Liu Y and Fan X *Mater. Chem. Phys.* **210** 2018 pp 136-45
[5] Chen L, Wei R, Tang K, Zhang J, Jiang F, He L and Sun J *Mater. Sci. Eng. A.* 2018
[6] Ko J and Hong S *J. Al. Comp.* **743** 2018 pp 115-25
[7] Wang Z, Baker I, Cai Z, Chen S, Poplawsky J and Guo W *A. Mater.* 2016
[8] Stepanov N, Yurchenko N, Tikhonovsky M and Salishchev G *J. Al. Comp.* **687** 2016 pp 59-71
[9] Stepanov N, Shaysultanov D, Chernichenko R, Ikornikov D, Sanin V and Zherebtsov S *Mater. Sci. Eng.* A. **728** 2018 pp 54-62
[10] Liu K, Komarasamy M, Gwalani B, Shukla S and Mishra R *Scr. Mater.* **158** 2019 pp 116-20

IOP Conf. Series: Materials Science and Engineering 1014 (2021) 012051 doi:10.1088/1757-899X/1014/1/012051

Studying Thermal Stability of Cast and Microcrystalline Alloys Al-(2.5, 4)%Mg-Sc-Zr

I S Shadrina[1*], A V Nokhrin[1], V I Kopylov[1,2], V N Chuvil'deev[1], A A Bobrov[1], and C V Likhnitskii[1]

[1] Lobachevsky State University of Nizhni Novgorod, 23 Gagarina ave., Nizhny Novgorod, 603950, Russian Federation

[2] Physical and Technical Institute, National Academy of Sciences of Belarus, 10 Kuprevichst., Minsk, 220141, Belarus

*Corresponding author: janashadr@gmail.com

Abstract. The article presents the results of experimental studies of the structure and mechanical properties of cast and microcrystalline (MC) aluminum alloys Al-(2.5, 4)%Mg-Sc-Zr with a total content of Sc+Zr = 0.32%. The Sc content was varied from 0.10% to 0.22% at an interval of 0.02%, with the Zr concentration in the alloy being changed proportionally. In their initial state, cast alloys are characterized by a homogeneous coarse-grained macrostructure. The structure of the alloy structure presents single submicron particles Al_3 (Sc,Zr) that are formed during crystallization of the alloys. It has been demonstrated that the recrystallization onset temperature in MC alloys depends on the scandium-to-zirconium ratio.

1. Introduction

Currently, the stability of the microcrystalline (MC) structure of Al-Mg aluminum alloys is ensured by the controlled breakdown of an oversaturated solid solution of Sc in Al with the formation of nanoparticles Al_3Sc [1-6]. The downside of Al_3Sc particles is their low thermal stability (high growth rate). This leads to an increase in the grain size during annealing of MC alloys where the intensity of the release and growth of Al_3Sc particles can significantly exceed similar characteristics in coarse-grained materials. This problem can be solved by partial replacing of Sc with rare earth elements (REEs) and transition metals (TM) with the diffusion rate in aluminum lower than that of scandium and high solubility in the Al_3Sc phase [2, 7-11]. This will allow the breakdown of a solid solution to form particles with an Al_3M (M – REEs, TM) structure, as well as particles with a "core Al_3Sc - shell Al_3M" structure with low growth (coalescence) rate.

The purpose of this study is to research thermal stability and mechanical properties of Al-Mg-Sc-Zr MC aluminum alloys where expensive scandium is partially replaced by less costly zirconium.

2. Methods and materials

Al-2.5%Mg-Sc-Zr and Al-4%Mg-Sc-Zr alloys are the objects of the study. The total content of Sc+Zr = 0.32%. The Sc content was varied from 0.10% to 0.22% at an interval of 0.02%, with the Zr concentration in the alloy being changed proportionally. Bars of 20×20×160 mm were produced by induction casting. The alloys were not tempered or homogenized after casting. The MC structure was formed by the Equal Channel Angular Pressing (ECAP) method at a temperature of 225 °C.

Microhardness measurements were taken with an HVS-1000 microhardness tester (50 g load, 10 s loading time). Specific electrical resistivity (SER) was measured using the eddy current method with a SIGMATEST 2.069 instrument. Microstructure was researched using a JSM-6490 scanning electron microscope with an Oxford Instruments INCA 350 energy-dispersive microanalyzer. Macrostructure was researched using a Leica IM DRM metallographic microscope. Samples were annealed using a SNOL air furnace with a controllable heating system.

3. Experimental results

In their initial state, Al-2.5%Mg-Sc-Zr cast alloys have a coarse-grained homogeneous structure in the central part of the bar and a columnar crystal structure at the edges of the bar. 4%Mg cast alloys have a coarse-grained homogeneous structure with agrain size of ~30-50 μm on the scandium-to-zirconium ratio in the alloy. The structure of alloys with a total content Sc + Zr> 0.30% demonstrates primary particles with an average size of ~0.35-1.3 μm containing scandium. No primary particles were detected in alloys with an Sc content < 0.30%. The research results suggest that SER experimental values are quite close to their theoretical values, calculated using the additivity rule [12] (see Table 1). This signals that most of scandium and zirconium is in a solid solution.

Table 1. SER values (μΩ·cm) for cast alloys.

Sc and Zr content	0.22Sc+ 0.10Zr	0.20Sc+ 0.12Zr	0.18Sc+ 0.14Zr	0.16Sc+ 0.16Zr	0.14Sc+ 0.18Zr	0.12Sc+ 0.20Zr	0.10Sc+ 0.22Zr
			Al-2.5%Mg-Sc-Zr alloys				
Experimental	4.50	4.59	4.37	4.62	4.55	4.54	4.50
Calculated	4.62	4.62	4.61	4.61	4.60	4.60	4.59
			Al-4%Mg-Sc-Zr alloys				
Experimental	5.43	5.24	5.37	5.39	5.34	5.37	5.25
Calculated	5.43	5.43	5.42	5.42	5.41	5.41	5.40

Figure 1 shows SER and microhardness dependences on the temperature of a 30-minute annealing of Al-2.5%Mg-Sc-Zr cast alloys. The analysis of the results demonstrates that substituting scandium with zirconium leads to an increase in thermal stability of the solid solution of scandium and zirconium in aluminum. The onset temperature of solid solution breakdown in alloys is 275 °C. At the same time, the rate of Al_3Sc particle precipitation is decreasing. The EDS microanalysis results demonstrate that most of the released particles contain both scandium and zirconium.

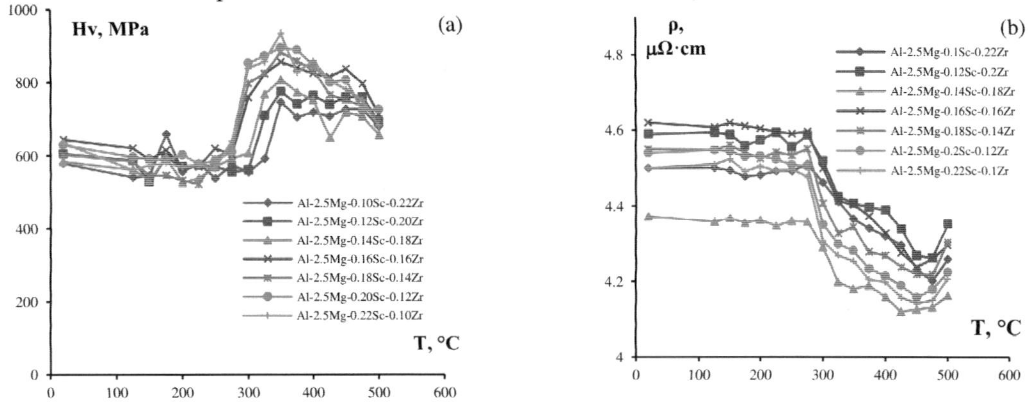

Figure 1. Dependences of microhardness (a) and SER (b) on the temperature of a 30-minute annealing of Al-2.5%Mg-Sc-Zr cast alloys with different contents of scandium and zirconium.

Figure 2 shows SER and microhardness dependences on the temperature of a 30-minute annealing of 4%Mg cast alloys. The analysis of the results demonstrates that substituting scandium with zirconium leads to an increase in thermal stability of the solid solution of scandium and zirconium in aluminum. The onset temperature of solid solution breakdown in alloys is ~200 °C. The EDS microanalysis results suggest that most of the formed particles contain both scandium and zirconium.

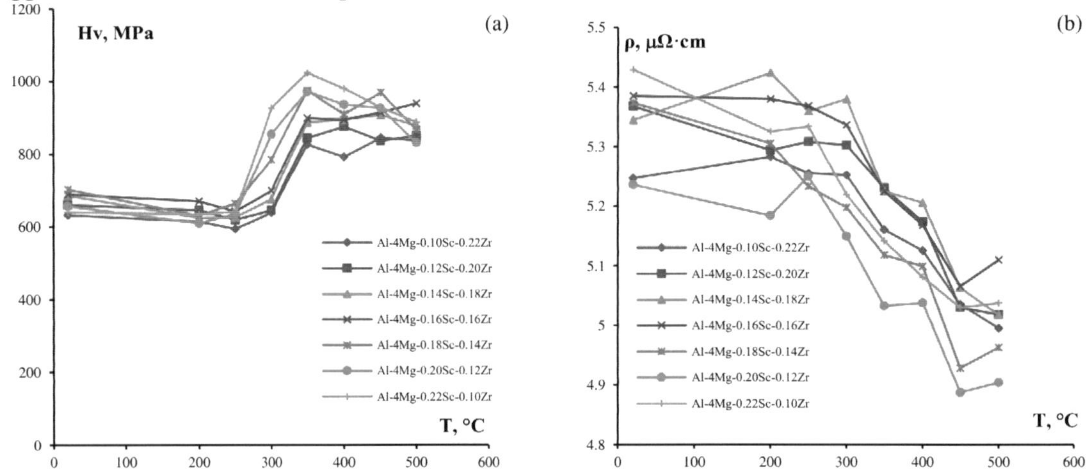

Figure 2. Dependences of microhardness (a) and SER (b) on the temperature of a 30-minute annealing of Al-4%Mg-Sc-Zr cast alloys with different contents of scandium and zirconium.

Analysis of $H_v(T)$ dependences suggests that the greatest hardening effect during annealing of cast alloys is observed for the alloys with an increased content of scandium (0.18-0.22%Sc). It should be noted that annealing at 475-500 °C causes SER to increase in 2.5Mg and 4Mg cast alloys. This is due to the dissolution of previously released Al_3Sc particles. This result indirectly indicates that during the breakdown of the solid solution, Al_3Sc and Al_3Zr particles were released separately.

The result was then confirmed while researching SER dependences on the duration of cast alloys isothermal annealing. Mechanisms of solid solution decomposition during annealing of aluminum alloys were identified from the analysis of researching specific electro resistance using the Johnson-Mehl-Avrami-Kolmogorov (JMAK) model [13]. It was demonstrated that breakdown of solid solution in cast alloys is two-staged, which is caused by a "separate" release of Al_3Sc particles at lower temperatures at the cores of lattice dislocations, and then, at higher temperatures, a release of Al_3Zr particles at the boundaries of dendritic crystals (grain boundaries).

Studies of MC aluminum alloys demonstrate that there is a post-ECAP increase in microhardness of aluminum alloys by ~200 MPa while SER remains practically the same.

Studies of the microstructure of annealed microcrystalline Al-2.5%Mg-Sc-Zr alloys demonstrate that the recrystallization onset temperature is 375-400 °C. It was found that SER dependences on the annealing temperature for all MC alloys are two-staged – a slight reduction of SER during annealing down to 275 °C and a rapid reduction of SER at higher temperatures occur (Figure 3). It should be noted that the onset temperature of solid solution breakdown in Al-2.5%Mg-Sc-Zr cast and MC alloys is 250-275 °C and is quite weakly dependent on the scandium-to-zirconium ratio (see Figure 3b). It should also be noted that, despite high microhardness values of post-ECAP MC alloys, the microhardness of Al-2.5%Mg-Sc-Zr MC alloys after annealing at 500 °C is lower than that of similarly-composed cast alloys after equivalent annealing. This result suggests that the second-phase particles are predominantly released at the MC alloy grain boundaries during annealing. This leads to faster growth of the released particles and, as per the Orowan equation, to a decrease in their contribution to the aluminum alloy hardness.

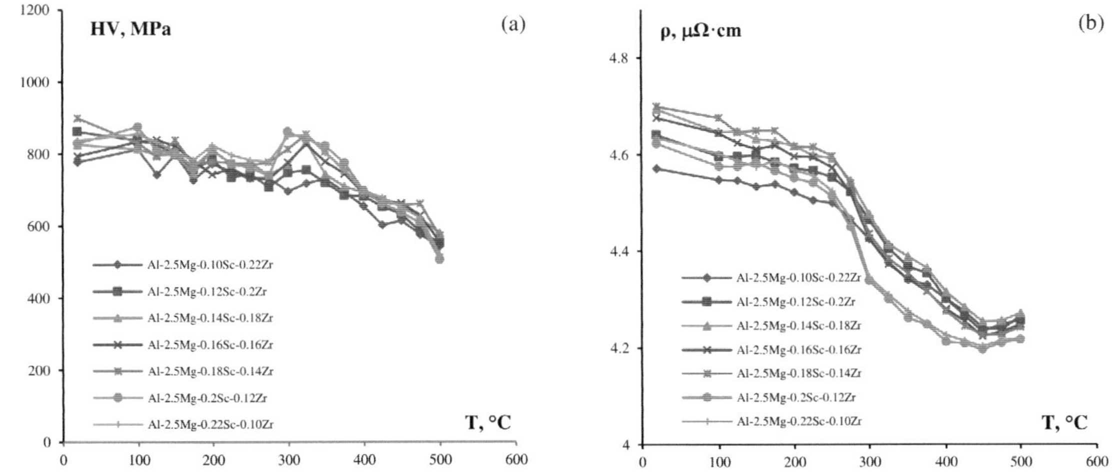

Figure 3. Dependences of microhardness (a) and SER (b) on the temperature of a 30-minute annealing of Al-2.5%Mg-Sc-Zr MC alloys with different contents of scandium and zirconium.

4. Conclusions

The specifics of solid solution breakdown in Al-2.5%Mg-Sc-Zr and Al-4%Mg-Sc-Zr cast and microcrystalline aluminum alloys have been studied. Studies have shown that partial substitution of Sc with Zr leads to a monotonic increase in thermal stability of the Sc and Zr solid solution in aluminum.

Acknowledgments

This study has been supported by the Russian Science Foundation (grant No. 18-13-00306).

The authors wish to thank V.V. Zakharov (All-Russian Institute of Light Alloys, Moscow) for his recommendations on the composition of aluminum alloys and their casting modes.

References

[1] Filatov Yu A, Elagin V I, Zakharov V V 2020 *Mater. Sci. Eng. A* **280** 97-101.
[2] Buranova Y, Kulitskiy V, Peterlechner M, Mogucheva A, Kaibyshev R, Divinski S V, Wilde G 2017 *Acta Materialia* **2017** 210-224.
[3] Mochugovskiy A, Mikhaylovskaya A V 2020 *Materials Letters* **275** 128096.
[4] Davydov V G, Rostova T D, Zakharov V V, Filatov Yu A, Yelagin V I 2020 *Mater. Sci. Eng. A* **280** 30-36.
[5] Chuvil'deev V N., Shadrina Ya S, Nokhrin A V, Kopylov V I, Bobrov A A, Gryaznov M Yu, Shotin S V, Tabachkova N Yu, Chegurov M K, Melekhin N V 2020 *Journal of Alloys and Compounds* **831** 154805.
[6] Dobatkin S, Estrin Y, Zakharov V V, Rostova T, Ukolova O, Chirkova A 2009 *International Journal of Materials Research* **100** 1697-1704.
[7] Zakharov V V 1997 *Metal Science and Heat Treatment* **39** 61-66.
[8] Zakharov V V 2015 *Metal Science and Heat Treatment* **57** 410-414.
[9] Chuvil'deev V N, Nokhrin A V, Makarov I M, Lopatin Yu G, Sakharov N V, Melekhin N V, Piskunov A V, Smirnova E S, Kopylov V I 2012 *Russ. Metall.* **2012** 415-427.
[10] Barkov R Y, Yakovtseva O A, Mamzurina O I, Loginova I S, Medvedeva S V, Proviryakov A S, Mikhaylovskaya A V, Pozdniakov A V 2020 *Physics of Metals and Metallography* **121** 604-609.
[11] Nokhrin A, Shadrina I, Chuvil'deev V, Kopylov V 2019 *Materials* **12** 316.
[12] Chuvil'deev V N, Nokhrin A V, Smirnova E S, Kopylov V I 2012 *Russ. Metall.* **2012** 985-993.
[13] Chuvil'deev V N, Nokhrin A V, Smirnova E S, Kopylov V I 2013 *Russ. Metall.* **2013** 676-690.

IOP Conf. Series: Materials Science and Engineering 1014 (2021) 012052 doi:10.1088/1757-899X/1014/1/012052

Effect of carbon content, deformation and annealing on the structure and properties of interstitial TRIP high-entropy alloys

D G Shaysultanov*, K Sh Raimov and N D Stepanov

Belgorod State National Research University, Pobeda, 85, Belgorod, 308015, Russia

*Corresponding author: shaysultanov@bsu.edu.ru

Abstract. In this work, the effect of carbon content on the structure and mechanical properties of $Fe_{(50-x)}Mn_{30}Co_{10}Cr_{10}C_x$ (x=0; 0.5; 1.0 at.%) interstitial high entropy alloys was studied. In the as-cast condition, the alloys were composed of the face centred cubic (fcc) and hexagonal close packed (hcp) phases; the amount of the hcp phase decreased from 46% to 20% as the carbon concentration increased from 0 to 1 at.%. The carbon content had a limited effect on the strength of the alloys, yet the ductility was enhanced significantly as the result of interstitial alloying. Furthermore, $Fe_{49}Mn_{30}Co_{10}Cr_{10}C_1$ alloy was cold rolled to a reduction of 56% reduction and annealed at 700-900°C for 1 hour. After cold working, the alloy has attained almost fully the hcp structure. Annealing resulted in (i) transformation of the deformation-induced hcp phase into fcc; (ii) development of recovery and recrystallization; (iii) precipitation of Cr-rich $M_{23}C_6$ carbides. The cold-rolled alloy was very strong but brittle; annealing resulted in softening and increase of the ductility. The relationships between the chemical composition, structure and mechanical properties of the alloys are briefly discussed.

1. Introduction

The so-called high-entropy alloys (HEAs) have recently emerged as promising materials with unprecedent mechanical characteristics, for example, a combination of strength and ductility [1]. Excellent properties of the alloys often come from the synergy of different deformation mechanisms. For instance, Z. Li et al. introduced $Fe_{50}Mn_{30}Co_{10}Cr_{10}$ (in at.%) alloy composed from two (face centered cubic, fcc and hexagonal close packed, hcp) phases, exhibiting transformation-induced plasticity (TRIP) effect [2]. The mechanical properties of HEAs can be significantly enhanced by the addition of interstitial elements like carbon or nitrogen [3,4], as was demonstrated in $Fe_{49.5}Mn_{30}Co_{10}Cr_{10}C_{0.5}$ alloy [5]. In this case, the alloy additionally benefited from interstitial solid solution hardening and precipitate (carbide) hardening. However, many aspects of the effect of carbon on the structure and properties of interstitial TRIP HEAs remain unclear. In addition, it is not completely clear how the structure and properties of such alloys evolve during thermomechanical processing [6]. Therefore, in this work, we (i) have examined the effect of carbon content (0-1 at.%) on the structure of the as-cast $Fe_{50}Mn_{30}Co_{10}Cr_{10}$-based alloys and (ii) studied the effect of thermomechanical processing on the structure and properties of the alloy with 1 at.% C.

2. Materials and methods

A series of the $Fe_{(50-x)}Mn_{30}Co_{10}Cr_{10}C_x$ (x=0; 0.5; 1.0 at.%) alloys were prepared by vacuum arc melting technique. High-purity (≥99.9 wt.%) elements were used as starting materials. The alloys were examined in the as-cast condition. In addition, the alloy with x=1 at.% C ($Fe_{49}Mn_{30}Co_{10}Cr_{10}C_1$) was subjected to cold rolling with a reduction of 56% and subsequent annealing at 700-900°C for 1 hour.

A scanning electron microscope with an electron backscattered diffraction (EBSD) attachment was primarily used for the microstructure characterization. Careful mechanical polishing on grinding papers with continuously decreasing grain size and final polishing with OP-S suspension was performed to prepare the specimens for microstructure characterization. In addition, transmission electron microscopy (TEM) method was used to get better understanding of the structure of the alloys in some conditions.

The mechanical properties of the alloys were accessed by tensile tests at room temperature. Dog-bone specimens with the gauge dimensions of 12×3×1.5 mm^3 were pulled to fracture with the initial strain rate of 10^{-3} s^{-1}.

3. Results and Discussion

The as-cast microstructures of the $Fe_{(50-x)}Mn_{30}Co_{10}Cr_{10}C_x$ (x=0; 0.5; 1.0 at.%) alloys are shown in figure 1. All alloys had a similar coarse grain size of 300-400 μm (not shown). Inside the grains, a dual-phase structure, composed of fcc and hcp phases, was observed. The addition of carbon has resulted in a gradual decrease in the fraction of the hcp phase: from 46% in the x=0 alloy to 36% in the x=0.5 alloy and, finally, to 20% on the x=1.0 alloy. The decrease in the hcp fraction is most likely associated with an increase of the stacking fault energy of the alloy due to carbon alloying. Also, the morphology of the hcp particles was affected by carbon alloying. In the carbon-free alloy, the particles had regular, plate-like shape with sharp straight boundaries. An increase in the carbon concentration resulted in more equiaxed particles with irregular shape and curved boundaries. Note that the TEM studies (the results are not shown) did not reveal the presence of any other phases than fcc and hcp.

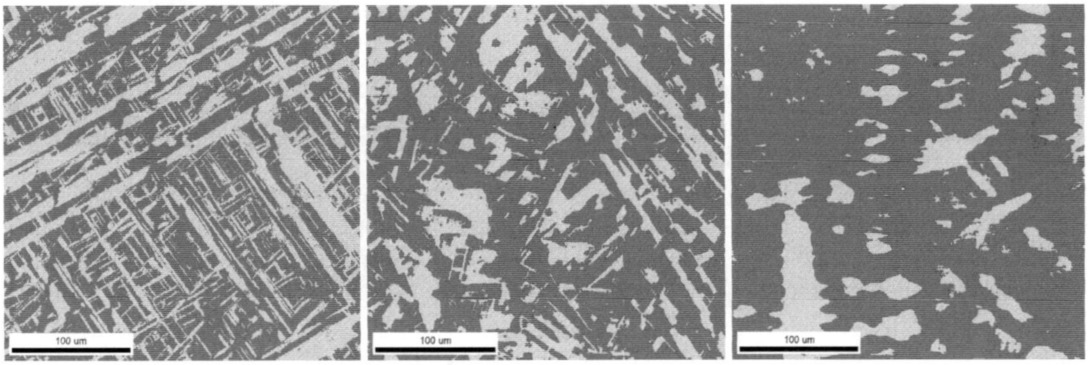

a b c

Figure 1. EBSD phase maps of the $Fe_{(50-x)}Mn_{30}Co_{10}Cr_{10}C_x$ alloys: a - x=0 at.%; b – x=0.5 at.%; c – x=1.0 at.%. The hcp phase is shown in green color and the fcc phase – in red color.

The x=1 ($Fe_{49}Mn_{30}Co_{10}Cr_{10}C_1$) alloy was further subjected to cold rolling and subsequent annealing at 700-900°C (figure 2). After cold working, the alloy had mostly hcp structure (figure 2a), which was a product of the deformation-induced martensitic transformation. The fraction of the fcc phase was only 13%; fcc particles had a characteristic plate-like morphology. Note the presence of black areas on the EBSD phase map; they indicate the low quality of the obtained Kikuchi patterns and the high density of defects in the alloy after rolling. Annealing at 700°C results in drastic changes in the microstructure (figure 2b, c). First, the deformation-induced hcp phase was mostly converted to the fcc phase; the fraction of the hcp phase was 15%. Partial recrystallization also occurred; the fraction and the size of the recrystallized grains were 62% and 1.8 μm, respectively. In the uncrystallized areas, the recovery took place as evidenced from the absence of black areas with low image quality. A further increase in the annealing temperature to 900°C resulted in the formation of a fully recrystallized microstructure with a mean grain size of 7.0 μm (figure 2d, e). The fraction of the hcp phase decreased further to 5%. In addition, TEM studies revealed the presence of Cr-rich $M_{23}C_6$ type carbides in the fcc matrix (figure 2f). The fraction and the size of the carbides were 4% and 140 nm, respectively. Note that similar carbides were detected already after annealing to 700°C (not shown).

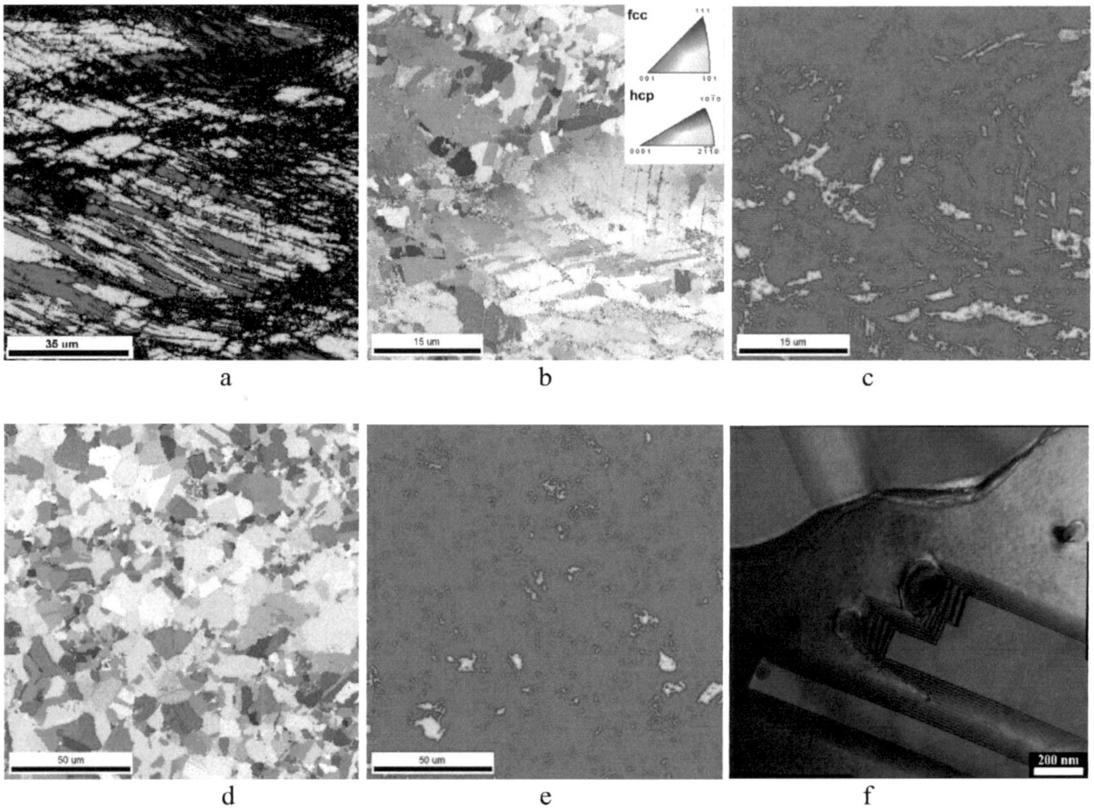

Figure 2. Microstructure of the $Fe_{49}Mn_{30}Co_{10}Cr_{10}C_1$ alloy after thermomechanical processing: a – cold rolling to a reduction of 56%; b, c – subsequent annealing at 700°C; d, e, f – subsequent annealing at 900°C. a, c, e – EBSD phase maps (color coding is the same as in figure 1); b, d – EBSD IPF maps; f – TEM bright-field image.

The effect of chemical composition and processing on the mechanical properties of the program $Fe_{(50-x)}Mn_{30}Co_{10}Cr_{10}C_x$ alloys was evaluated using tensile testing (figure 3). The carbon content had a strong and not readily anticipated effect on the mechanical properties of the as-cast alloys. The $Fe_{50}Mn_{30}Co_{10}Cr_{10}$ and $Fe_{49}Mn_{30}Co_{10}Cr_{10}C_1$ alloy had rather close strength – yield strength of 245 MPa and 240 MPa, respectively, and ultimate tensile strength of 705 MPa and 680 MPa (figure 3a). However, the carbon-doped alloy had more than 2 times better ductility – the respective values of uniform elongation were 38% and 86%, respectively. Such a huge difference in ductility can be potentially attributed to a higher fraction of the softer fcc phase (figure 1) that can transform to a harder hcp phase during plastic deformation and thus provide effective strain hardening, i.e. the enhanced TRIP effect. Yet, this assumption is contrary to the recent results on similar $Fe_{50}Mn_{30}Co_{10}Cr_{10}$ alloy [6]. Thus, additional work is required to elucidate the reasons of improved ductility of the carbon-doped alloys.

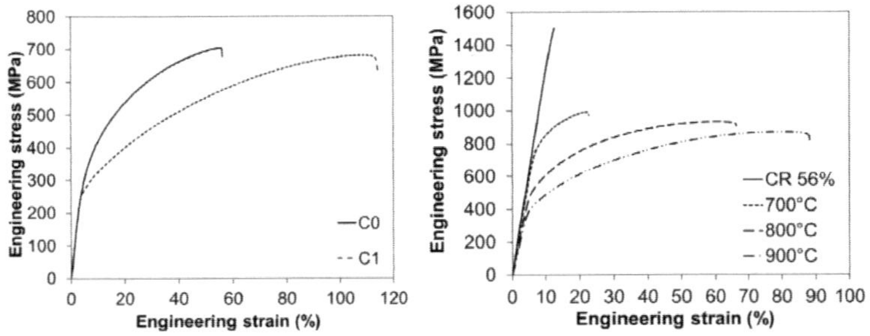

a b

Figure 3. Stress-strain curves obtained during tensile testing of: a – as-cast $Fe_{(50-x)}Mn_{30}Co_{10}Cr_{10}C_x$ (x=0 and x=1) alloys; b - $Fe_{49}Mn_{30}Co_{10}Cr_{10}C_1$ after thermomechanical processing.

Thermomechanical processing had a more anticipated effect on the mechanical properties of the $Fe_{49}Mn_{30}Co_{10}Cr_{10}C_1$ alloy (figure 3b). The cold-worked alloy had high strength (yield strength of 1415 MPa), but very limited ductility (0.5%), probably due to the mostly hcp structure (figure 2a). Annealing at 700°C resulted in considerable softening (yield strength of 680) due to the transition of hcp to fcc (figure 2c), but the ductility remained rather low (10%) because of the partially recrystallized microstructure. Excellent ductility was obtained in the fully recrystallized alloy after annealing at 900°C (68%). The yield strength of the annealed alloy was 370 MPa. The obtained results clearly suggest typical strength-ductility trade-off in the $Fe_{49}Mn_{30}Co_{10}Cr_{10}C_1$ alloy.

4. Conclusions

1) The increase in the carbon percentage from 0 to 1 at.% resulted in a decrease in the hcp phase fraction from 46% to 20% in the as-cast $Fe_{(50-x)}Mn_{30}Co_{10}Cr_{10}C_x$ alloys. The strength of the alloys showed some decrease with carbon addition, however, the ductility increased substantially.

2) After cold rolling, the $Fe_{49}Mn_{30}Co_{10}Cr_{10}C_1$ alloy had almost fully the hcp structure. Annealing at 700-900°C resulted in (i) transformation of deformation-induced hcp phase into fcc; (ii) development of recovery and recrystallization; (iii) precipitation of Cr-rich $M_{23}C_6$ carbides. The cold-rolled alloy was very strong but brittle; annealing resulted in softening and an increase of the ductility.

Acknowledgements

The authors gracefully acknowledge the financial support from the Russian Science Foundation, grant № 20-79-10093.

References

[1] Miracle D B and Senkov O N 2017 A critical review of high entropy alloys and related concepts *Acta Mater.* **122** 448–511

[2] Li Z, Pradeep K G, Deng Y, Raabe D and Tasan C C 2016 Metastable high-entropy dual-phase alloys overcome the strength–ductility trade-off *Nature* **534**

[3] Klimova M V, Semenyuk A O, Shaysultanov D G, Salishchev G A, Zherebtsov S V and Stepanov N D 2019 Effect of carbon on cryogenic tensile behavior of CoCrFeMnNi-type high entropy alloys *J. Alloys Compd.* **811** 152000

[4] Klimova M, Shaysultanov D, Semenyuk A, Zherebtsov S, Salishchev G and Stepanov N 2020 Effect of nitrogen on mechanical properties of CoCrFeMnNi high entropy alloy at room and cryogenic temperatures *J. Alloys Compd.* 156633

[5] Li Z, Tasan C C, Springer H, Gault B and Raabe D 2017 Interstitial atoms enable joint twinning and transformation induced plasticity in strong and ductile high-entropy alloys *Sci. Rep.* **7** 40704

[6] Li Z, Tasan C C, Pradeep K G and Raabe D 2017 A TRIP-assisted dual-phase high-entropy alloy: Grain size and phase fraction effects on deformation behavior *Acta Mater.* **131** 323–35

IOP Conf. Series: Materials Science and Engineering 1014 (2021) 012053 doi:10.1088/1757-899X/1014/1/012053

Structural-phase features of WC-based ceramics obtained by the spark plasma sintering method

K E Smetanina*, P V Andreev, E A Lantsev, M M Vostokov, N V Malekhonova, A A Murashov

Lobachevsky State University of Nizhny Novgorod, Nizhny Novgorod, Russia

*Corresponding author: smetanina-ksenia@mail.ru

Abstract. Ceramics obtained by spark plasma sintering of tungsten carbide powder were studied by X-ray diffraction layer-by-layer analysis. The surface of the ceramic samples was mechanical grinded and polished in several stages. The distribution of crystal phases by the depth of ceramics was investigated.

1. Introduction

WC-based hard alloys produced by sintering [1] of powders are promising materials for use as metal cutting tool. It is characterized by high hardness, high wear resistance, high melting point, chemical resistance to corrosion and oxidation. At the same time, pure tungsten carbide ceramics are characterized by low flexural strength [2].

Original WC powders are sintered with various additives to improve the physical and mechanical properties of hard alloys. The addition of cobalt increases the crack resistance and strength of WC-based hard alloy [1].

Improvement of the hard alloy physical properties is also associated with the formation of a homogeneous high-density fine-grained structure in them [3]. In this regard, the most promising method for manufacturing of hard alloys is spark plasma sintering (SPS) of powders. The method consists in heating the powder at a high rate up to the sintering temperature in a vacuum under pressure [4]. The powder is poured into a hollow cylindrical graphite mold and fixed with graphite punches.

Previously, an X-ray diffraction layer-by-layer analysis of samples of the hard alloy WC + 10% Co revealed the heterogeneity of the phase composition over the depth of the sample [5]. It was established that the brittle η-phase (Co_3W_3C) appears in the studied samples at a depth of ≥ 100 μm. The observed effect indirectly confirms the hypothesis of carbon diffusion from graphite punches contacting the surface of sintered samples. Thus, the carbon deficiency was replenished in the surface layers of the samples with preventing the formation of the η-phase.

It is known that carbon not only prevents the formation of brittle phases in the sintering process (W_2C in WC ceramics and the η-phase ($Co_xW_yC_z$) in WC – Co hard alloys), but also dissolves in cobalt. Therefore, for a detailed study of the discovered phenomenon, it was necessary to consider the original system – WC.

The purpose of the work is to determine the degree of homogeneity of WC ceramics obtained by the SPS method by the approach of X-ray diffraction layer-by-layer analysis.

2. Materials and techniques

Industrial WC powder with an average particle size of 3 μm (produced by JSC "KZTS") was used as a raw material for the manufacture of ceramic samples. The content of carbon in the original powder was 6.13 ± 0.01% wt., where free carbon was 0.05 ± 0.01% wt.

SPS of WC powder was performed on the "Dr. Sinter model SPS-625" (SPS Syntex, Japan). Sintering was carried out in a vacuum in a graphite mold with an internal diameter of 12 mm.

The objects of the study were 2 ceramic samples. Sample 1 was obtained in the following mode: heating speed was 50°C/min, pressure was 70 MPa, sintering temperature was 1690°C.

Before sintering sample 2, punches and a mold were coated with BN to isolate the sintering powder from the surfaces of carbon parts. Otherwise, the sintering conditions for samples 1 and 2 were the same.

The surface of the sintered ceramic samples was mechanical grinded with diamond discs and sequentially mechanical polished with diamond paste to a roughness level of 1 μm. The operation of mechanical grinding was carried out on a "Struers Secotom-10", polishing – on a "Buehler Automet 250".

There were conducted 5 stages, each consisted of grinding and subsequent polishing. The thickness subtracted at each stage was 60, 70, 40, 140, 460 μm for sample 1 and 100, 100, 120, 180, 380 μm for sample 2. The height of the sample was controlled using a micrometer.

Phase composition of the original powder and sintered ceramic samples was studied by X-ray powder diffraction. X-ray diffraction experiments were performed on a diffractometer "XRD-7000" (Shimadzu, Japan) in the "wide slit mode" using CuKα-radiation ($\lambda = 1.54$ Å) in the range of angles $2\theta = 30 - 80°$ with a scanning step of 0.04° and an exposure time of 2 s.

Quantitative phase analysis was carried out by the reference intensity ratio method using integral intensities of the analytical diffraction peaks of crystal phases detected in the studied samples [6]. The analytical peak of α-WC was peak (101) and W_2C was peak (111).

Vickers hardness (H_v) was measured using a "Struers Duramin-5" with 2 kg load.

3. Description of experimental results

Quantitative X-ray phase analysis of the original WC powder demonstrated the presence of 2.5 ± 0.2% wt. of W_2C. This explains the introduction of free carbon into the WC powder by the manufacturer.

Figure 1 shows the results of X-ray diffraction experiments of ceramic samples 1 and 2. The numbers of X-ray diffraction patterns correspond to the number of the layer-by-layer analysis stage. Thus, 1 – diffraction pattern of the surface layer of the sample subjected to primary surface treatment, 5 – diffraction pattern of the layer closest to the center of the sample (at a depth of ~ 900 μm from the surface layer).

Significantly lower content of W_2C – 0.8 ± 0.1% wt. is fixed in both samples after the first stage of surface treatment (up to 100 μm depth) than in the original powder. The recovery of W_2C to α-WC can take place only due to carbon [7].

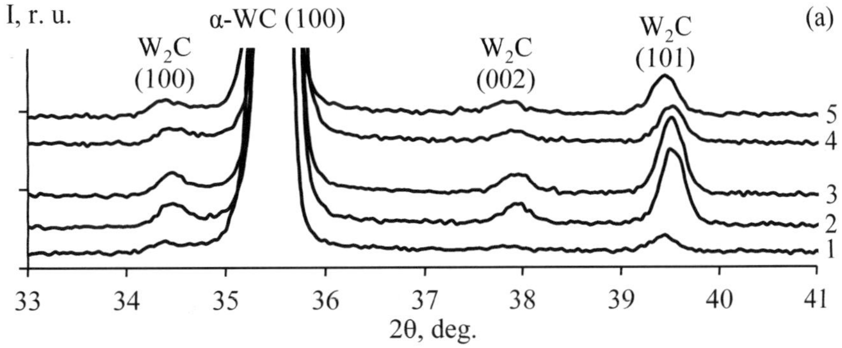

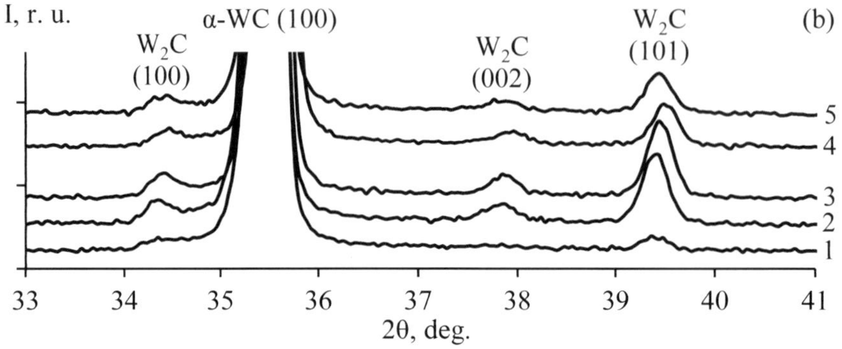

Figure 1. X-ray diffraction patterns of WC ceramic samples at the different stages of surface treatment: a) sample 1 (without BN covering of graphite parts of the mold), b) sample 2 (with BN covering of graphite parts of the mold)

The similar character of the diffraction patterns of both samples after the first stage of their surface treatment, and also the mentioned results on WC + 10% Co, allow us to draw a conclusion that the recovery of W_2C to WC on the surface occurs most likely due to the carbon diffusion from graphite punches.

Subsequent layers of both samples (at a depth of ≈ 200 μm) contain 3.0 ± 0.2% wt. of W_2C that slightly exceeding the content of W_2C in the original powder. The constant content of W_2C – 1.7 ± 0.2% wt. is fixed from a depth of 300-400 μm (see figure 2). An increasing in hardness H_V is also observed with increasing degree of sample homogeneity in deeper layers.

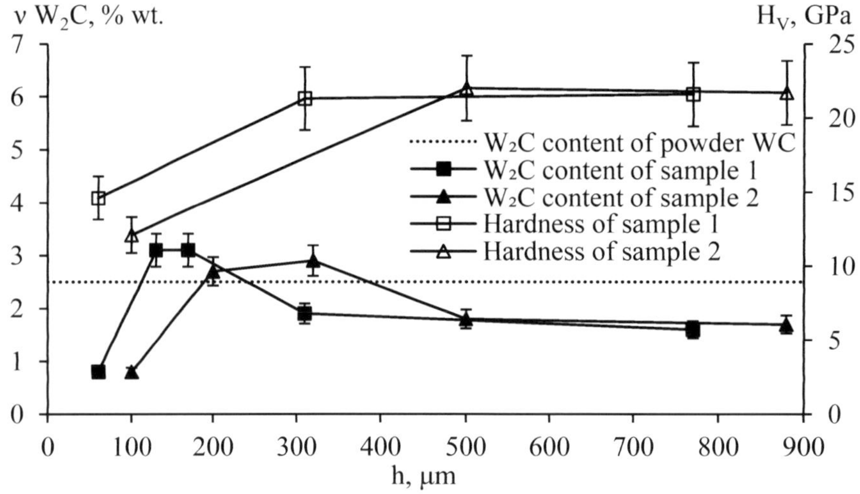

Figure 2. Distribution of W_2C and H_V by depth of studied ceramic samples: sample 1 (without BN covering of graphite parts of the mold) and sample 2 (with BN covering of graphite parts of the mold)

The detected heterogeneity of the phase composition over the sample depth may be related to the features of the WC powder sintering process. So, adsorbed oxygen on the WC powder surface at heating reacts with WC, which leads to partial decarburization of WC particles with the formation of gases CO and CO_2, and therefore with the formation of new W_2C particles.

But the free carbon introduced into the original WC powder to achieve the equilibrium carbon value (6.13% wt.) is insufficient to suppress both the existing W_2C and the W_2C formed by decarburization during heating.

It should also be noted that the sample is heated from the mold. This leads to an uneven sintering process and prevents the formation of a homogeneous structure in the sintered sample.

4. Conclusion

The result of this study is the detection of the phase composition heterogeneity of WC-based ceramics obtained using a promising technology – spark plasma sintering of powder. It was found that at a depth of 100-300 μm, the content of W_2C reaches its maximum. This maximum is closer to a depth of 200 μm for a sample without BN and for a sample with BN – around 300 μm. In this case, the coating of the mold with a layer of BN did not affect the characteristic values of the W_2C phase content of the ceramic sample within the error limits. Thus, it was shown that in order to further study the sintering mechanisms of WC-based ceramics it is necessary to take into account the temperature distribution inside the sample.

Acknowledgments

The reported study was funded by Russian Science Foundation (RSF), project №18-73-10177.

References

[1] Panov V and Chuvilin A 2001 *Technology and properties of sintered carbide alloys and products from them* (Moscow: MISIS) p 428
[2] Kurlov A and Gusev A 2013 *Tungsten Carbides: Structure, Properties and Application in Hardmetals* (Springer International Publishing) p 242
[3] Chuvildeev V et al 2013 *Nanotechnologies in Russia* **8 (1-2)** 108-122
[4] Tokita M 2013 Spark Plasma Sintering (SPS) Method, Systems, and Applications *Handbook of Advanced Ceramics* (2nd edition) ed Shigeyuki Somiya (Academic Press) chapter 11.2.3 pp 1149-77
[5] Smetanina K, Andreev P, Lantsev E, Vostokov M and Malekhonova N 2020 *Industrial laboratory. Diagnostics of materials* **86(8)** 38-42
[6] Andreev P, Smetanina K and Lantsev E 2019 *Industrial laboratory. Diagnostics of materials* **85(8)** 37-42.
[7] Cao S, Cai Z, Zhou J and Li J 2006 *Journal of Central South University of Technology* **13** 603-607

Effect of microstructure on the oxidation behavior of a β-solidified γ-TiAl based alloy after nitrogen ion implantation

D O Panov[1], V S Sokolovsky[1*], N D Stepanov[1], S V Zherebtsov[1], V Y Novikov[1], N A Nochovnaya[2], P V Panin[2], G A Salishchev[1]
[1] Laboratory of Bulk Nanostructured Materials, Belgorod National Research University, Belgorod 308015, Russia
[2] Federal State Unitary Enterprise "All-Russian scientific research institute of aviation materials" (FSUE "VIAM"), 17 Radio str., Moscow 105005, Russia

*Corresponding author: sokolovskiy@bsu.edu.ru

Abstract. The microstructure of the oxide layer of a β-solidifying γ-TiAl alloy (Ti-43.2Al-1.9V-1.1Nb-1.0Zr-0.2Gd-0.2B) after nitrogen ion implantation was studied after cyclic oxidation tests at 800 °C. Oxidation resulted in the formation of the continuous oxide scale, consisting of thin inner and outer TiO_2, dense Al_2O_3 and mixed Al_2O_3 + TiO_2 layers. It was revealed that the nitrogen ion implantation considerably inhibited the growth of the oxide scale, internal oxidation and Al-depleted zones. Thick Al-depleted zones and a mixture of Al_2O_3 and TiO_2 were formed along the colony boundaries. Tooth-shaped Al-depleted zones propagated along the lamella boundaries. In addition, the presence of Gd-rich particles in the oxide scale promoted the formation of Al_2O_3 particles in surrounding areas and increased the depth of the oxide scale around.

1. Introduction

β-solidified gamma titanium aluminide (γ-TiAl) based alloys have attracted tremendous attention as potential high-temperature materials due to their good creep resistance and superior strength-to-weight ratio [1]. These alloys consist mainly of (α_2+γ)-lamellae colonies and have the most promising mechanical properties for aerospace engine components. However, the γ-TiAl based alloys are prone to rapid oxidation at temperatures above 600 °C, resulting in the formation of protective Al_2O_3 layer, TiO_2, a mixture of Al_2O_3 and TiO_2, inner oxide and Al-depleted zones on the surface [1-4].

It is well established that plenty of factors affect the oxide scale growth, namely the chemical composition and structure of the base material, surface finish, environment etc. [1, 5-8]. For instance, the oxidation resistance of Ti-45Al-8Nb (at.%) was considerably enhanced by the Y addition of 0.3 at.% that promoted the protective Al_2O_3 layer formation [9]. On the other hand, coarse Y_2O_3-particles, found along the grain boundaries in the alloys with high Y content (>0.3 at.%), provided fast path for the inward oxygen diffusion and significantly increased the oxidation rate [9]. Although the beneficial effects of the addition of rare-earth elements on the oxidation resistance of various alloys were known [10], there is no comprehensive understanding of the effect of Gd on the oxidation behavior of a γ-TiAl based alloy. Apart from adjusting the bulk chemical composition, surface modification, for example, by ion implantation of different non-metallic or metallic species (F, I, N, Cl, Nb, Al, Si, Mo) can effectively improve the oxidation resistance of the alloys [11-13].

The microstructure of the base material can also affect the oxidation behavior. For example, refinement of α_2-phase particles in the Ti–46Al–1Cr–0.2Si alloy inhibits oxidation dramatically [8]. Meanwhile, the orientation of lamellar colonies relative to the surface can influence the oxidation behavior of the γ-TiAl based alloys because lamellar boundaries can serve as fast diffusion pathways. However, this effect has never been properly studied. In this paper, the effect of microstructure on the oxide layer formation of a β-solidifying γ-TiAl alloy is considered.

2. Materials and methods

The β-solidifying γ-TiAl alloy with the actual composition Ti-43.2Al-1.9V-1.1Nb-1.0Zr-0.2Gd-0.2B (at. %) was used as the program material. The samples with a size of 20×20×2 mm^3 were cut from the ingots using a Sodick AQ300L wire electrical discharge machine. The samples surface was ground with emery paper up to № 2000, and then mechanically polished with an OP-S colloidal silicon suspension. The nitrogen ion implantation was produced at temperature of 400 °C in a vacuum chamber with a pressure in the range of 10^{-4} - 10^{-5} Pa. The radiation dose and ion energy were 10^{17} cm^{-2} and 40 keV, respectively.

Cyclic oxidation tests was performed in open air using a Nabertherm furnace. The samples were soaked for 24 hours at 800 °C with subsequent air cooling. After that the samples were weighed and then could be returned to the furnace again for the next test cycle. Up to 5 cycles of oxidation tests were performed.

The structure of the alloy was studied mostly using scanning electron microscopy (SEM). The following tapes of samples were tested by SEM: a beveled sample with a polished surface inclined ~5° – for the surface and material structure after nitrogen ion implantation; a cross-section of samples - for structure and element distribution of oxidation scale. The samples were examined using a FEI Quanta 600 electron microscope equipped with a back-scattered electron (BSE) and an energy-dispersive X-ray spectroscopy (EDS) detector with an accelerating voltage of 30 kV.

3. Results and discussion

The microstructure of the β-solidifying Ti-43.2Al-1.9V-1.1Nb-1.0Zr-0.2Gd-0.2B alloy consisted of (α_2+γ) lamellae colonies surrounded by γ- and (β+ω)-phase particles (Figure 1a, b). Particles of γ- and (β+ω)- phases were mainly found between the colonies. The volume fraction of γ- and (β+ω)- phase's particles was 1%. Some Gd-rich phases and a small amount of borides were found in the microstructure (Figure 1a). The volume fraction of Gd-rich particles was ≈2%. A more detailed investigation of the as-cast microstructure of the program material was presented in the previous paper [14]. The microstructure after ion implantation was presented in figure 1c. The modified layer had light grey color and depth of 1 μm; the boundary between the base material and modified layer was curved. No difference in the microstructure between the base material and modified layer was detected by SEM-BSE.

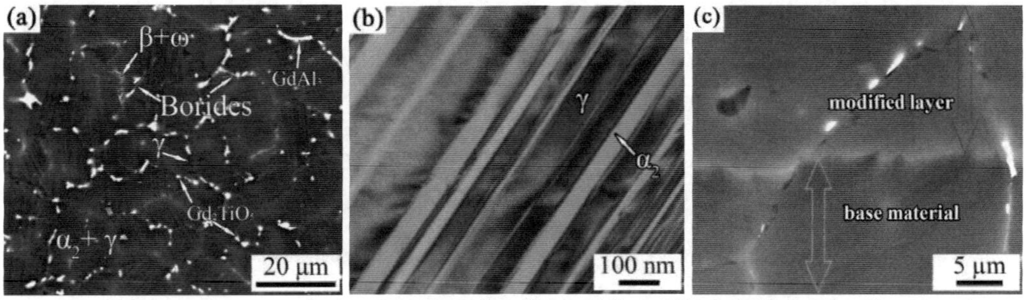

Figure 1. Microstructure of the Ti-43.2Al-1.9V-1.1Nb-1.0Zr-0.2Gd-0.2B alloy in the as-cast condition (a, b) and after ion implantation (c): (a, c) SEM-BSE images; (b) TEM image.

Figure 2 shows the growth of oxide scale and the internal oxidation and Al-depleted zones (inner zones) depth for the as-cast and nitrogen ion implanted alloy depending on the number of oxidation cycles. In this case, the internal oxidation zone referred to the oxygen-enriched aria under the oxide scale, which might also include some oxide particles. A typical example of an oxidized microstructure with identification of all layers is shown below in Figure 3. Two stages were distinguished in both conditions. At the first stage, parabolic kinetics was observed up to 3 cycles which is associated with the formation of a protective oxide layer. The formation of a protective layer led to slowing the oxide scale and inner oxide and Al-depleted zones growth at the second stage (3-5 cycles) in case of the nitrogen ion implanted alloy. After 5 cycles, the depth of the oxide scale and inner zones of the ion implanted alloy was 14 μm and 2 μm, respectively. Note that these values were considerably larger than the depth of the modified layer (~1 μm). However, almost linear growth of the oxide scale and inner oxide and Al-depleted zones for the as-cast alloy was detected at the second stage.

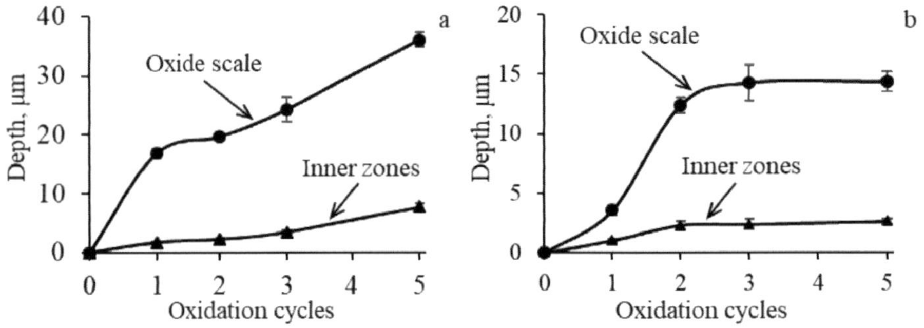

Figure 2. Dependence the depth of the oxide scale and inner zones (internal oxidation area and Al-depleted zones) of the Ti-43.2Al-1.9V-1.1Nb-1.0Zr-0.2Gd-0.2B alloy in the as-cast condition (a) and after ion implantation (b) on number of oxidation cycles at 800 °C.

The cross-section structure after 2 cycles of oxidation test is shown in Figure 3. The oxide scale mostly consisted of two oxides phases (Ti and O-rich light gray areas - TiO_2; Al and O-rich dark gray areas - Al_2O_3) [9] (Figure 2). The following layers were found: I – outward-growing TiO_2; II – Al_2O_3; III – inner TiO2; IV – a mixture of Al2O3 + TiO2; V – internal oxidation zone; VI – Al-depleted zone. The structure of the oxide scale was typical for γ-TiAl based alloys [9]. The sequence of layers depended on the ratio of outward-diffusion of Al and Ti and inward-diffusion of O on the one hand, and the ratio of the growth rate of TiO_2 and Al_2O_3, on the other hand. Obviously, TiO_2 and Al_2O_3 formed and grew simultaneously because of the similar free energy of these formation, but the growth rate of TiO_2 was drastically higher in comparison with Al_2O_3 [15]. Thus, the outward-growing TiO_2 layer quickly covered the surface of the material. The Al_2O_3 layer was found underneath the TiO_2 layer because of the diffusion of Ti toward the surface. The inner TiO_2 layer was likely formed because of the subsequent depletion of Al beneath the Al_2O_3 layer. The mixture of Al_2O_3+TiO_2 grew by parallel outward- and inward-diffusion [16]. As a result of the higher diffusion coefficient of Al in comparison with Ti [4], the Al-depleted zone appeared.

Thick Al-depleted zones and a mixture of Al_2O_3 and TiO_2 oxides formed along the colony boundaries. Closer look has revealed that tooth-shaped Al-depleted zones propagated along the lamella boundaries (see high resolution insert in figure 3). The presence of Gd-rich particles enhanced the formation of Al_2O_3 in surrounding areas and increased the depth of the oxide scale. Rapid diffusion of oxygen anions along colony boundaries, γ-lamellas, and Gd-rich particles can therefore increase the effective oxidation rate. It should be suggested that the noticeable positive effect of the nitrogen implantation on suppressing the oxide scale growth can be probably associated with enhancing the formation of the protective Al_2O_3 layer and, as a result, slowing the oxygen propagation. However, further studies are required to establish the oxidation mechanisms in detail.

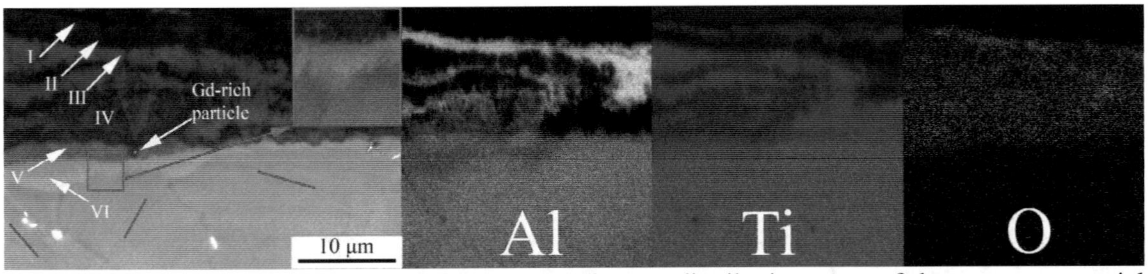

Figure 3. Cross-section SEM-BSE images and element distribution maps of the program material after nitrogen ion implantation after 2 cycles of oxidation tests. I – outward-growing TiO_2 layer; II – Al_2O_3 layer; III – inner TiO_2 layer; IV – a mixture of Al_2O_3 + TiO_2; V – internal oxidation zone; VI – Al-depleted zone; red lines indicate lamellae direction.

4. Conclusions

The dependence of the oxide scale formation on the microstructure of the base layer was studied using the example of a β-solidifying γ-TiAl alloy (Ti-43.2Al-1.9V-1.1Nb-1.0Zr-0.2Gd-0.2B) after nitrogen ion implantation. The nitrogen ion implantation improved the oxidation resistance of the alloy considerably, as was evidenced by the dependence of oxide scale thickness on number of oxidation cycles. Thick Al-depleted zones and a mixture of Al_2O_3 and TiO_2 were formed along the colony boundaries. Tooth-shaped Al-depleted zones propagated along the lamella boundaries. In addition, the presence of Gd-rich particles in the oxide scale promoted the formation of Al_2O_3 particles in surrounding areas and increased the depth of the oxide scale around.

Acknowledgment

The work was supported by the Russian Science Foundation under Grant № 19-79-30066. The authors are also grateful to Joint Research Center "Materials and Technologies", Belgorod State University, for the assistance with instrumental analysis.

References

[1] Leyens C, Peters M 2003 *Titanium and Titanium Alloys: Fundamentals and Applications* (Weinheim: Wiley-VCH)
[2] Fergus J W 2002 *Mater. Sci. and Eng. A* **338** 108-125
[3] Tetsui T, Ono S 1999 *Intermetallics* **7** 689–697
[4] Rahmel A, Quadakkers W J, Schutze M 1995 *Mater. Corros.* **46** 271–285
[5] Zheng N, Quadakkers W J, Gil A and Nicke H 1995 *Oxidation of Metals* **44** 477-499
[6] Taniguchi S, Hongawara N, Shibata T 2001 *Materials Science and Engineering A* **307** 107–112
[7] Rakowski J M, Meier G H and Pettit F S 1996 *Scripte Materialia* **Vol. 35** 1417-1422
[8] Perez P, Jimenez J A, Frommeyer G and Adeva P 2000 *Oxidation of Metals* **Vol. 53** 138-147
[9] Zhao L L, Li G Y , Zhang L Q, Lin J P, Song X P, Ye F, Chen G L 2010 *Intermetallics* **18** 1586–1596
[10] Cueff R, Buscail H, Caudron E, Issartel C, Riffard F 2003 *Corros Sci* **45** 1815-1831
[11] Taniguchi S, Uesaki K, Zhu Y C, Fujita K, Iwamoto N, Matsunaga Y, Nakagawa K 1999 *Mater. Sci. Eng. A* **266** 267-275
[12] Hui S, Zhang Z, Xiao J, Yuan G 1999 *Rare Met.* **18 (3)** 162-167
[13] Donchev A, Schütze M 2008 *Mater Corros* **59** 489-493
[14] Sokolovsky V S, Stepanov N D, Zherebtsov S V, Nochovnaya N A, Panin P V, Zhilyakova M A, Popov A A, Salishchev G A 2018 *Intermetallics* **94** 138–151
[15] Dai J, Zhu J, Chen C, Weng F 2016 *J. Alloys Compd* **685** 784 798
[16] Becker S, Rahmel A, Schorr M, Schütze M 1992 *Oxid Met* **38** 425–464

IOP Conf. Series: Materials Science and Engineering 1014 (2021) 012055 doi:10.1088/1757-899X/1014/1/012055

Creep behavior and microstructure of a Ta-added 9%Cr steel with high B and low N contents

E Tkachev[*], **A Belyakov**

Belgorod State National Research University, Pobeda, 85, Belgorod, 308015, Russia

[*]Corresponding author: tkachev_e@bsu.edu.ru

Abstract. The creep behavior and microstructure of an advanced a Ta-added 9%Cr steel were investigated. The steel is characterized by increased strength in the long-term creep regime, which can be associated with fine dispersion of (Nb,Ta)(C,N) particles. Strengthening of the studied steel due to the fine MX particles was estimated using the Orowan particle strengthening model.

1. Introduction

High-chromium martensitic steels are the primary material for the manufacture of high-temperature elements of modern coal-fired power units. The increasing demands to the energy efficiency of thermal power plants require increasing the temperature and pressure of the steam entering the steam turbine, thus such elements as superheater and reheater tubes have to operate at more stressed conditions.

It is known that the creep strength of martensitic steels can be significantly increased by reducing the initial size and coarsening rate of the second phase particles [1-3]. The $M_{23}C_6$ and MX (where M is a metal and X is C and/or N) precipitates are the most important in terms of increasing the stability of the tempered martensite lath structure and long-term creep properties of these steels [1-3]. In recent works, it was shown that an increase in the boron content and additional alloying with tantalum can increase the dispersion of $M_{23}C_6$ and MX particles, and thereby increase the creep strength of the steel [3]. However, so far no studies have been carried out on the creep behavior and the distribution of secondary particles in Ta-alloyed steels with a low N content.

This study examines the microstructure and creep behavior of a 9% Cr martensitic steel with an additional amount of boron and tantalum.

2. Materials and methods

Steel Fe-0.1C-9Cr-1.8W-0.6Mo-3Co-Nb-V-0.0012B-0.007N-0.085Ta was produced by vacuum induction melting and subjected to solution treatment at 1200 °C. Then the steel was austenized at 1050 °C for 0.5 h and tempered at 780 °C for 3 h.

The round-bar creep specimens of 6-mm in diameter and 60-mm in gauge length were subjected to tensile creep tests in air under initial applied stresses of 160, 120, and 110 MPa at 650 °C using ATS2330 lever arm machines.

Crept microstructures and dispersed particles of the studied steel were analyzed by transmission electron microscopy (TEM) using a Jeol JEM-2100 microscope operating at 200kV, equipped with an INCA energy-dispersive X-ray (EDX) spectrometer. The dislocation densities were estimated by counting individual dislocations in the lath interiors per unit area on arbitrarily selected TEM images. Extraction carbon replicas were prepared to determine the size and chemical composition of different types of precipitates. The volume fractions of secondary (Nb,Ta)(C,N) and V(C,N) particles were estimated using ThermoCalc TCFE7 software.

3. Results and discussion

The creep curves at 650 °C (Figure 1) show that the studied steel is characterized by increased strength in the long-term creep regime, while the creep strength at a relatively high stress of 160 MPa is close to that for standard P92 steel [4]. So that the creep test with an applied stress of 110 MPa is still in progress after about 10 000 hours without rupture.

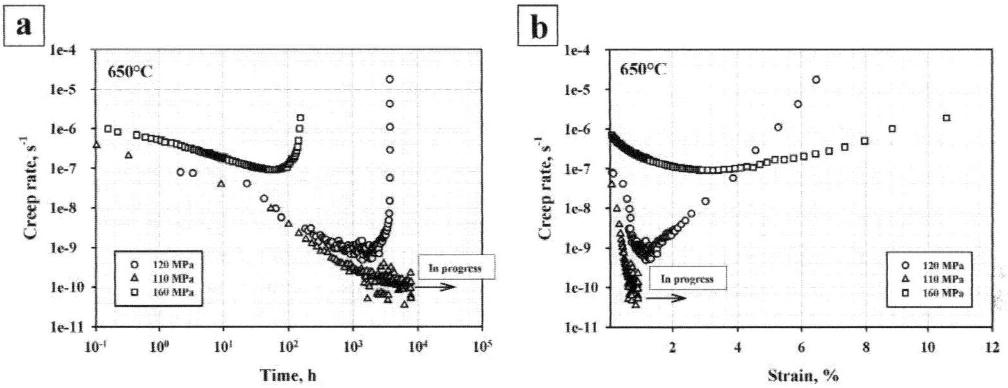

Figure 1. Creep rate versus creep time (a) and creep rate versus strain (b) curves for the studied steel subjected to creep tests at 650 °C.

It is seen that a decrease in the applied stress leads to a significant decrease in the minimum creep rate from 9.0×10^{-8} s^{-1} at 160 MPa to about 1.3×10^{-10} s^{-1} at 110 MPa. Such a rapid decrease in the creep rate with decreasing the applied stress is inherent to the threshold creep behavior and is associated with the presence of dispersed particles.

The microstructure of the studied steel after tempering was described in detail elsewhere [5]. A typical tempered martensite lath structure with particles of $M_{23}C_6$ and MX precipitates was observed. The examination of crept microstructures (Figure 2) reveals precipitation of chains of relatively coarse $(Fe,Cr)_2(W,Mo)$ Laves phase particles.

Figure 2. TEM microstructures of the studied steel after creep under 160 MPa for 158 h (a) and under 120 MPa for 3738 h (b).

The microstructural parameters are summarized in Table 1. The creep for 3738 h at 650°C is accompanied by an increase in the lath/subgrain width by ~80%. It is seen in Figure 1 that small subgrains remain in the grain interiors after creep, but there are also large recovered regions without sub-boundaries that have been formed between the packet/prior austenite grain boundaries. In such recovered regions, the uniformly distributed particles of MX carbonitrides are the main obstacles for the moving dislocations.

Table 1. Parameters of tempered and crept microstructures of the studied steel.

	Tempered at 780°C	160 MPa / 158 h	120 MPa / 3738 h
Lath/subgrain width, nm	355 ± 35	460 ± 20	630 ± 45
Dislocation density, $\times 10^{14}$ m^{-2}	2.4 ± 0.2	0.9 ± 0.2	0.6 ± 0.1
Mean size of $M_{23}C_6$, nm	75 ± 4	78 ± 4	85 ± 5
Mean size of $(Fe,Cr)_2(W,Mo)$, nm	–	184 ± 20	239 ± 20
Mean size of MX, nm	23 ± 3	35 ± 2	39 ± 3

It should be noted that the MX carbonitrides formed in the studied steel are very stable against coarsening, and their mean size after 3,738 h of creep is less than 40 nm (Table 1, Figure 3). Increasing the dispersion of carbonitride particles in high-chromium martensitic steels with a low nitrogen content is particularly important, since it compensates the decreased volume fraction of VN particles.

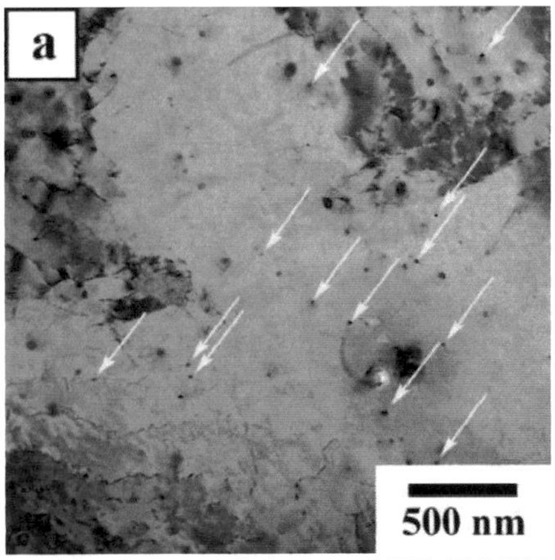

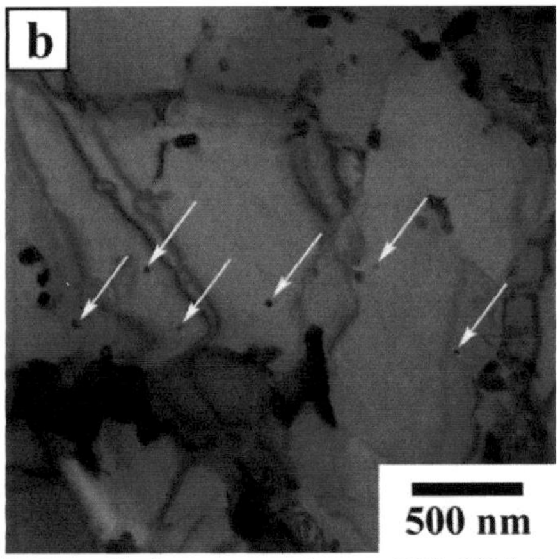

Figure 3. Fine precipitates of (Nb,Ta)(C,N) in the studied steel after tempering at 780 °C (a) and after creep for 3738 h (b).

To determine the strengthening contribution from the MX particles in the as-tempered conditions, the well-known Orowan particle strengthening model was used [4]:

$$\sigma_{Or} = 0{,}3MGb/2(1+\nu)\lambda\,, \quad (1)$$

where λ_i – interparticle spacing, M - Taylor's factor (= 3), G - shear modulus (~59 GPa at 650 °C), ν – Poisson's ratio, b - interatomic distance in slip direction (= 0,25 nm). The following equation correlates λ to the number density N_S:

$$\lambda_{MX} = 0{,}5\, N_S^{-1/2} \quad (2)$$

The number density N_S was calculated as

$$N_S = \frac{3f_v}{2\pi r^2} \quad (3)$$

where f_v is the volume fraction and r - mean radius of MX precipitates. The obtained value of strengthening from the MX carbonitrides particles in the studied steel was compared with that in steel P92 and 9%Cr steel with high B and low N contents (Table 2).

Table 2. Nitrogen content, volume fraction, diameter and spacing of MX carbonitrides in 9%Cr steels, together with Orowan stress estimated form the values of interparticle spacing.

	Nitrogen content, wt.%	f_v of MX	Mean radius of MX, nm	λ_{MX}, nm	σ_{Or}, MPa
P92 steel [4]	0.05	0.00310	20	260	148
9%Cr steel with high B and low N contents [6]	0.007	0.00086	17.5	431	89
Studied Ta-alloyed 9%Cr steel with high B and low N contents	0.007	0.00146	12.5	237	162

It can be seen that the estimated strengthening from MX particles for the studied Ta-alloyed steel is almost twice that in 9%Cr steel with the same nitrogen content. This indicates that addition of Ta to high-chromium martensitic steels with a low N content is a promising approach to improve precipitation hardening due to fine MX carbonitrides particles.

4. Conclusions

The creep behavior and microstructure of advanced Ta-added 9% Cr steel with high B and low N contents have been studied. The main results are as follows:

1. Fine precipitation of (Nb,Ta)(C,N) particles during tempering effectively decreases the minimum creep rate at the long-term creep regime.
2. Microstructural observations of the crept specimens showed additional precipitation of relatively coarse $(Fe,Cr)_2(W,Mo)$ particles while $M_{23}C_6$ and MX particles exhibit a low coarsening rate during creep.
3. The Orowan strengthening model suggests a significant increase in precipitation hardening of Ta-alloyed steel due to the precipitation of fine (Nb,Ta)(C,N) particles.

Acknowledgments
This study was supported by the Russian Science Foundation, under grant No. 19-73-10089.

References

[1] Gao J, Dong Z 2020 *Appl. Phys. A Mater. Sci. Process.* **126**.
[2] Tkachev E, Belyakov A, Kaibyshev R, 2020, *Mater. Sci. Eng. A.* **772** 138821.
[3] Fedoseeva A E, Nikitin I S, Fedoseev A E, Tkachev E S and Kaibyshev R O 2020, *IOP Conf. Ser. Mater. Sci. Eng.*, **848**.
[4] Maruyama K, Sawada K, Koike J I 2001 *ISIJ Int.* **41** 641–653.
[5] Tkachev E, Belyakov A, 2019, *IOP Conf. Ser. Mater. Sci. Eng.*, **525**.
[6] Fedorova I, Kostka A, Tkachev E, Belyakov A, Kaibyshev R 2016 *Mater. Sci. Eng. A.* **662** 443–455.

IOP Conf. Series: Materials Science and Engineering 1014 (2021) 012056 doi:10.1088/1757-899X/1014/1/012056

Creep strength and microstructure of a modified P911-type steel weld joint

E Tkachev*, **A Belyakov**

State National Research University, Pobeda, 85, Belgorod, 308015, Russia

*Corresponding author: tkachev_e@bsu.edu.ru

Abstract. The creep strength and microstructure of the weld joint of the modified P911-type steel has been studied. The creep rupture time of the welded joint at 650° of 1375 h is close to that of the base metal. The heat affected zone -is found to be the weakest area due to the increased size and relatively high coarsening rate of precipitates. The increased boron content in the weld steel effectively stabilizes the $M_{23}(C,B)_6$ particles and is beneficial for the creep strength of the weld joint in the fusion zone.

1. Introduction

Modified 9-12%Cr martensitic steels are promising materials for high temperature components of modern power plants. Gas tungsten arc welding (GTAW) of these steels is widely used to join different components of the power units. However, the creep strength of welded joints is significantly lower as compared to the base metal (BM) due to the microstructural changes caused by heat input during the welding process. Thus, the design of appropriate welding consumables to prevent degradation of the creep properties of high-chromium martensitic steel weldments is important. The main reason for the microstructural degradation during creep is the coarsening of precipitates, leading to a decrease in pinning forces, lath widening and the formation of a coarse subgrain structure [1]. It has been observed that an increased boron content in 9-12%Cr steels is beneficial for improving the dispersion and stability of $M_{23}(C,B)_6$ precipitates [2-4]. However, the effect of the addition of boron to the filler material of the 9%Cr steel welds on the overall creep strength and microstructure in different weld zones is not well established.

In this work, we investigate the creep strength of a weld joint of grade P911 steel obtained with a modified 9%Cr steel consumable with high B and low N contents.

2. Materials and methods

Plates of 9Cr-1Mo-1W (P911) steel with a double-V butt joint configuration were preheated to ~250°C and welded by the gas tungsten arc welding (GTAW) method using a 9% Cr steel with high boron and low nitrogen contents. The chemical compositions of the base material and the welding wire are shown in Table 1.

Table 1. Chemical compositions of base and filler materials (wt.%).

	Fe	C	Si	Mn	Cr	Co	Mo	W	V	Nb	B	N
Base metal	bal.	0.11	0.06	0.36	9.09	-	1.02	1.11	0.22	0.09	0.005	0.050
Filler metal	bal.	0.10	0.12	0.41	9.05	2.86	0.58	1.8	0.20	0.05	0.012	0.007

Then the welded joint was cooled to ~100 °C and subjected to post-weld heat treatment (PWHT), which was carried out at 760 °C for 2 h.

Flat creep specimens with a length of 25 mm and a cross section of 7×3 mm^2 were subjected to tensile creep tests in air under initial applied stresses of 140, 120, and 100 MPa at 650 °C using ATS2330 lever arm machines.

The microstructures and dispersed particles in different zones of the weld joint were investigated by transmission electron microscopy (TEM) using a Jeol JEM-2100 microscope operating at 200kV, equipped with an INCA energy-dispersive X-ray (EDX) spectrometer. The dislocation densities were estimated by counting individual dislocations in the lath interiors per unit area on arbitrarily selected TEM images. Extraction carbon replicas were prepared to determine the size and chemical composition of precipitates. The volume fractions of precipitates were determined using the Thermo-Calc software with the TCFE7 database for Fe-based alloys.

3. Results and discussion

Figure 1 shows the stress vs. rupture time relation for the studied weld joint at 650 °C in comparison with that for the base metal of grade P911.The creep fracture of the specimens occurred in the softened heat affected zone (HAZ) near the fusion zone (FZ). The estimated creep rupture strength of the weld joint at 100,000 hours and T=650 °C is lower by 10 MPa as compared to that of P911 steel.

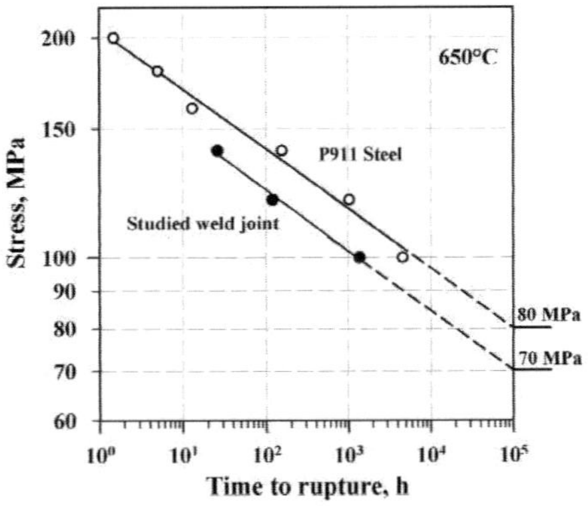

Figure 1. Creep rupture properties of the studied weld joint and aP911 steel.

All studied zones of the weld joint after PWHT had a tempered martensite lath structure with a high density of $M_{23}C_6$ and MX precipitates [5]. The crept microstructures of the studied weld joint in HAZ and FZ are shown in Figure 2 together with base metal in the grip portion, aged for 1375 h.

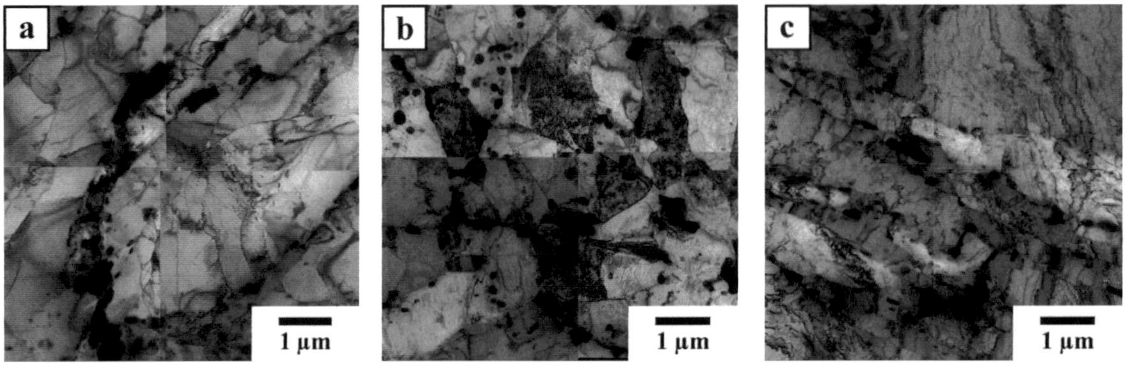

Figure 2. Microstructure of the studied weld joint after creep at 100 MPa in FZ (a); in HAZ (b) and base metal in the grip portion of the sample aged for 1375 h (c) at 650 °C.

The microstructure parameters and the mean size of precipitates in different portions of the crept specimens are summarized in Table 2. It is seen that the lath structure of the tempered martensite evolves into a subgrain structure during creep. The formation of the intermetallic Laves phase particles was observed in the base and weld metals, but their size in HAZ is almost twice that found in the fusion zone. An increase in the creep duration leads to a significant drop in the dislocation density in the FZ from 2.15×10^{14} m^{-2} after creep for 26 h to 0.45×10^{14} m^{-2} after creep for 1375 h.

Table 2. The microstructure parameters in the different portions of crept specimens of the studied weld joint.

	140 MPa, 26 h			100 MPa, 1375 h		
	Base metal in grip portion	Heat affected zone	Fusion zone	Base metal in grip portion	Heat affected zone	Fusion zone
Lath width, μm	0.61±0.02	0.69±0.03	0.65±0.01	0.69±0.02	0.94±0.10	0.90±0.04
Dislocation density, $\times10^{14}$ m^{-2}	1.05±0.20	0.80±0.20	2.15±0.25	0.80±0.20	0.45±0.10	0.45±0.15
Average size of $M_{23}C_6$, nm	140±9	162±14	118±7	171±13	190±14	130±8
Average size of MX, nm	55±4	60±7	43±8	59±6	67±8	46±5
Average size of $Fe_2(W,Mo)$, nm	207±9	242±11	127±5	380±14	394±18	234±9

It is suggested that the increased creep strength of the weld metal in the FZ is due to the increased stability of the $M_{23}C_6$ particles and the Laves phase particles. These particles were found to effectively prevent migration of (sub)grain boundaries [2,3]. The determined eequilibrium volume fractions of different precipitates are listed in Table 3.

Table 3. Equilibrium volume fractions of precipitates in the base and filler steels calculated by Thermo-Calc.

	Base metal	Filler metal
Volume fraction of $M_{23}(C,B)_6$, %	2.69	2.16
Volume fraction of (Nb,V)(C,N), %	0.39	0.09

Volume fraction of $Fe_2(W,Mo)$, %	0.57	0.96

The overall equilibrium volume fraction of precipitates at 650 °C in the base and the weld steel is nearly the same. The volume fraction of $M_{23}(C,B)_6$ in the base steel is somewhat higher than that in the weld metal. The relatively low volume fraction of (Nb,V)(C,N) in the weld metal is attributed to the low N content in the consumable steel filler wire. The well-known Zener equation was used to estimate the pinning pressure from precipitates at the (sub)grain boundaries in HAZ and FZ of the weld joint [6]:

$$P_Z = \frac{3\gamma F_V}{d} \tag{1}$$

where γ is the boundary energy per unit area, Fv is the volume fraction of precipitates and d is the mean diameter of particles. Figure 3 displays the changes of the pinning pressures exerted by precipitates in HAZ and FZ during creep exposure. After PWHT, the steel in HAZ is characterized by increased stability as compared to the weld metal due to pinning from (Nb,V)(C,N) particles. However, the precipitation of $Fe_2(W,Mo)$ particles in the weld metal during creep significantly increases the pinning pressures in the FZ. Since $Fe_2(W,Mo)$ and $M_{23}(C,B)_6$ particles in FZ are characterized by relatively low coarsening rate, they effectively stabilize the (sub)grain boundaries during creep exposure.

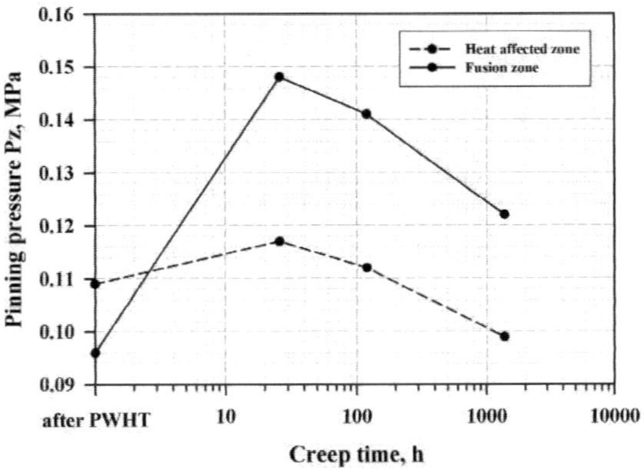

Figure 3. The evolution of Zener pinning pressures exerted by precipitates in HAZ and FZ of the studied weld joint during creep exposure.

4. Conclusions

The creep strength and microstructure of the modified P911-type steel weld joint has been studied. The main results can be summarized as follows:

1. The creep strength of the modified P911-type steel weld joint at T=650 °C is lower by ~10 MPa as compared to that of the base P911 steel due to the softening of the base metal in the HAZ.
2. The decreased strength of the base steel in HAZ is attributed to the increased size and relatively high coarsening rate of the $M_{23}(C,B)_6$ and (Nb,V)(C,N) particles.
3. The formation of relatively small particles of $Fe_2(W,Mo)$ and reduced coarsening rate of $M_{23}(C,B)_6$ particles in the weld metal results in the increased creep strength of the studied weld joint in the FZ.

References

[1] Fedoseeva A, Dudova N, Kaibyshev R, 2017 *J.Mater. Sci.* **52** 2974
[2] Gao J, Dong Z, 2020 *Appl. Phys. A Mater. Sci. Process.* **126**.
[3] Tkachev E, Belyakov A, Kaibyshev R, 2020 *Mater. Sci. Eng. A.* **772** 138821.
[4] Abstoss K G, Schmigalla S, Schultze S, Mayr P, 2019 *Mater. Sci. Eng. A.* **743** 233–242.
[5] Tkachev E, Belyakov A, 2018 *AIP Conf. Proc.* **2051** 020307.
[6] Manohar P A, Ferry M, Chandra T, 1998 *ISIJ Int.* **38** 913–924.

IOP Conf. Series: Materials Science and Engineering 1014 (2021) 012057 doi:10.1088/1757-899X/1014/1/012057

Synthesis of the high-entropy $(BiSbSe_{1.5}Te_{1.5})_{1-x}Cu_x$ compound promising for thermoelectric applications

A E Vasil'ev[1], M N Yaprintsev[1], O N Ivanov[1,2*], M V Zhezhu[1], E N Yapryntseva[1]

[1]Belgorod State University, Belgorod, 308015, Russian Federation
[2]Belgorod State Technological University named after V.G. Shukhov, Belgorod, 308012, Russian Federation

*Corresponding author: ivanov.oleg@bsu.edu.ru

Abstract. Single phase and nanosized powders of high-entropy $(BiSbSe_{1.5}Te_{1.5})_{1-x}Cu_x$ compound with x=0; 0.3, 0.6 were synthesized by using soft chemistry approaches. Owing to its low thermal conductivity, this compound is promising for thermoelectric applications. The synthesized powders were used to prepare bulk samples totally suitable for testing thermoelectric properties.

1. Introduction

It is known that large-scale commercial application of thermoelectric materials has been remarkably limited by their low energy conversion efficiency [1]. The efficiency is expressed by the dimensionless figure of merit, $ZT=S^2T/\rho k$, where S, T, ρ and k are the Seebeck coefficient, absolute temperature, specific electrical resistivity and total thermal conductivity, respectively. To enhance the thermoelectric conversion efficiency, many efforts based on various modern scientific and technological approaches, focusing on optimal S, ρ and k combination, have been made in recent decades [2-4]. The concept of high-entropy alloy is a new strategy to improve the thermoelectric efficiency of thermoelectric materials [5, 6]. High-entropy alloys are usually considered as solid solutions in which more than five principal elements each in a 5-35% molar ratio compete for the same crystallographic site, yielding a high entropy of mixing, that results in unusual and exciting properties. As for developing of thermoelectricity, the lattice thermal conductivity in high-entropy alloys can be remarkably reduced via severe lattice distortion effects, resulting in the enhancement of ZT. An important technological challenge in the development of high-entropy alloys, is to prepare single phase compound with a desired structure and elemental composition.

The aim of this work is to find a synthesis method based on soft chemistry approaches, which is suitable to prepare the high-entropy $(BiSbSe_{1.5}Te_{1.5})_{1-x}Cu_x$ compound. This compound is a new high-entropy one, which seems to be promising for thermoelectric applications. Besides, the compound can be considered as a derivate from the Bi_2Te_3 compound, which is a typical low-temperature thermoelectric. It should be also underlined that, in contrast to common methods used to prepare high-entropy alloys, soft chemistry synthesis is implemented at rather low temperatures.

2. Materials and methods

A soft chemistry method was developed to synthesize starting $(BiSbSe_{1.5}Te_{1.5})_{1-x}Cu_x$ powders with x=0; 0.3, 0.6. It is known that reducing the lattice thermal conductivity in high-entropy alloys is often accompanied by a degradation of the carrier mobility [5, 6]. To reduce the degradation, Cu doping was used. All the synthesis manipulations were carried out in a conic flask with reflux and constant stirring. In a typical procedure (for 2 g powder), high pure ($Bi(NO_3)_3*5H_2O$, SeO_2, TeO_2, $Cu(CH_3COO)_2*H_2O$) precursors were dissolved in 400 ml ethylene glycol in stoichiometric proportions, with 10 g of NaOH addition for controlling *pH*-value and heated to 373 K for 1 h . The obtained solution was cooled to room temperature. 10 ml of hydrazine hydrate was slowly added with stirring into the solution and heated to 433 K. The resulting dark solution was maintained at this temperature for 3 h. After the reaction process, black powder was precipitated. The powder was purified by centrifugation and washing for 3 times with ethanol and acetone. Then the powder was dried in an argon atmosphere at 523 K for 2 hours.

The powder was further used to compact bulk samples by cold pressing in a steel mold at 100 MPa. The compacted samples were finally annealed at 673 K for 2 h, as a result, samples suitable for the examination of the thermoelectric properties were obtained. The density of the bulk materials, measured by the Archimed method, was estimated to be ~90%.

The morphology and size of the powder particles were identified by transmission electron microscopy (Jeol JEM 2100 microscope).

The crystal structure was determined by the X-Ray diffraction method (Rigaku Ultima IV diffractometer).

The elemental composition and microstructure of the samples were examined by scanning electron microscopy (FEI Quanta 600 FEG microscope) with an energy dispersive spectroscopy (EDS) detector.

3. Results and discussion

XRD patterns for the starting $(BiSbSe_{1.5}Te_{1.5})_{1-x}Cu_x$ powders with x=0; 0.3, 0.6 are shown in Figure 1.

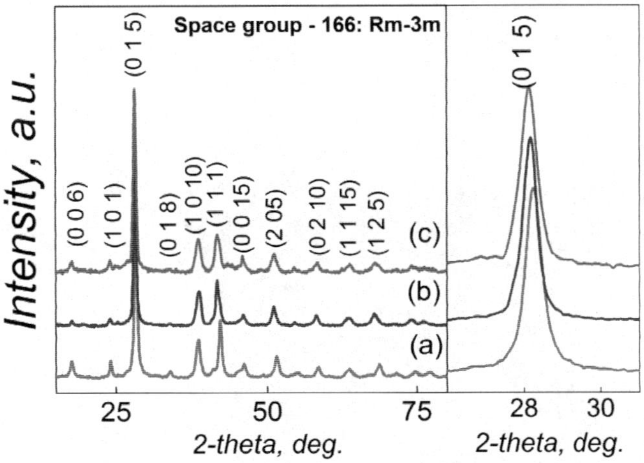

Figure 1. The XRD patterns of the $BiSbSe_{1.5}Te_{1.5}$ (a), $(BiSbSe_{1.5}Te_{1.5})_{0.7}Cu_{0.3}$ (b), $(BiSbSe_{1.5}Te_{1.5})_{0.4}Cu_{0.6}$ (c) powders.

According the XRD data, all the powders are single phase with tetragonal symmetry corresponding to the space *R*3*m* group. The lattice parameters were found to be slightly increased with increasing *x*, as shown, for instance, for the (015) peak. The incorporation of Cu atoms in the crystal structure can lead to this lattice expansion. As result, the diffraction peaks are shifted to low angles. No significant structure changes of the powders and bulk samples were observed while drying and annealing.

The use of $N_2H_4{\cdot}H_2O$ as a common reducing agent results in simultaneously reducing the temperature and duration synthesis. The decomposition of hydrazine hydrate during synthesis to non-reactive gases (NH_3, N_2) and water does not pollute the resulted nanopowder. Ethylene glycol has the coordinating properties inducing surface functionalization and colloidal stabilization of nanosized particles. This method of synthesis summarizes favorable properties of hydrazine hydrate and ethylene glycol, that, in turn, allows preparing single phase, highly-crystalline and nanosized ($(BiSbSe_{1.5}Te_{1.5})_{1-x}Cu_x$) powders for a short time and at low temperatures. The content of Bi, Sb, Se, Te, Cu in the powders with various x was determined by the EDS method. The element composition of all the powders was found to be corresponding to their nominal composition. The results of TEM-examination of the powders with various x demonstrated similar pattern for all the compositions. A typical morphology of the $(BiSbSe_{1.5}Te_{1.5})_{0.7}Cu_{0.3}$ powder is presented in Figure 2 (a). The powder consisted of irregularly shaped agglomerates of nanoplates.

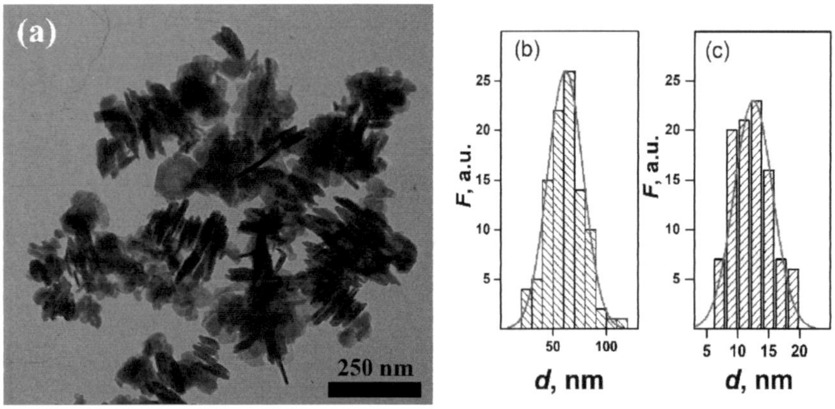

Figure 2. The TEM-image of the $(BiSbSe_{1.5}Te_{1.5})_{0.7}Cu_{0.3}$ powder (a) and histograms of the nanoparticles diameter size (b) and thickness (c) distributions.

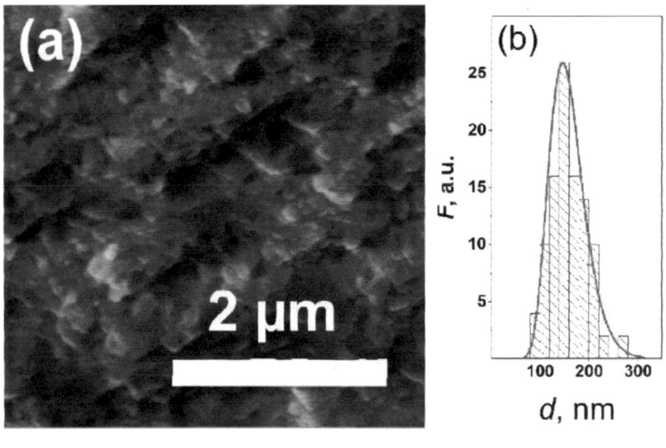

Figure 3. Typical micrmisctructure of the fractured surface for the bulk $(BiSbSe_{1.5}Te_{1.5})_{0.7}Cu_{0.3}$ sample (a) and the histogram of grain size distribution (b).

It was found that the probability density function, $P(d)$ for the histograms of the nanoparticles size distribution in Figure 2 (b) and (c) can be described by an unimodal lognormal distribution

$$P(d) = \frac{1}{\sqrt{2\pi}\sigma d} exp\left(\frac{-(lnd - lnd')^2}{2w^2}\right) \quad (1)$$

where *d* is the current nanoparticles size, *d'* is the average nanoparticles size, and *w* is the standard deviation of the logarithms of the particle sizes.

The analysis based on expression (1) of the experimental histograms of the nanoparticles size distribution in the powders material allowed estimating the average diameter size as ~ 60 nm and thickness size as ~ 12 nm.

It is known that nanocrystals of compounds based on Bi_2Te_3 usually form two-dimensional plates because of their crystal structure [7]. Bulk Bi_2Te_3 crystal has a layered structure that is composed of stacked two-dimensional layers of Te(1)–Bi–Te(2)–Bi–Te(1) with a Te(1)–Te(1) weak van der Waals interaction, leading to the ease of cleavage along the planes perpendicular to the *c*-axis.

The SEM-image taken from the fractured surface of the bulk $(BiSbSe_{1.5}Te_{1.5})_{0.7}Cu_{0.3}$ sample is shown in Figure 3 (a). All the bulk samples have a disordered grain structure with an average grain size of ~ 150 nm. The size was extracted from the analysis of histograms of the grain size distribution (figure 3 (b)). One can see that the the growth of grains occurred during the fabrication of material consisting of the starting nanosized powder. This phenomenon destroying the initial nanostructural state is characteristic of high-temperature sintering of nanomaterials [8].

4. Conclusion

Thus, a simple method based on soft chemistry approaches was developed to prepare the $(BiSbSe_{1.5}Te_{1.5})_{1-x}Cu_x$ powders with x=0; 0.3, 0.6. The method allowed synthesizing single phase and nanosized starting powders. Then, bulk materials, suitable for examination of the thermoelectric properties, were prepared.

Acknowledgments

This work was also financially supported by the Ministry of Education and Science of the Russian Federation under project No FZWN-2020-0015.

References

[1] Nolas G S, Sharp J, and Goldsmid H J, Thermoelectrics Basic Principles and New Materials Developments, Springer, Berlin, 2001.
[2] Chen Z G, Han G, Yang L, Cheng L, and Zou J, "Nanostructured thermoelectric materials: current research and future challenge", Progress in Natural Science: Materials International, Vol. 22, pp. 535-549, 2012.
[3] Yaprintsev M, Lyubushkin R, Soklakova O and Ivanov O, "Effects of Lu and Tm doping on thermoelectric properties of Bi_2Te_3", Journal of Electronic Materials, Vol. 47, pp. 1362-1370 2018.
[4] Hu L, Zhu T, Liu X, and Zhao X, "Point defect engineering of high-performance bismuth-telluride-based thermoelectric materials", Advanced Functional Materials, Vol. 24, pp. 5211-5218, 2014.
[5] Fan Z, Wang H, Wu Y, Liu X J, Lu Z P, "Thermoelectric high-entropy alloys with low lattice thermal conductivity," RSC Advances, Vol. 6, pp. 52164-52170, 2016
[6] Fan Z, Wang H, Wu Y, Liu X J, Lu Z P "Thermoelectric perfomance of PbSnTeSe high-entropy alloys," Materials Research Letters, Vol. 5 No. 3, pp. 187-194, 2017.
[7] Goldsmid H J, "Bismuth telluride and its alloys as materials for thermoelectric generation", Materials, Vol. 7, pp. 2577-2592, 2014.
[8] Scheele M, Oeschler N, Meier K, Kornowski A, Klinke C, Weller H, "Synthesis and thermoelectric characterization of Bi_2Te_3 nanoparticles", Advanced Functional Materials, Vol. 19, pp. 3476-3483, 2009.

Structure and mechanical properties of near-eutectic refractory Al-Cr-Nb-Ti-Zr high entropy alloys

N Yurchenko*, E Panina, G Salishchev and N Stepanov

Laboratory of Bulk Nanostructured Materials, Belgorod National Research University, Belgorod 308015, Russia

*Corresponding author: yurchenko_nikita@bsu.edu.ru

Abstract. In this study, the structure and mechanical properties of the near-eutectic refractory $Al_{24}Cr_{11}Nb_{11}Ti_{37}Zr_{17}$, $Al_{20}Cr_{20}Nb_{25}Ti_{25}Zr_{10}$, and $Al_{28}Cr_{15}Nb_{15}Ti_{20}Zr_{22}$ (at.%) high entropy alloys were investigated. Both the $Al_{24}Cr_{11}Nb_{11}Ti_{37}Zr_{17}$ and $Al_{20}Cr_{20}Nb_{25}Ti_{25}Zr_{10}$ alloys had a hypoeutectic structure, while the $Al_{28}Cr_{15}Nb_{15}Ti_{20}Zr_{22}$ alloy had a hypereutectic one. The proeutectic phase in the $Al_{24}Cr_{11}Nb_{11}Ti_{37}Zr_{17}$ and $Al_{20}Cr_{20}Nb_{25}Ti_{25}Zr_{10}$ alloys was the ordered B2 phase and in the $Al_{28}Cr_{15}Nb_{15}Ti_{20}Zr_{22}$ alloy – the C14 Laves phase. In all three alloys, the eutectic regions possessed a fine lamellar structure composed of a mixture of the C14 Laves and B2 phases. The densities of the alloys were 5.20-5.89 g/cm^3. The $Al_{24}Cr_{11}Nb_{11}Ti_{37}Zr_{17}$ and $Al_{28}Cr_{15}Nb_{15}Ti_{20}Zr_{22}$ alloys were extremely brittle at 22°C, while the $Al_{20}Cr_{20}Nb_{25}Ti_{25}Zr_{10}$ alloy showed some ductility ($\varepsilon = 0.7\%$) and high yield strength of 1830 MPa. At 800°C, all alloys became ductile. The highest yield strength (1010 MPa) was demonstrated by the $Al_{20}Cr_{20}Nb_{25}Ti_{25}Zr_{10}$ alloy and the lowest one (235 MPa) – by the $Al_{24}Cr_{11}Nb_{11}Ti_{37}Zr_{17}$ alloy. The comparison between the experimental as-solidified structures and the CALPHAD predictions was made.

1. Introduction

Refractory high entropy alloys (RHEAs) demonstrate exceptional resistance to softening at temperatures up to 1600°C that makes them attractive candidates for high-temperature applications [1]. RHEAs are usually based on single-phase bcc or ordered B2 solid solution and can also contain intermetallic secondary phases. However, the selection of a proper type of structure for the best mechanical properties is a complex task due to the variety of alloys' compositions.

Recent studies showed an intriguing option for the development of the RHEAs with a ductile bcc matrix and embedded coherent B2 nanoparticles that resembled the γ/γ' structure of commercial nickel-based superalloys [2]. Such a microstructure design enhances room-temperature plasticity but does not prevent the rapid strength loss at $T > 600°C$ due to undesired phase transformations [3].

Higher performance at elevated temperatures can be achieved in RHEAs with a composite-like structure [4]. One of the notable examples of RHEAs composites is the $Al_{20}Cr_{10}Nb_{15}Ti_{20}V_{25}Zr_{10}$ alloy [5]. Owing to a hypoeutectic structure comprised of relatively soft B2 and hard C14 Laves phases, and low density (5.55 g/cm^3), the alloy was found to be one of the strongest RHEAs at 800°C. Unfortunately, the presence of V, which is harmful to oxidation resistance [6], can limit potential structural applications of this RHEA. In the current study, we reported several V-free variants of the $Al_{20}Cr_{10}Nb_{15}Ti_{20}V_{25}Zr_{10}$ alloy and investigated their structure and mechanical properties.

2. Materials and methods

The program alloys with nominal compositions of $Al_{24}Cr_{11}Nb_{11}Ti_{37}Zr_{17}$, $Al_{20}Cr_{20}Nb_{25}Ti_{25}Zr_{10}$, and $Al_{28}Cr_{15}Nb_{15}Ti_{20}Zr_{22}$ (at.%) were produced by vacuum arc melting. The phase composition and microstructure of the alloys were studied for the as-cast state using X-ray diffraction (XRD) and scanning electron microscopy (SEM). The densities of the $Al_{24}Cr_{11}Nb_{11}Ti_{37}Zr_{17}$, $Al_{20}Cr_{20}Nb_{25}Ti_{25}Zr_{10}$, and $Al_{28}Cr_{15}Nb_{15}Ti_{20}Zr_{22}$ alloys, determined by the hydrostatic weighing method, were 5.20 ± 0.02, 5.89 ± 0.03, and 5.50 ± 0.03 g/cm^3, respectively. Isothermal compression of rectangular samples measured 6×4×4 mm^3 was carried out at 22 °C or 800 °C. The initial strain rate was 10^{-4} s^{-1}. Sheil solidification diagrams were constructed using a Thermo-Calc (version 2020a) software and a TCHEA3 (high-entropy alloys) database.

3. Results and discussion

According to XRD analysis (Figure 1a), all program alloys had a dual-phase structure composed of B2 and C14 (hcp) Laves phases. The EBSD phase maps had shown that the $Al_{24}Cr_{11}Nb_{11}Ti_{37}Zr_{17}$ alloy had a hypoeutectic microstructure consisted of a proeutectic B2 (red color) phase separated by eutectic regions composed of the lamellar C14 Laves (green color) and B2 phases (Figure 1b). The $Al_{20}Cr_{20}Nb_{25}Ti_{25}Zr_{10}$ alloy also possessed the hypoeutectic microstructure (Figure 1c), similar to the $Al_{24}Cr_{11}Nb_{11}Ti_{37}Zr_{17}$ alloy. The $Al_{28}Cr_{15}Nb_{15}Ti_{20}Zr_{22}$ alloy showed the hypereutectic microstructure comprised of the proeutectic C14 Laves phase surrounded by eutectic (C14 Laves+B2) domains. According to EDS analysis (Table 1), the B2 phase, in all program alloys, was enriched with Ti and Nb, while the C14 Laves phase was enriched with Cr, Zr, and (slightly) Al.

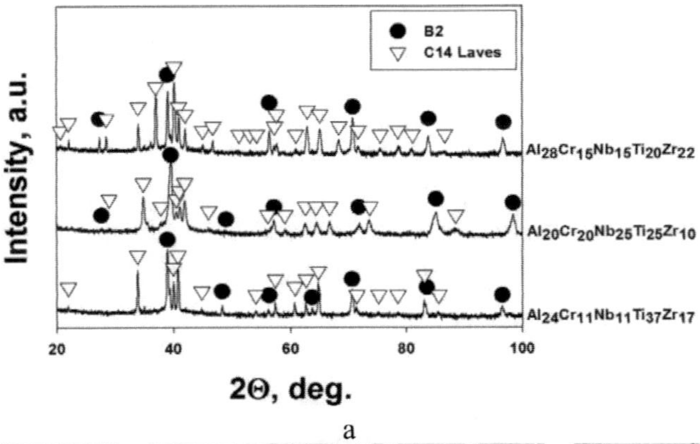

a

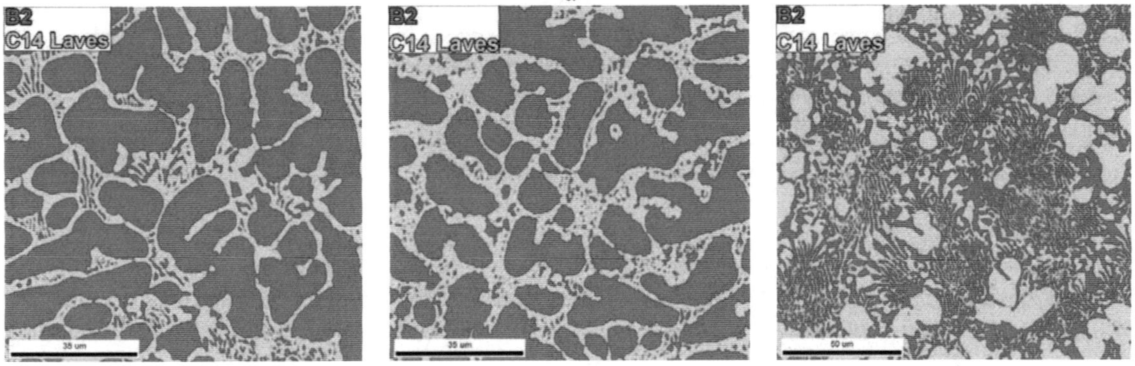

b c d

Figure 1. Structure of the $Al_{24}Cr_{11}Nb_{11}Ti_{37}Zr_{17}$ (a, b), $Al_{20}Cr_{20}Nb_{25}Ti_{25}Zr_{10}$ (a, c), and $Al_{28}Cr_{15}Nb_{15}Ti_{20}Zr_{22}$ (a, d) alloys; XRD pattern – a; EBSD phase maps – b-d.

Table 1. Chemical compositions of structural constituents of the $Al_{24}Cr_{11}Nb_{11}Ti_{37}Zr_{17}$, $Al_{20}Cr_{20}Nb_{25}Ti_{25}Zr_{10}$, and $Al_{28}Cr_{15}Nb_{15}Ti_{20}Zr_{22}$ alloys.

Element, at.%	Al	Cr	Nb	Ti	Zr	Volume fraction, %
			$Al_{24}Cr_{11}Nb_{11}Ti_{37}Zr_{17}$			
B2 phase	22.2	8.2	14.0	41.4	14.2	69±2
C14 Laves phase	25.2	14.7	9.1	28.7	22.3	31±2
Alloy composition	23.7	10.7	11.9	36.2	17.5	-
			$Al_{20}Cr_{20}Nb_{25}Ti_{25}Zr_{10}$			
B2 phase	19.5	14.9	30.2	29.5	5.9	64±1
C14 Laves phase	22.3	27.8	16.8	17.8	15.3	36±1
Alloy composition	20.8	20.0	25.1	25.2	8.9	-
			$Al_{28}Cr_{15}Nb_{15}Ti_{20}Zr_{22}$			
C14 Laves phase	30.3	20.8	11.5	11.4	26.0	58±2
B2 phase	23.8	6.9	20.8	31.3	17.2	42±2
Alloy composition	27.5	14.6	15.8	19.8	22.3	-

Figure 2 presents the engineering stress-strain curves of the $Al_{24}Cr_{11}Nb_{11}Ti_{37}Zr_{17}$, $Al_{20}Cr_{20}Nb_{25}Ti_{25}Zr_{10}$, and $Al_{28}Cr_{15}Nb_{15}Ti_{20}Zr_{22}$ alloys obtained during compression at 22 and 800 °C. At 22°C, the $Al_{24}Cr_{11}Nb_{11}Ti_{37}Zr_{17}$ and $Al_{28}Cr_{15}Nb_{15}Ti_{20}Zr_{22}$ alloys fractured in the elastic region at 1570 and 795 MPa, respectively (Figure 2a). In turn, the $Al_{20}Cr_{20}Nb_{25}Ti_{25}Zr_{10}$ alloy showed some ductility ($\varepsilon = 0.7\%$) and a high yield strength of 1830 MPa. At 800 °C, the program alloys became ductile (Figure 2b). The $Al_{24}Cr_{11}Nb_{11}Ti_{37}Zr_{17}$ alloy had the lowest yield strength of 285 MPa and, after attaining the yield point, showed a prolonged steady-state flow stage. In the $Al_{28}Cr_{15}Nb_{15}Ti_{20}Zr_{22}$ alloy, after yielding at 685 MPa, a notable strengthening stage followed by softening, and then a steady-state flow stage was observed. The $Al_{20}Cr_{20}Nb_{25}Ti_{25}Zr_{10}$ alloy was the strongest with a yield strength of 1010 MPa. Despite the impressive work hardening capacity resulting in a peak stress value of 1225 MPa, the alloy then gradually softened until reaching ~26-27% of plastic deformation; flow stress stabilized at a level of ~780-800 MPa.

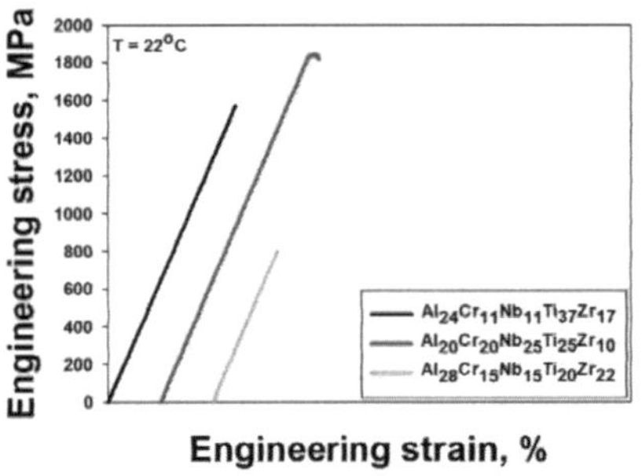

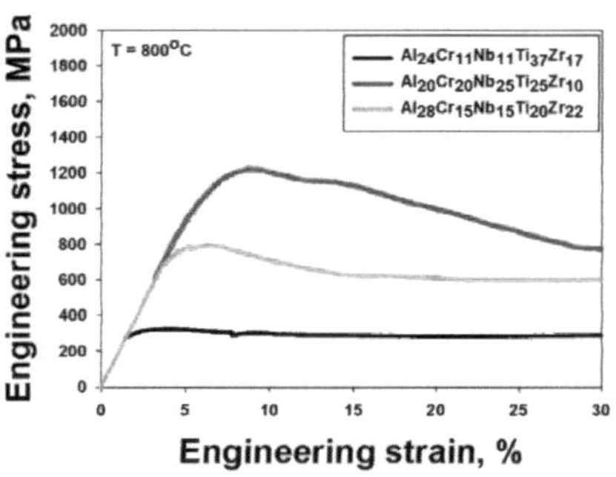

Figure 2. Engineering stress-strain curves of the program alloys obtained during compression at 22 °C (a) and 800 °C (b).

The CALPHAD method was used to analyze the experimental structures of the program alloys (Figure 3). According to the Scheil solidification diagrams, crystallization of all three alloys should start at ~1700 °C, but terminate differently: at ~1450 °C ($Al_{24}Cr_{11}Nb_{11}Ti_{37}Zr_{17}$), ~1500 °C ($Al_{20}Cr_{20}Nb_{25}Ti_{25}Zr_{10}$), and 1350 °C ($Al_{28}Cr_{15}Nb_{15}Ti_{20}Zr_{22}$). The solidification of the $Al_{24}Cr_{11}Nb_{11}Ti_{37}Zr_{17}$ alloy is expected to follow the sequence of L→L+bcc→L+bcc+C14 Laves (Figure 3a). At the same time, in the $Al_{20}Cr_{20}Nb_{25}Ti_{25}Zr_{10}$ and $Al_{28}Cr_{15}Nb_{15}Ti_{20}Zr_{22}$ alloys, the sequence of L→L+C14 Laves→L+bcc+C14 Laves is predicted (Figures 3b, c). Note that the L+C14

Laves→L+bcc+C14 Laves transition in the $Al_{20}Cr_{20}Nb_{25}Ti_{25}Zr_{10}$ alloy should occur in a very narrow temperature interval, thereby suggesting a "eutectic-like" reaction of L→L+bcc+C14 Laves (Figure 3b).

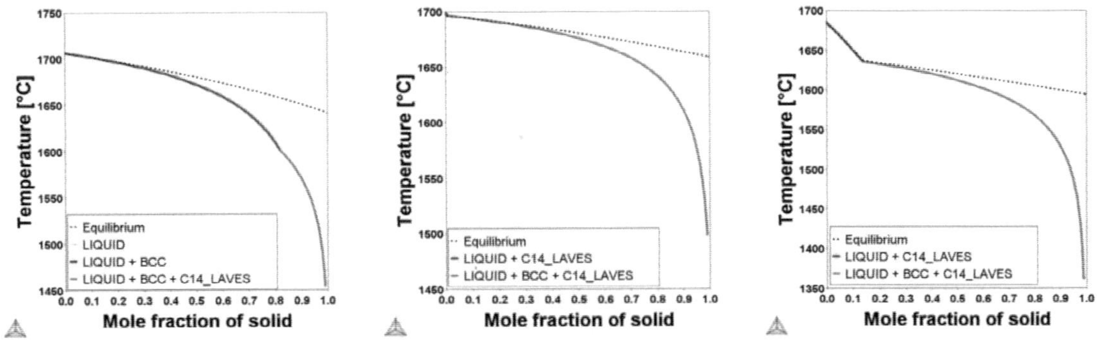

a b c

Figure 3. Scheil solidification diagrams of the $Al_{24}Cr_{11}Nb_{11}Ti_{37}Zr_{17}$ (a), $Al_{20}Cr_{20}Nb_{25}Ti_{25}Zr_{10}$ (b), and $Al_{28}Cr_{15}Nb_{15}Ti_{20}Zr_{22}$ (c) alloys.

To sum up, we introduced three new RHEAs with composite-like near-eutectic structures composed of B2 and C14 Laves phases and a rather low density of 5.20-5.89 g/cm^3. Two alloys, $Al_{24}Cr_{11}Nb_{11}Ti_{37}Zr_{17}$ and $Al_{20}Cr_{20}Nb_{25}Ti_{25}Zr_{10}$, had the hypoeutectic microstructure, meanwhile the $Al_{28}Cr_{15}Nb_{15}Ti_{20}Zr_{22}$ alloy – the hypereutectic one. Among the program alloys, the $Al_{20}Cr_{20}Nb_{25}Ti_{25}Zr_{10}$ showed the best combination of mechanical properties: modest ductility at 22 °C and the highest yield strength at 22 and 800 °C. It was found that CALPHAD correctly predicted the as-solidified microstructures of the $Al_{24}Cr_{11}Nb_{11}Ti_{37}Zr_{17}$ and $Al_{28}Cr_{15}Nb_{15}Ti_{20}Zr_{22}$ alloys, but failed in the case of the $Al_{20}Cr_{20}Nb_{25}Ti_{25}Zr_{10}$ alloy; the B2 ordering of the bcc phase was not anticipated in any of the alloys. Such inaccuracies must be taken into account during the design of novel (near)-eutectic RHEAs with advanced properties based on the Al-Cr-Nb-Ti-Zr alloy system.

Acknowledgments
The authors gratefully acknowledge the financial support from the Russian Science Foundation Grant no. 19-79-30066.

References

[1] Senkov O N, Miracle D B and Chaput K J 2018 Development and exploration of refractory high entropy alloys — A review *J. Mater. Res.*

[2] Soni V, Senkov O N, Gwalani B, Miracle D B and Banerjee R 2018 Microstructural Design for Improving Ductility of An Initially Brittle Refractory High Entropy Alloy *Sci. Rep.* 1–10

[3] Senkov O N, Couzinie J-P, Rao S I, Soni V and Banerjee R 2020 Temperature dependent deformation behavior and strengthening mechanisms in a low density refractory high entropy alloy Al10Nb15Ta5Ti30Zr40 *Materialia* 100627

[4] Liu Y, Zhang Y, Zhang H, Wang N, Chen X, Zhang H and Li Y 2017 Microstructure and mechanical properties of refractory HfMo0.5NbTiV0.5Six high-entropy composites *J. Alloys Compd.* **694** 869–76

[5] Yurchenko N, Panina E, Tikhonovsky M, Salishchev G, Zherebtsov S and Stepanov N 2020 Structure and mechanical properties of an in-situ refractory Al20Cr10Nb15Ti20V25Zr10 high entropy alloy composite *Mater. Lett.* 127372

[6] Butler T M, Chaput K J, Dietrich J R and Senkov O N 2017 High temperature oxidation behaviors of equimolar NbTiZrV and NbTiZrCr refractory complex concentrated alloys (RCCAs) *J. Alloys Compd.* **729** 1004–19

Synthesis and electric transportation properties of high-entropy oxides based on cerium oxide

M N Yapryntsev[1*], I V Sudzhanskaya[1], R A Lyubushkin[2], E N Yapryntseva[1]

[1]Belgorod State University, Belgorod, 308015, Russian Federation
[2]Belgorod State Technological University Named After V.G. Shukhov, Belgorod, 308012, Russian Federation

*Corresponding author: yaprintsev@bsu.edu.ru

Abstract. $CeGdRLaPrO_{2-\delta}$ (where R: La, Y and Dy) rare-earth high-entropy oxides (HEOx) with the Ia-3 (206) structure were obtained. The structural, microstructural and transport properties of the compounds were studied. All samples exhibiting good homogeneity and electrical conductivity typical of materials for the manufacture of solid electrolytes.

1. Introduction

The idea of developing high-entropy oxides was borrowed from the accumulated research experience in the design of high-entropy alloys [1]. In the literature, the first works in this area have been found since 2015. For the first time, an entropy-stabilized (CoCuMgNiZn)O solid solution with the structure Fm-3m was synthesized [2]. Since then, a number of studies have been published on both the properties of these oxides [3] and the search for new high-entropy oxide systems. As a result, the range of known high-entropy oxides has expanded significantly. Despite a very early stage of their development, oxides with high entropy have already shown great promise as functional materials, demonstrating, for example, colossal dielectric constant, high ionic conductivity [4] and low thermal conductivity. Some of the high-entropy oxides have been successfully tested as anode materials for lithium-ion batteries or cathode materials for Na-ion cells [1-4].

In this work, single-phase high-entropy oxides of the composition $(CeGdDySmPr)O_{2-\delta}$, $(CeGdYSmPr)O_{2-\delta}$ and $CeGdLaSmPrO_{2-\delta}$ were obtained for the first time. These materials were synthesized by the sol-gel method followed by cold uniaxial compaction and free sintering at 1500 °C in air. Comprehensive studies of structural, microstructural and electrical transport properties were carried out. The results of studying the electrical transport properties demonstrate a similarity with the properties of classical systems based on cerium oxides ($Ce_{0.8}Pr_{0.2}O_{2-\delta}$).

2. Materials and methods

Cerium(III) nitrate hexahydrate ($Ce(NO_3)_3 \cdot 6H_2O$, 99.9%), gadolinium (III) acetate tetrahydrate ($Gd(CH_3COO)_3 \cdot 4H_2O$, 99.9%), dysprosium (III) acetate tetrahydrate ($Dy(CH_3COO)_3 \cdot 4H_2O$, 99.9%), samarium (III) nitrate hexahydrate ($Sm(NO_3)_3 \cdot 6H_2O$, 99.9%), praseodymium (III) acetate sesquihydrate ($Pr(NO_3)_3 \cdot 6H_2O$, 99.9%), lanthanum (III) nitrate hexahydrate ($La(NO_3)_3 \cdot 6H_2O$, 99.9%), ytrium (III) acetate tetrahydrate ($Y(CH_3COO)_3 \cdot 4H_2O$, 99.9%), 28–30% ammonium hydroxide (NH_4OH) and deionized water were used for the synthesis. Starting materials were weighted in a way to maintain equimolar proportions of the cations and then they were dissolved in 100 ml of deionized water. The pH of the solution was adjusted by adding 28% NH_4OH dropwise until the pH reached ~11. The solution was stirred for 1 h, then the resulting gel was evaporated at a temperature of 120 °C until complete drying. In the second step, the obtained precursors were annealed at a temperature of 900 °C for 2 hours in an air atmosphere. Next, all samples were pressed into pellets of 20 mm diameter, under pressure of 200 MPa, using a uniaxial hydraulic press. The pellets were then free-sintered in a chamber furnace for 4 h at 1500 °C, and subsequently cooled down to room temperature with the furnace.

The morphology and size of the powder particles were identified by transmission electron microscopy (TEM Jeol 2100). The crystal structure was examined by x-ray diffraction (UltimaIV). The element composition and microstructure of specimens was controlled by scanning electron microscopy Quanta 200 3D with EDS (energy dispersive spectroscopy) detector. The specific electrical conductivity (σ) of the AC was measured on a Novocontrol Concept 43 impedance meter using a four-probe method.

3. Results and discussion

Figure 1 (a) presents the XRD diffractograms of the studied HEO ($CeGdDyLaPrO_{2-\delta}$, $CeGdDyLaPrO_{2-\delta}$ and $CeGdDyLaPrO_{2-\delta}$) samples after annealing at a temperature of 900 °C. Already at this stage of obtaining powdery materials, all the samples under study are single-phase, (body-centered cubic) and the peaks corresponding to space symmetry group Ia-3 (206).

The changes in the lattice parameters for the samples are reflected in the corresponding changes in the position of the diffraction peaks, as shown, for example, for the peak (2 2 2) (Figure 1 (a)).. The lattice parameters calculated by the Rietveld refinement for all the samples are presented in Figure 1 (b). The parameters increase with increasing ionic radius ($r_{ion}(R^{+3})$) of the variable element. Significant structural changes of all materials during sintering were not observed, the values of the lattice parameters were constant.

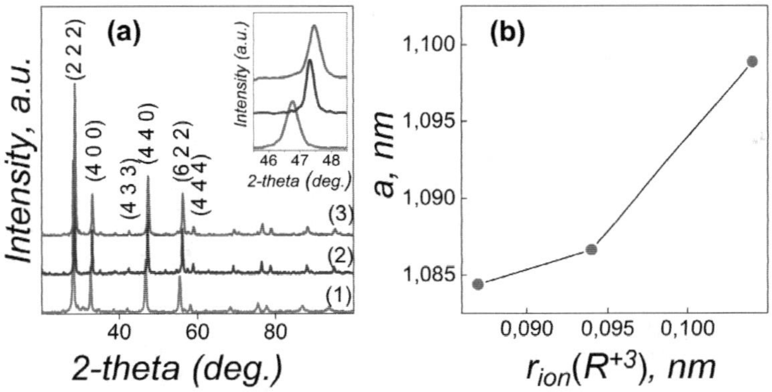

Figure 1. The XRD patterns (a) of $CeGdLaSmPrO_{2-\delta}$ (1), $CeGdDySmPrO_{2-\delta}$ (2), and $CeGdYSmPrO_{2-\delta}$ (3) and ionic radius of the variable element effect on the lattice parameters (b)

Typical morphology of powders annealed at 900 °C are presented at Figure 2. The results of TEM investigation of the synthesized powders demonstrate a similar pattern for all compositions. The typical morphology of $CeGdLaSmPrO_{2-\delta}$ powders are presented in Figure 2. After heat treatment, the nanoparticles of $CeGdLaSmPrO_{2-\delta}$ have an irregular shape. To estimate the average size of nanoparticles in the powder of $CeGdLaSmPrO_{2-\delta}$, histograms of the particles size distribution were plotted using the TEM images in Figure 2 (b). Next, these histograms were analyzed by lognormal unimodal distribution fittings. It is shown that the nanoparticles have an average size of 28 nm, with a standard deviation of ±7 nm.

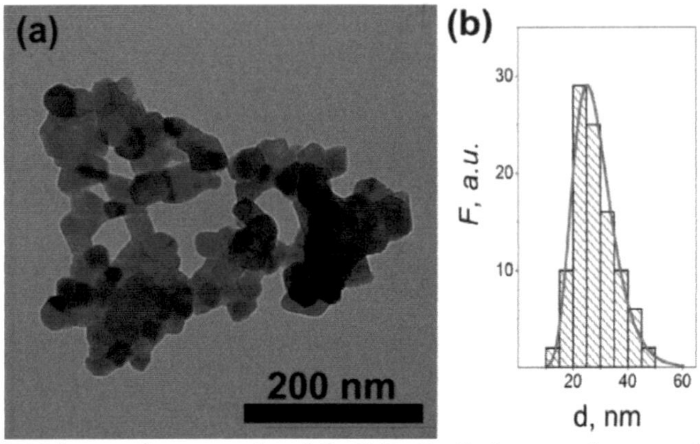

Figure 2. Typical TEM image for example of $CeGdLaSmPrO_{2-\delta}$ starting powders (a) and histogram of the nanoparticles size distribution (b)

The elemental analysis of $CeGdRLaPrO_{2-\delta}$ (where R: La, Y and Dy) was measured by scanning electron microscopy-energy dispersive spectroscopy (SEM-EDS). EDS - analysis shows that the element composition ofthe tested powders corresponded to the nominal composition. This reveals that the nanoparticles are composed of the stoichiometric phase of $CeGdRLaPrO_{2-\delta}$ (where R: La, Y and Dy).

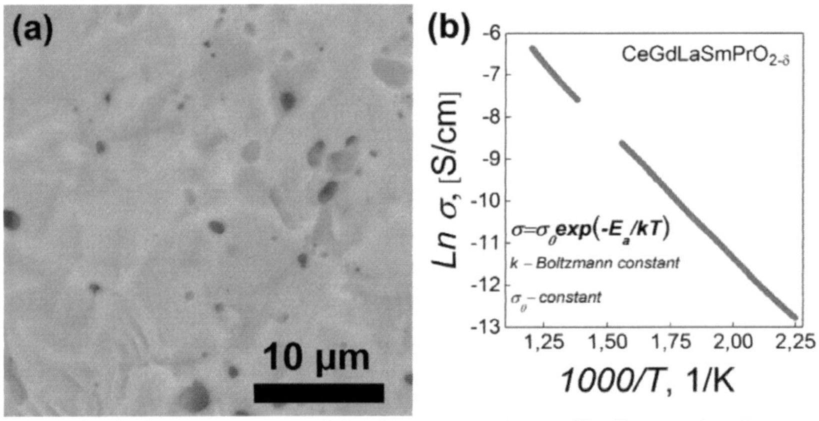

Figure 3. Typical microsctructure of the fracture surface of bulk samples for example of $CeGdLaSmPrO_{2-\delta}$ (a) and temperature dependence of the electrical conductivity in lnσ (1/T) coordinates (b)

A SEM image of the fractured surface sow of the exemplary samples are presented in Figure 3 (a). The SEM image obtained in a phase contrast mode (detector-BSD) show that the bulk samples are phase homogeneous and have a disordered grain structure with an average size of grains ~ 3,3 μm. Closed pores are observed on the fractured surface. As can be seen from the presented EDS data, the obtained samples are close to the assumed equimolar composition, with the only exception related to the lower content of Pr (Table 1). The presence of pores and the violation of stoichiometry most probably resulted from the evaporation of Pr during thermal treatment, as indicated by the bluish residues visible in the alumina crucibles. According to the results of the EDS analysis, all samples on average, the concentration of praseodymium decreases about 1.5 at.% relative to the pre assigned stoichiometric composition.

Electrical conductivity (σ) was measured in the temperature range from room temperature to 800 °C in air. Figure 3 (b) shows the temperature dependence of the electrical conductivity of the

$CeGdLaSmPrO_{2-\delta}$ ceramic system in lnσ(1/T) coordinates according to the Arrhenius law (see the corresponding equation in figure 3 (b)). [5]

Table 1. Average composition of the studied materials derived from EDS area analysis of bulk samples

Sample	Element (At. %)							
	Ce	Gd	La	Dy	Y	Sm	Pr	O
$CeGdLaSmPrO_{2-\delta}$	8.0	7.9	7.7	-	-	7.8	6.0	62.6
$CeGdDySmPrO_{2-\delta}$	8.3	7.7	-	7.8	-	7.4	5.9	62.9
$CeGdYSmPrO_{2-\delta}$	7.9	7.8	-	-	7.6	7.7	6.1	62.9

The values of the activation energy (E_a) of the conduction process, calculated for two temperature (T) range, from 170 to 400 °C and from 460 to 560 °C. It was found that both in the low-temperature region (170 °C - 400 °C) and in the middle-temperature region (450 ° - 560 °), the activation energy of the $CeGdLaSmPrO_{2-\delta}$ ceramic system remained practically unchanged, amounting to 0.52 eV and 0.53 eV, which is associated with a constant mechanism of oxygen-ionic conductivity caused by the drift motion of oxygen ions under the influence of an external electric field in a given temperature range [6-8]. The electrical conductivity of the obtained ceramics at a temperature of 510 °C reaches $0.1 \cdot 10^{-2}$ S/cm, which meets the requirements for the ionic conductivity of materials for the manufacture of solid electrolytes [6], and corresponds to materials based on cerium oxide.

4. Conclusion

In this work, three different compositions from the $CeGdRLaPrO_{2-\delta}$ (where R: La, Y and Dy) systems were prepared using the method of sol-gel synthesis and sintering at 1500 °C. All the samples under study are single-phase, (body-centered cubic), and the peaks correspond to that space symmetry group Ia-3 (206). Based on the evolution of the lattice constants, as well as the EDS-spectroscopy data, it can be stated that the received materials are composed of the stoichiometric phase of $CeGdRLaPrO_{2-\delta}$ (where R: La, Y and Dy). The synthesized initial powders are nanosized, with an average particle size of 28 nm. When nanopowders are sintered, bulk materials are obtained. Sintering at 1500 °C leads to partial evaporation of praseodymium, which is confirmed by the EDS method. In the low-temperature region (170 °C - 400 °C) and in the middle-temperature region (450 °C - 560 °C), the activation energy of electrical conductivity of the $CeGdLaSmPrO_{2-\delta}$ ceramic system remained practically unchanged, amounting to 0.52 eV and 0.53 eV. The electrical conductivity of the obtained ceramics at a temperature of 510 °C reaches $0.1 \cdot 10^{-2}$ S/cm, which meets the requirements for the ionic conductivity of materials for the manufacture of solid electrolytes.

References

[1] Cantor B, Chang I T, Knight P and Vincent A J, 2004 *Sci. Eng: A* **375-377** 213–218.
[2] Rost C M, Schaet R, Borman T, Moballegh A, Dickey E C, Hou D, Jones J L, Curtarolo S and Maria J P, 2015 *Nat. Commun.* **6** 8485.
[3] Rost C M, Rak Z, Brenner D W and Mari J P, 2017 *J. Am. Ceram. Soc.* **100** 2732–2738.
[4] Bérardan D, Franger S, Meena A K and Dragoe D, 2016 *J. Mater. Chem. A.* **4** 9536–9541.
[5] Panigrahi S, Biswa R C, Anwar S, Besra L and Bhattacharjee S, 2013 *J. Am. Ceram. Soc.* **96** 2846–2851.
[6] Skinner S J and Kilner J A, 2003 *Materials Today* **6** 30-37
[7] Dąbrowa J, Szymczak M, Zajusz M, Mikuła A, Moździerz M, Berent K, Wytrwal-Sarna M, Bernasik A, Stygar M and Świerczek K, 2020 *Journal of the European Ceramic Society* **40** 5870-5881.
[8] Yahiro H, Eguchi Y, Eguchi K and Arai H, 1988 *J.of Applied Electrochemistry* **18** 527-531.

IOP Conf. Series: Materials Science and Engineering 1014 (2021) 012060 doi:10.1088/1757-899X/1014/1/012060

New high-entropy oxide phases with the perovskite structure

O V Zaitseva*, S A Gudkova, E A Trofimov, V E Zhivulin, A Yu Punda, A Yu Starikov and D A Vinnik

Department of Materials Science and Physical Chemistry of Materials, South Ural State University, Chelyabinsk, Russia

*Corresponding author: zaitcevaov@susu.ru

Abstract. Results of studying the structure and chemical composition of perovskite phases synthesized by solid-state sintering in multicomponent oxide systems: $(Ba,Sr,Ca,Mg,Pb)(Ti,Zr)O_3$, $(Ba,Sr,Ca,Mg,Pb)(Ti,Zr,Hf)O_3$, $(Ba,Sr,Ca,Mg,Pb)(Ti,Zr,Hf,Sn)O_3$, and $(Ba,Sr,Ca,Mg,Pb)(Ti,Zr,Hf,Sn,Mn)O_3$ are presented. The chemical composition of the obtained samples was investigated using a JEOL JSM7001F scanning electron microscope equipped with an Oxford INCA X-max 80 energy-dispersive X-ray fluorescence spectrometer. The structure was studied with a Rigaku Ultima IV powder X-ray diffractometer. It was demonstrated the possibility of obtaining a high-entropy perovskite phase. Elements (Ba, Sr, Ca, Ti, Zr, Hf, Sn), which can form the high-entropy perovskites have been determined, as well as the elements (Pb, Mg, Mn), which, for various reasons, are hardly incorporated in the high-entropy perovskites.

1. Introduction

The main goal of this research was to obtain new high-entropy oxides with a perovskite structure by solid-state sintering in systems that were not explored before as high-entropy phases [1-6], and to study the composition and structure of them. Such materials are of potential interest for further physical studies. It is known that a number of perovskites have pronounced magnetic properties, others can participate in the processes of energy conversion and storage, some exhibit superconductivity or ferroelectric properties. In contrast to the known works on partially substituted perovskites, in the course of present work there were studied systems in the concentration area where the entropy of mixing reaches its maximum, i.e. in the region where the concentrations of the main components of one type (A or B for the formula ABO_3) are close to be equimolar.

The results presented include the preparation of high-entropy perovskites in the 10 metals oxide system $(Ba,Sr,Ca,Mg,Pb)(Ti,Zr,Hf,Sn,Mn)O_3$, as well as in more simple systems – $(Ba,Sr,Ca,Mg,Pb)(Ti,Zr)O_3$, $(Ba,Sr,Ca,Mg,Pb)(Ti,Zr,Hf)O_3$ and $(Ba,Sr,Ca,Mg,Pb)(Ti,Zr,Hf,Sn)O_3$.

2. Experiment

The analytical grade $BaCO_3$, $SrCO_3$, $CaCO_3$, $MgCO_3$, PbO, TiO_2, ZrO_2, HfO_2, SnO_2 and MnO_2 were used as precursors. In preliminary experiments the optimum temperature for solid-state sintering was determined. In these experiments, samples of a given composition were heat-treated in air for 5 hours at temperatures from 1000 to 1400 °C with an interval of 50 °C. Then the samples were examined with an electron microscope. The criterion for the optimal synthesis temperature was the large size and well-formed shape of the crystals of high-entropy perovskite phases.

The found optimum temperature (1400 °C) was used for heat treatment of four multi-component compositions. The batch compositions for solid-state sintering were calculated based on the following hypothetical formulas:

$(Ba_{0.2}Sr_{0.2}Ca_{0.2}Mg_{0.2}Pb_{0.2})(Ti_{0.5}Zr_{0.5})O_3$
$(Ba_{0.2}Sr_{0.2}Ca_{0.2}Mg_{0.2}Pb_{0.2})(Ti_{0.33}Zr_{0.33}Hf_{0.33})O_3$
$(Ba_{0.2}Sr_{0.2}Ca_{0.2}Mg_{0.2}Pb_{0.2})(Ti_{0.25}Zr_{0.25}Hf_{0.25}Sn_{0.25})O_3$
$(Ba_{0.2}Sr_{0.2}Ca_{0.2}Mg_{0.2}Pb_{0.2})(Ti_{0.2}Zr_{0.2}Hf_{0.2}Sn_{0.2}Mn_{0.2})O_3$.

These elements were chosen according to the ionic radii of the elements (in the context of the Goldschmidt's rule), their coordination number in the structures related to the perovskite-type, as well as the literature data on the composition of minerals and artificial multicomponent phases with the perovskite-like structure, which are prone to isomorphism (perovskite, celanite, snaps, loparite, dysanalite, rhinitis, tausonite, macedonite, etc.).

Pellets, pressed from a thoroughly ground batch mixture, were sintered for 5 hours in an air atmosphere at a temperature of 1400 °C on a platinum plate placed in a tubular furnace with silicon carbide heaters.

After heat treatment, the fractured surface of the pellets was investigated using a Jeol JSM7001F scanning electron microscope equipped with an Oxford INCA X-max 80 X-ray spectrometer for elemental analysis of samples. In order to study the structure, the obtained samples were ground and investigated by X-ray phase analysis using a Rigaku Ultima IV X-ray powder diffractometer with Cu Kα radiation.

3. Results and discussions

The powder X-ray diffraction study showed that in all four samples the perovskite phase was formed in significant quantities. The other oxide phases were also found in noticeable quantities. The study of the morphology of the sample by electron microscopy made it possible to detect uniform equiaxial or rectangular microcrystals (figures 1-4).

Figure 1. Morphology of the $(Ba,Sr,Ca,Mg,Pb)(Ti,Zr)O_3$ sample.

The electron microscopy study of the fracture of the samples, combined with micro X-ray spectral analysis, made it possible to determine the compositions of the microcrystals. Data on the composition of equiaxial or rectangular microcrystals, that dominate in the sample, are presented in table 1. Here, the content of elements A and B is normalized to 100 atomic % for each group of elements separately. Since several measurements were made for each sample, the table shows average, maximum and minimum results (to demonstrate the range of possible values).

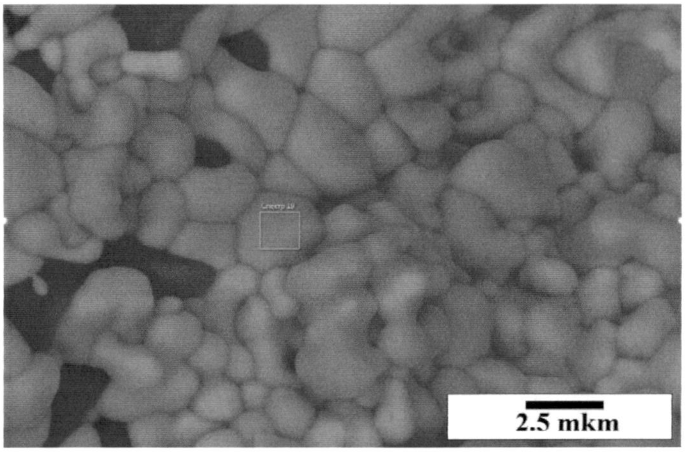

Figure 2. Morphology of the (Ba,Sr,Ca,Mg,Pb)(Ti,Zr,Hf)O_3 sample.

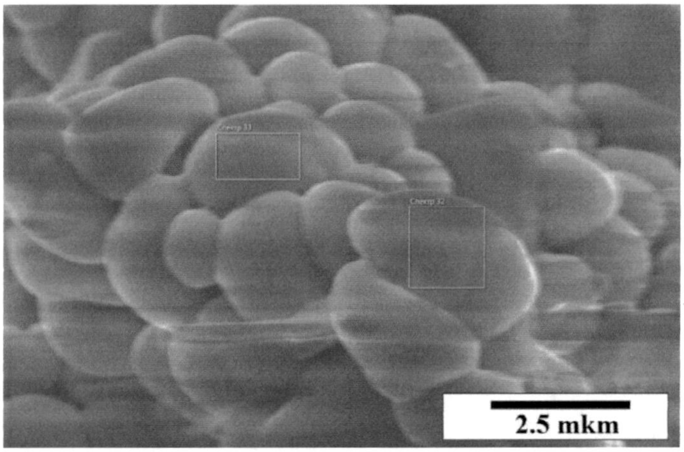

Figure 3. Morphology of the (Ba,Sr,Ca,Mg,Pb)(Ti,Zr,Hf,Sn)O_3 sample.

Figure 4. Morphology of the (Ba,Sr,Ca,Mg,Pb)(Ti,Zr,Hf,Sn,Mn)O_3 sample.

Table 1. Composition of equiaxial or rectangular microcrystals.

Sample	Composition level	[A], at.% of elements of type "A"					[B], at.% of elements of type "B"				
		Ba	Sr	Ca	Mg	Pb	Ti	Zr	Hf	Sn	Mn
(Ba,Sr,Ca,Mg,Pb)(Ti,Zr)O_3	min	24	16	10	3	4	33	27	-	-	-
	max	45	31	30	11	23	73	67	-	-	-
	aver	**37**	**23**	**18**	**6**	**16**	**56**	**44**	**-**	**-**	**-**
(Ba,Sr,Ca,Mg,Pb)(Ti,Zr,Hf)O_3	min	47	18	12	3	1	25	14	44	-	-
	max	57	27	12	21	2	30	26	60	-	-
	aver	**52**	**22**	**12**	**12**	**2**	**28**	**20**	**52**	**-**	**-**
(Ba,Sr,Ca,Mg,Pb)(Ti,Zr,Hf,Sn)O_3	min	25	27	19	1	0	33	15	11	30	-
	max	43	49	24	8	3	36	25	17	32	-
	aver	**37**	**36**	**20**	**5**	**2**	**35**	**20**	**14**	**31**	**-**
(Ba,Sr,Ca,Mg,Pb)(Ti,Zr,Hf,Sn,Mn)O_3	min	39	23	20	0	0	18	20	22	19	1
	max	52	36	24	1	5	21	31	39	23	4
	aver	**46**	**30**	**22**	**0**	**2**	**20**	**26**	**30**	**21**	**3**

An analysis of the composition shows that in the ABO_3 perovskite formula, Ca, Sr, and Ba can certainly act as element "A". Mg in the equiaxial or rectangular microcrystals was found in quantities noticeably smaller than expected from the batch composition. On the other hand, Mg is a component of other phases formed, primarily of phases with octahedral crystals. Pb in the studied samples either is not found or is present in quantities significantly less than expected. This is probably due to the evaporation of Pb oxide during heat treatment.

From table 1 follows that in the perovskite formula, Ti, Zr, Hf, and Sn can certainly act as "B" elements. Mn in the samples is found in quantities noticeably smaller than expected, so, as well as Mg, it may be a component of the formed side phases.

4. Conclusion

Investigation of the structure and chemical composition of the samples belonging to the oxide system (Ba,Sr,Ca,Mg,Pb)(Ti,Zr,Hf,Sn,Mn)O_3, showed the possibility of obtaining high-entropy perovskite phases. The elements that easily form the high-entropy phase are Ca, Sr, Ba, Ti, Zr, Hf and Sn. Pb, Mg, Mn for various reasons practically do not enter the high-entropy phase. Further studies in this direction will make it possible to obtain phase pure samples suitable for studying its electromagnetic characteristics.

Acknowledgments

The research was done with the support of the Russian Science Foundation (project No. 19-73-10046).

References

[1] Vinnik D, Trofimov E, Zhivulin V, Gudkova S, Zaitseva O, Zherebtsov D, Starikov A, Sherstyuk D, Amirov A, Kalgin A, Trukhanov S and Podgornov F 2020 *Nanomaterials* **10** 268

[2] Jiang S, Hu T, Gild J, Zhou N, Nie J, Qin M, Harrington T, Vecchio K and Luo J 2018 *Scr. Mater.* **142** pp 116–120

[3] Sarkar A, Djenadic R, Wang D, Hein C, Kautenburger R, Clemens O and Hahn H 2018 *J. Eur. Ceram. Soc.* **38** pp 2318–27

[4] Biesuz M, Fu S, Dong J, Jiang A, Ke D, Xu Q, Zhu D, Bortolotti M, Reece M J and Hu C 2019 *J. Asian Ceram. Soc.* **7** pp 127–32

[5] Sharma Y, Musico B, Gao X, Hua C, May A, Herklotz A, Rastogi A, Mandrus D, Yan J and Lee H 2018 *Phys. Rev. Mater.* **2** 60404

[6] Zhong Y, Sabarou H, Yan X, Yang M, Gao M, Liu X and Sisson R 2019 *Mater. Des.* **182** 108060

IOP Conf. Series: Materials Science and Engineering 1014 (2021) 012061 doi:10.1088/1757-899X/1014/1/012061

Structure and corrosion properties of stainless steel after high-power ion beam processing

M V Zhidkov*, M Y Gazizova[1], A E Ligachev[2], S K Pavlov[3] and G E Remnev[3]

[1]Belgorod State National Research University, 85 Pobedy Str., 308015, Belgorod, Russia
[2]Prokhorov General Physics Institute of the Russian Academy of Sciences, 38 Vavilova St., 119991 Moscow, Russia
[3]Tomsk Polytechnic University, 30 Lenina St., 634050 Tomsk, Russia

*Corresponding author: zhidkov@bsu.edu.ru

Abstract. In this study, the surfaces of AISI 321 stainless steel samples were irradiated with one shot of a high-power ion beam (HPIB) at pulse energy densities of 1 and 3 J/cm^2. The surface morphology and structural-phase state in the near-surface layers of the treated samples were analyzed using scanning electron microscopy and electron backscatter diffraction. Additionally, the influence of HPIB processing on the resistance to intergranular corrosion was investigated using electrochemical experiments.

1. Introduction

High-power ion beam (HPIB) processing is one of the promising methods used for the surface modification of metallic materials. The high-energy impact of powerful ion beams with an intensity of 10^7 W/cm^2 and more can be accompanied by a number of processes, such as heating, melting and evaporation, excitation of compression waves in the target because of the recoil pulse and rapid cooling of the heated surface layer of the target at a rate of 10^9-10^{10} K/s [1]. These processes lead to structural-phase transformations and changes in the surface topography of materials. Moreover, materials exposed to HPIB exhibit appreciable surface properties, such as resistance to wear and corrosion, microhardness, and fatigue strength [2-4].

The main features of the formation of craters on high-speed steel surfaces after ion irradiation have been extensively studied [5, 6]. Additionally, another study investigated the topography and properties of 316L stainless steel after HPIB processing [7].

In this study, we elucidated the topographical and structural changes in AISI 321 stainless steel irradiated by a HPIB and investigated its subsequent susceptibility to intergranular corrosion.

2. Experimental details

AISI 321 stainless steel (Fe–0.12C–18Cr–10Ni–0.5Ti) was used in our experiments after warm rolling ($\varepsilon = \sim 1$, T = 400 °C). This material was prepared in the form of rectangular slabs, and its surfaces were mechanically polished using a rotary polishing machine LaboPol-5 (Struers). Ion irradiation of the samples was performed using a TEMP-4M pulse accelerator [8]. The composition of the ion beam was as follows: carbon ions (C^+, C^{n+}) and protons (H^+) at a ratio of 85:15. The accelerating voltage was 250 kV, and the pulse duration (at half maximum of the accelerating voltage) was 100 ns. The residual pressure in the chamber was 2×10^{-2} Pa. The pulse energy densities (F) were 1 and 3 J/cm^2.

The structure and topography of the ion-beam-processed surface were studied using a scanning electron microscope (SEM, FEI Quanta 600) equipped with energy dispersive spectrometry (EDX, resolution: 0.2.–0.25%) and electron backscatter diffraction (EBSD).

Electrochemical experiments were performed using an IPC-Pro potentiostat and a standard three-electrode system. Silver chloride was used as the reference electrode with a carbon counter electrode. The corrosion resistance was examined in a solution of 0.01 mol/L KSCN and 0.5 mol/L H_2SO_4 using the double loop electrochemical potentiodynamic reactivation (DL-EPR) method at 30 °C.

3. Results and discussion

3.1. Surface topography

Figure 1 illustrates the SEM images of the irradiated stainless steel surfaces at different values of F. Rounded craters with concave centers were formed on the steel surfaces after ion beam processing. Craters were uniformly distributed over the sample surface. The average crater size was approximately 20 μm for F = 1 J/cm^2, and increased to 25 μm for F = 3 J/cm^2. The surface density of the craters increased from 2.7 x 10^4 to 4 x 10^4.

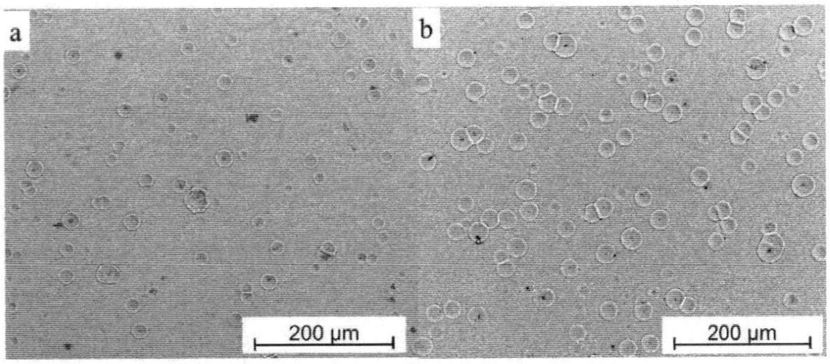

Figure 1. Top-view topography of the surface of samples of AISI 321 steel irradiated by HPIB at F=1 J/cm^2 (a) and F=3 J/cm^2 (b).

Figure 2 illustrates the EDX maps of the craters on the irradiated surface of AISI 321 stainless steel. Figure 2 (b) corresponds to the map distribution of chemical elements: Ti, S, Fe, Cr, Ni. In our experiments, the centers of the craters were rich in Ti and S, but poor in Ni, Cr and Fe. Craters appear to be formed on the inclusions of TiS or $Ti_4C_2S_2$ [9] that were present in the steel prior to processing.

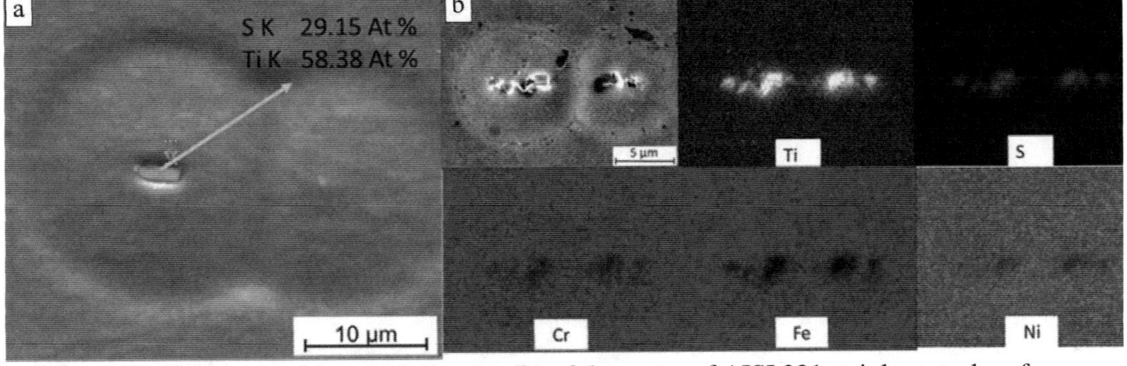

Figure 2. SEM image (a) and EDX maps (b) of the crater of AISI 321 stainless steel surface irradiated by HPIB.

The mechanism of crater formation has been proposed [5]. If the energy density exceeds the ablation threshold, the inclusions vaporize prior to matrix. Local ablation and eruption induced by partial vaporization drive the melt flow, leading to the formation of craters [5]. In [6], it was reported that the inclusion can also generate a crater during ion processing, despite the lack of vaporization of the inclusions on the target surface [6]. In this case, the difference in surface tension between the molten inclusion and the matrix can drive the melt to flow, generating a crater on the molten target surface [6].

3.2. Structure and phase composition of the near-surface layers

Figure 3 illustrates the structure of the steel surface before and after ion beam processing. In the initial state, the steel has a fiber-like structure with a mean grain size of approximately 500 nm and an axial inequality coefficient of approximately 5. The share of high-angle grain boundaries (HAGBs) was 57%. The ion irradiation altered the near-surface layer structure. The steel near-surface layers obtained at F = 1 J/cm^2 consist of a structure close to the equiaxed ultrafine-grains, with an average grain size of ~450 nm. In this case, the share of HAGB increased to 90%. Meanwhile, after irradiation with F = 3 J/cm^2, the average grain size increased to 600 nm (Figure 3c).

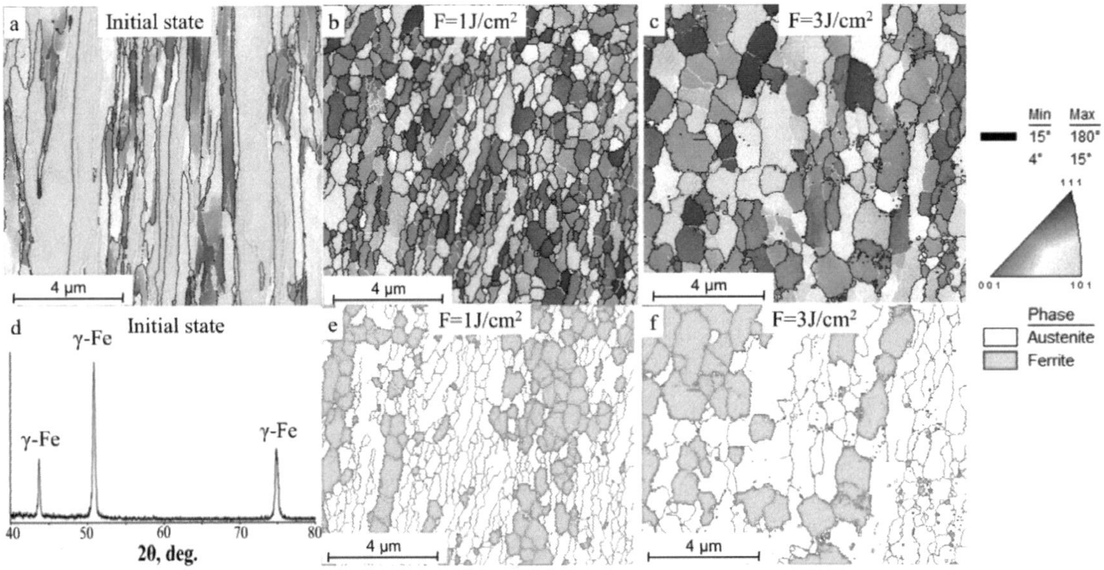

Figure 3. Microstructure and phase composition of AISI 321 steel before (a, d) and after HPIB processing at F = 1 J/cm^2 (b, e) and F = 3 J/cm^2 (c, f). OIM (a-c), phase distribution (e, f), EBSD maps, and XRD pattern (d).

In the initial state, the steel exhibits a single austenite (fcc) phase structure (Figure 3d). Ion irradiation promotes the formation of the ferrite (bcc) phase that coexists with the fcc phase (Figure 3e,f). According to the EBSD data, the fraction of ferrite is 35% for F = 1 J/cm^2 and 23% for F = 3 J/cm^2.

3.3. Corrosion properties

The DL-EPR polarization curves for AISI 321 stainless steel are shown in Figure 4. The current density of the first (passivation) peak increases after ion beam processing (No.1-3 in Figure 4a). The increase in the current density indicates more intense dissolution. This can be attributed to the formation of a secondary phase (ferrite) in the near-surface layers after irradiation. Simultaneously, the intensity of the reverse (reactivation) peak decreases compared with that of the peak of the original untreated surface (No.1-3 in Figure 4b). This may be associated with the redistribution of chromium due to HPIB processing, which facilitates the formation of a more stable protective oxide film on the steel surface.

The degree of sensitization is characterized by the reactivation coefficient R_r; it is the ratio of the areas of the reactivation peak (Figure 4b) to the passivation peak (Figure 4a). According to the method used, the steel with a R_r of less than 0.11 is resistant to intergranular corrosion. The smaller the R_r value, the greater the intergranular corrosion resistance. The original state with the R_r = 0.06 is resistant to intergranular corrosion, as can be observed from Figure 4c. R_r decreased to 0.009 after HPIB processing with F = 1 J/cm^2. The steel had the same resistance to intergranular corrosion (R_r = 0.007) as the coarse-grained austenitic structure after processing with F = 3 J/cm^2.

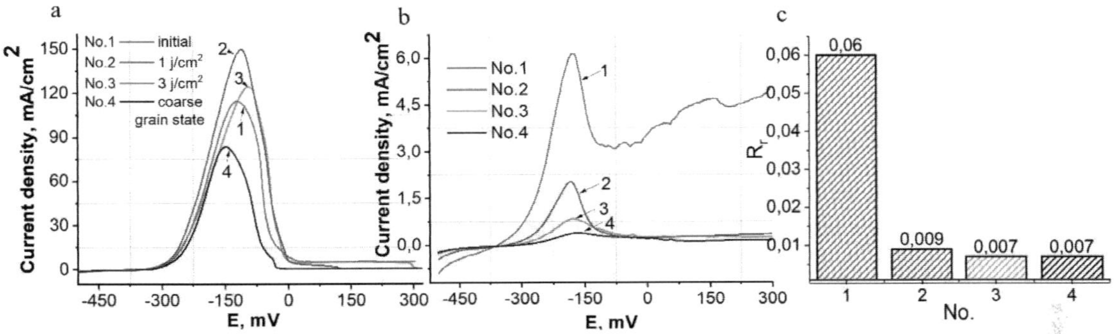

Figure 4. Potentiodynamic polarization curves [forward (a) and reverse (b)] and reactivation ratios (c) of AISI 321 stainless steel before and after HPIB processing.

4. Conclusion

1. The surface irradiation of AISI 321 stainless steel by a high-power ion beam leads to surface defects in the form of rounded craters. It was demonstrated that inclusions based on Ti and S are the preferred sites for crater formation.
2. HPIB processing of AISI 321 stainless steel causes a variation in the structure of the surface layers, involving the formation of microstructure grains that are more uniform in size. This results in a high degree of equiaxiality and an increased fraction of high-angle grain boundaries in the modified layer.
3. The resistance of steel to intergranular corrosion improves significantly after HPIB processing.

Acknowledgments
The reported study was funded by RFBR, project number 20-08-00907.

References

[1] Korotaev A D, Tyumentsev A N, Pinzhin Yu P, Remnev G E, 2004 Surface & Coatings Technology **185** 38–49
[2] Shulov V A, Nochovnaya N A, Remnev G E, 1998 Materials Science and Engineering A **243** 290 –3
[3] Ma X, Zhang G, Wang G, Zhu G, Zhou W, Wang J, Sun B, 2014 Applied Surface Science **311** 567-73
[4] Wang X, Lei M K, Zhang J S, 2007 Surface & Coatings Technology **201** 5884–90
[5] Zhong H, Zhang J, Shen J, Yu X, Liang G, Cui X, Zhang X, Zhang G, Yan S, Le X, 2017 Nuclear Instruments and Methods in Physics Research B **409** 298-301
[6] Zhong H, Zhang J, Shen J, Liang G, Zhang Sh, Xu M, Yu X, Yan S, Remnev G E, Le X, 2020 Vacuum **179** 109541
[7] Wang X, Han X G, Lei M K, Zhang J S, 2007 Materials Science and Engineering A **457** 84–89
[8] Remnev G E, Shulov V A 1993 Laser and Particle Beams **11** 707–731
[9] Wilson P R and Chen Z, 2007 Scripta Materialia **56** 753–756

New high-entropy oxide phases with the magnetoplumbite structure

V E Zhivulin*, E A Trofimov, A Yu Starikov, S A Gudkova, A Yu Punda, D A Zherebtsov, O V Zaitseva and D A Vinnik

Department of Materials Science and Physical Chemistry of Materials, South Ural State University, Chelyabinsk, Russia

*Corresponding author: splav.zlat@mail.ru

Abstract. In the studied BaO–SrO–CaO–La_2O_3–Fe_2O_3–TiO_2–Al_2O_3–Mn_2O_3–CoO–In_2O_3–Ga_2O_3–Cr_2O_3 system there were synthesized three new high-entropy magnetoplumbite phases by solid-state sintering in air at 1300, 1350, and 1400 °C.
The average composition of the high-entropy phases obtained at 1350 °C can be described by the formulas:
$Ba_{0.48}Sr_{0.30}Ca_{0.05}La_{0.15}Fe_{1.62}Ti_{1.78}Al_{1.55}Mn_{1.45}Co_{1.11}In_{1.29}Ga_{1.52}Cr_{1.71}O_{19}$;
$Ba_{0.39}Sr_{0.37}Ca_{0.05}La_{0.18}Fe_{2.88}Ti_{1.60}Al_{1.69}Mn_{1.09}Co_{0.92}In_{1.20}Ga_{1.27}Cr_{1.38}O_{19}$;
$Ba_{0.37}Sr_{0.29}Ca_{0.05}La_{0.28}Fe_{5.85}Ti_{1.15}Al_{0.97}Mn_{0.74}Co_{0.56}In_{0.85}Ga_{0.94}Cr_{0.97}O_{19}$.
The data obtained indicate that a significant portion of calcium was not incorporated into the high-entropy phase. This must be taken into account in the course of subsequent experimental work on the creation of homogeneous phases necessary for the study of their electrical and magnetic properties.

1. Introduction

In recent years, significant work has been carried out in obtaining novel non-metallic high-entropy systems [1-10]. The aim of our recent works is to study the possibility of obtaining multicomponent oxide phases with a magnetoplumbite structure, characterized by a high value of the configurational entropy of mixing [11-15]. Powder X-ray diffraction analysis of the samples allows us to confirm the phase purity of the created materials. Based on the obtained diffraction patterns, the parameters of the crystal lattice of the phases were calculated and a conclusion was made that these crystal structures belong to the magnetoplumbite structure (space group $P6_3/mmc$).

Such materials have potential applications in various areas of science and technology: in manufacturing permanent magnets, in magneto-optics, acoustoelectronics, microwave devices, and information storage devices. Studies carried out in recent decades have revealed that control of the properties of hexaferrites is possible by replacing some of the iron atoms with atoms of other elements [16-17]. Obtaining a high-entropy material with the magnetoplumbite structure opens up wide possibilities for smooth regulation of the entire complex of its electromagnetic characteristics, according to the demand of electronics manufacturers.

This work report the results of the synthesis of multicomponent phases with the magnetoplumbite structure in the BaO–SrO–CaO–La2O3–Fe_2O_3–TiO_2–Al_2O_3–Mn_2O_3–CoO–In_2O_3–Ga_2O_3–Cr_2O_3 system, characterized by a high configuration entropy of mixing. It is assumed that a high entropy of mixing can stabilize multicomponent solid solutions with the magnetoplumbite structure.

2. Experiment

The synthesis was carried out in a system having the formula $(Ba,Sr,Ca,La)Fe_x(Ti,Al,Mn,Co,In,Ga,Cr)_{12-x}O_{19}$, where x = 1.5, 3 and 6. It is obvious that the maximum configurational entropy in the sublattice formed by the elements Fe, Ti, Al, Mn, Co, In, Ga, and Cr is achieved when the concentrations of all these elements are equal to x = 1.5. However, in this work, we also investigated compositions having a larger iron fraction than the fraction of the other elements of B type (in a general formula $AB_{12}O_{19}$). An increase of the iron content will lead to a slight decrease in the configurational entropy of mixing ($\Delta S_{conf.}$); however, compositions with larger iron content are of higher interest from the point of view of potential applications of the materials obtained. The batch compositions for solid-state sintering were calculated in accordance with the above formula (table 1).

Table 1. Batch compositions of $(Ba,Sr,Ca,La)Fe_x(Ti,Al,Mn,Co,In,Ga,Cr)_{12-x}O_{19}$ samples.

	Sample number		
	1	2	3
x Fe	1.5	3	6
$BaCO_3$, wt. %	4.247	4.274	4.329
$SrCO_3$, wt. %	3.177	3.197	3.239
$CaCO_3$, wt. %	2.154	2.168	2.196
La_2O_3, wt. %	3.506	3.528	3.574
Fe_2O_3, wt. %	10.310	20.752	42.040
TiO_2, wt. %	10.313	8.896	6.007
Al_2O_3, wt. %	6.583	5.679	3.835
Mn_2O_3, wt. %	10.193	8.792	5.937
CoO, wt. %	9.676	8.346	5.636
In_2O_3, wt. %	17.925	15.463	10.442
Ga_2O_3, wt. %	12.102	10.439	7.049
Cr_2O_3, wt. %	9.813	8.465	5.716

The following reagents were used for the experiments: $BaCO_3$, $SrCO_3$, $CaCO_3$, La_2O_3, Fe_2O_3, TiO_2, Mn_2O_3, CoO, In_2O_3, Cr_2O_3, Ga_2O_3, Al_2O_3 of analytical grade purity.

Weighed portions of reagents were thoroughly mixed, pressed into tablets, and heated in air in a SiC furnace at 1300, 1350, or 1400 °C for 5 hours. After heat treatment, the sintered samples were crashed, and the fracture surfaces were examined using a Jeol JSM7001F scanning electron microscope equipped with an Oxford INCA X-max 80 X-ray spectrometer for elemental analysis. In order to confirm the structure, the obtained samples were investigated by X-ray phase analysis using a Rigaku Ultima IV X-ray powder diffractometer.

3. Results and discussions

Typical microcrystals found in the samples are presented in figure 1. The averaged results of X-ray spectral analysis of hexagonal microcrystals are listed in table 2. The compositions are presented as the average number of metal atoms per 13 metal atoms in the formula unit. In addition to the elemental analysis data, the table shows the target compositions (when the batch components would be completely incorporated into the magnetoplumbite phase).

From the data presented, it can be seen that the ratio between the number of atoms of type B and the number of atoms of type A in hexagonal crystals of all samples is quite close to 12. This ratio, as well as the data of X-ray phase analysis, indicate that solid-state synthesis enables obtaining high-entropy phases with the magnetoplumbite structure.

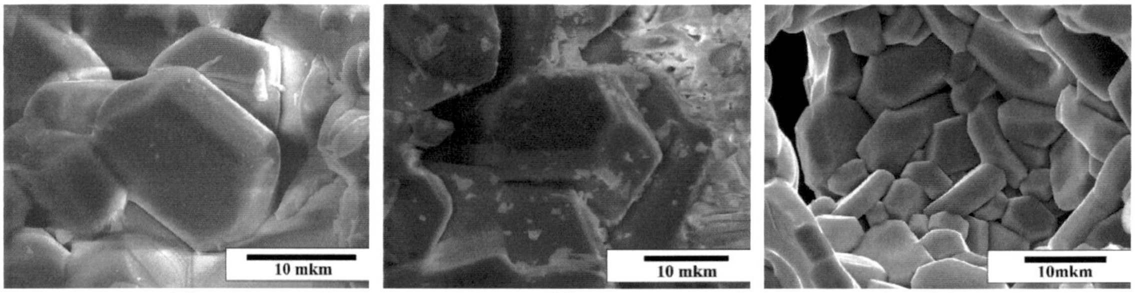

(a) (b) (c)

Figure 1. Morphology of crystals found in the samples: (a) composition 1, 1350 °C; (b) composition 2, 1400 °C; (c) composition 3, 1300 °C.

High-entropy magnetoplumbite phases were formed at all three temperatures used for synthesis. The obtained micrographs (figure 1) show that at higher temperatures, the average size of the formed hexagonal crystals increases. At the same time, an increase in temperature leads to the appearance of a larger number of molten areas, as well as particles of irregular shape.

The average composition of high-entropy phases in samples obtained at 1350 °C can be described by the formulas:

$Ba_{0.48}Sr_{0.30}Ca_{0.05}La_{0.15}Fe_{1.62}Ti_{1.78}Al_{1.55}Mn_{1.45}Co_{1.11}In_{1.29}Ga_{1.52}Cr_{1.71}O_{19}$;

$Ba_{0.39}Sr_{0.37}Ca_{0.05}La_{0.18}Fe_{2.88}Ti_{1.60}Al_{1.69}Mn_{1.09}Co_{0.92}In_{1.20}Ga_{1.27}Cr_{1.38}O_{19}$;

$Ba_{0.37}Sr_{0.29}Ca_{0.05}La_{0.28}Fe_{5.85}Ti_{1.15}Al_{0.97}Mn_{0.74}Co_{0.56}In_{0.85}Ga_{0.94}Cr_{0.97}O_{19}$.

The data obtained indicate that under the experimental conditions a large amount of calcium was not incorporated into the high-entropy phase. This fact must be taken into account considering subsequent syntheses of homogeneous phases necessary for the study of their electrical and magnetic characteristics. Probably, the optimal composition of the batch for the synthesis should be chosen from the compositions of the formed phases (table 2).

Table 2. Averaged results of hexagonal crystals elemental analysis (the atomic fractions are presented as its number normalized by 13 atoms).

Sample number	[A] (for $AB_{12}O_{19}$)				[B] (except [Fe] for $AB_{12}O_{19}$)							[Fe]
	[Ba]	[Sr]	[Ca]	[La]	[Al]	[Ti]	[Cr]	[Mn]	[Co]	[Ga]	[In]	
1 (batch)	0.25	0.25	0.25	0.25	1.5	1.5	1.5	1.5	1.5	1.5	1.5	1.5
1(experiment)	0.476	0.300	0.051	0.152	1.545	1.784	1.713	1.445	1.110	1.518	1.287	1.619
2 (batch)	0.25	0.25	0.25	0.25	1.286	1.286	1.286	1.286	1.286	1.286	1.286	3
2(experiment)	0.389	0.367	0.050	0.176	1.688	1.597	1.375	1.093	0.922	1.267	1.199	2.879

3 (batch)	0.25	0.25	0.25	0.25	0.857	0.857	0.857	0.857	0.857	0.857	0.857	6
3(experiment)	0.371	0.291	0.053	0.278	0.969	1.146	0.965	0.736	0.561	0.936	0.847	5.848

4. Conclusions

During the study of the $BaO–SrO–CaO–La_2O_3–Fe_2O_3–TiO_2–Al_2O_3–Mn_2O_3–CoO–In_2O_3–Ga_2O_3–Cr_2O_3$ system, the new high-entropy phases with the magnetoplumbite structure were obtained by solid-state sintering. The average composition of these phases obtained at a temperature of 1350 °C can be described by the formulas:

$Ba_{0.48}Sr_{0.30}Ca_{0.05}La_{0.15}Fe_{1.62}Ti_{1.78}Al_{1.55}Mn_{1.45}Co_{1.11}In_{1.29}Ga_{1.52}Cr_{1.71}O_{19}$;
$Ba_{0.39}Sr_{0.37}Ca_{0.05}La_{0.18}Fe_{2.88}Ti_{1.60}Al_{1.69}Mn_{1.09}Co_{0.92}In_{1.20}Ga_{1.27}Cr_{1.38}O_{19}$;
$Ba_{0.37}Sr_{0.29}Ca_{0.05}La_{0.28}Fe_{5.85}Ti_{1.15}Al_{0.97}Mn_{0.74}Co_{0.56}In_{0.85}Ga_{0.94}Cr_{0.97}O_{19}$.

Conclusions are drawn about the possible directions for obtaining homogeneous samples.

Acknowledgments

The work was supported by the Russian Science Foundation (project No. 18-73-10049).

References

[1] Rost C, Sachet E, Borman T, Moballegh A, Dickey E, Hou D, Jones J, Curtarolo S and Maria JP 2015 *Nat. Commun.* **6** 8485
[2] Bérardan D, Franger S, Dragoe D, Meena A and Dragoe N *Phys. Status Solidi RRL* 2016 **10** pp 328–33
[3] Sarkar A, Djenadic R, Usharani N, Sanghvi K, Chakravadhanula V, Gandhi A, Hahn H and Bhattacharya S 2017 *J. Eur. Ceram. Soc.* **37** pp 747–54
[4] Berardan D, Franger S, Meena A and Dragoe N 2016 *J. Mater. Chem. A* **4** pp 9536–41.
[5] Rak Z, Rost C, Lim M, Sarker P, Toher C, Curtarolo S, Maria J P and Brenner D 2016 *J. Appl. Phys.* **120** 9
[6] Rost C, Rak Z, Brenner D and Maria J P 2017 *J. Am. Ceram. Soc.* **100** pp 2732–38
[7] Berardan D, Meena A K, Franger S, Herrero C and Dragoe N 2017 *J. Alloys Compd.* **704** pp 693–700
[8] Sarkar A, Loho C, Velasco L, Thomas T, Bhattacharya S S, Hahn H and Djenadic R 2017 *Dalton Trans.* **46** pp 12167–76
[9] Djenadic R, Sarkar A, Clemens O, Loho C, Botros M, Chakravadhanula V S, Kübel C, Bhattacharya S, Gandhi A and Hahn H 2017 *Mater. Res. Lett.* **5** pp 102–09
[10] Lin M I, Tsai M H, Shen W J and Yeh J W 2019 *Thin Solid Films* **518** pp 2732–37
[11] Zaitseva O, Vinnik D and Trofimov E, 2019 *Mater. Sci. Forum* **946** pp 186–91
[12] Vinnik D, Zhivulin V, Trofimov E, Starikov A, Zherebtsov D, Zaitseva O, Gudkova S, Taskaev S, Klygach D, Vakhitov M, Sander E, Sherstyuk D and Trukhanov A 2019 *Nanomaterials* **9** 559
[13] Vinnik D, Trofimov E, Zhivulin V. Zaitseva O, Gudkova S, Starikov A, Zherebtsov D, Kirsanova A, Häßner M and Niewa R 2019 *Ceram. Int.* **45** pp 12942–48
[14] Vinnik D, Trofimov E, Zhivulin V, Zaitseva O, Zherebtsov D, Starikov A, Sherstyuk D, Gudkova S and Taskaev S 2020 *Ceram. Int.* **46** pp 9656–9660
[15] Vinnik D, Trukhanov A, Podgornov F, Trofimov E, Zhivulin V, Starikov A, Zaitseva O, Gudkova S, Kirsanova A, Taskaev S, Uchaev D, Trukhanov S, Almessiere M, Slimani Y. and Baykal A 2020 *J. Eur. Ceram. Soc.* **40** pp 4022–28
[16] Vinnik D, Ustinova I, Ustinov A, Gudkova S, Zherebtsov D, Trofimov E, Zabeivorota N, Mikhailov G and Niewa R 2017 *Ceram. Int.* **17** pp 15800–04
[17] Pullar R C 2012 Hexagonal ferrites: A review of the synthesis, properties and applications of hexaferrite ceramics *Prog. Mater. Sci.* **57** pp 1191–1334

IOP Conf. Series: Materials Science and Engineering 1014 (2021) 012063 doi:10.1088/1757-899X/1014/1/012063

Nanoscale structural evolution of Al2Cu precipitates in friction-stir welded Al–Cu–Mg alloy

I Zuiko*, S Malopheyev, S Mironov, R Kaibyshev

Laboratory of Mechanical Properties of Nanostructured Materials and Superalloys, Belgorod National Research University. Belgorod 308015, Pobeda str. 85, Russia

*Corresponding author: zuiko.ivan@gmail.com

Abstract. The work is devoted to FSW of cold rolled AA2519. The non-uniform distribution of all second phase precipitates (including coherent θ″ and non-coherent θ) was shown *via* the TEM-technique. The nucleation of the stable θ-phase, along with decreased dislocation density, led to a drop (~60 $HV_{0.2}$) in the microhardness of the nugget zone.

1. Introduction

Over the past hundred years, age hardening in alloys has been successfully exploited on an industrial scale to strengthen materials. For example, the most remarkable improvements in strength of Al–Cu–Mg alloys (i.e., 2XXX-series) are obtained through the formation of secondary precipitates in course of ageing. Previous studies [1-8] showed that precipitation scenario in Al–Cu–Mg alloys (with Cu/Mg ratio >5.6) could be represented as follows:

SSSS → {001} clusters → GP zones (Al_3Cu) → θ″ (Al_2Cu) → θ′ (Al_2Cu) → θ (Al_2Cu) (1)

{111} clusters → Ω (Al_2Cu) → θ′ (Al_2Cu) (2)

where SSSS stands for supersaturated solid solution and GP stands for Guinier-Preston zones. The first and second sequences occur on $\{100\}_\alpha$ and $\{111\}_\alpha$, respectively [2-5].

Now it is realized that to obtain maximum benefits in terms of strength, the age-hardened Al–Cu–Mg alloys need to be subjected to thermo-mechanical treatment which includes solid solution treatment, cold working and subsequent artificial ageing. Usually, a straighten procedure is applied to relieve the residual stress induced by quenching. But materials used in industry should demonstrate weldability. Experience has shown that this could be produced by applying a solid-state joining process called friction stir welding (FSW). However, in age hardening alloys, the heat generated during FSW causes the evolving of grains, secondary phases and crystalline defects (which affect the mechanical properties). Essentially, the weld microstructure is determined by the thermal cycle, which can be varied by changing the welding parameters (primarily, the tool travel speed and spindle rate). It should be mentioned that FSW generates a non-homogeneous distribution of temperature and, hence, a mixed (non-homogeneous) microstructure appears. Typically, it consists of a base material, a heat affected zone, a thermo-mechanically affected zone and a stir zone.

The deformation process associated with the welding procedure is very complicated and involves recovery and continuous dynamic recrystallization. The precipitation phenomenon in heat-treatable aluminium alloys is sophisticated and includes dissolution, coarsening and phase transformations. All aforementioned processes are complicated by the presence of a large quantity of crystalline defects (first of all deformation induced vacancies, dislocations and moving boundaries), which are unequivocally affected.

In the most cases, FSW is applied to completely treated (aged) material. It is thought that in the thermo-mechanically affected zone and the stir zone, the precipitates are dissolved, whereas in the heat-affected zone, particle coarsening occurs. Therefore, the demand for improving their properties is an urgent scientific challenge. The authors hypothesized that the production of high-quality welds of heat-treatable Al alloys may be possible through applying FSW after solution treatment and cold deformation. In this case, the thermal cycle will be similar to ageing. To further increase the strength characteristics, the use of post-weld ageing is proposed. As far as the authors know, this processing strategy has never been used before.

2. Methodology

The nominal chemical composition of the used AA2519 is Al–5.64Cu–0.33Mn–0.23Mg–0.15Zr-0.11Ti–0.09V–0.08Fe–0.08Zn–0.04Sn–0.01Si (in weight %). An ingot of the alloy with dimensions of $30\times100\times120$ mm^3 was homogenized (24 hours at 510°C) and hot rolled (T=425°C) to a thickness of 12 mm ($\varepsilon \sim 0.88$). Next, plates with a thickness of 3.75 mm, a width of 50 mm, and a length of 150 mm were cut, treated with a solid-solution at 535°C for 1h, quenched in water and rolled at room temperature to a total thickness reduction of 20% ($\varepsilon \sim 0.22$). Then the rolled material was friction-stir butt welded using an AccuStir 1004 FSW machine operating in two opposite regimes:

a) "cold" (spindle rate of 500 rpm and tool travel speed of 380 mm/min);
b) "hot" (spindle rate of 1100 rpm and tool travel speed of 760 mm/min)

This division is based on the statement that the FSW is a function of process variables (primary traverse and rotation speeds) [4]. The welding tool was fabricated from tool steel and consisted of a shoulder of 12.5 mm in diameter and an M5 cylindrical probe (pin) of 1.9 mm in length. A tilting angle of 2.5° was employed in all cases. The rolling and welding directions were parallel to each other. Finally, for imparting peak ageing the welded plates were annealed at 165°C for 6 hours. The choice of the regime is based on the analysis of ageing curves, which were shown in the preliminary work [6]. To view the microstructure distribution in the welds (i.e. to define various microstructural zones), the midthickness microhardness profiles were measured. TEM experiments were carried out in the longitudinal plane containing the welding/rolling and normal directions. Foils were cut directly from the centre of the welds (i.e. from the nugget region) and prepared in a standard manner. Importantly, all microstructural observations were conducted on the post-weld heat-treated samples. In the SAED patterns, reflections from the θ', θ'' and Ω phases are marked in green, blue and red colours, respectively. The schemes of indexed SAED patterns as well as other experimental details, could be found in Refs. [2-3, 5-6, 8].

3. Results and discussion

Figure 1 shows the microhardness profile measured across the welded joints in both the "cold-" and "hot-welded" conditions. Previously, we showed that after cold rolling with a reduction of 20% ($\varepsilon \sim 0.22$), the hardness of the alloy increased from 96±1 $HV_{0.2}$ (after solid solution treatment) to 155±1 $HV_{0.2}$. Subsequent age hardening (165°C for 6 hours) leads to a further increase in hardness to 167±1 $HV_{0.2}$. After both types of the welding process, the hardness-distance curves can roughly be divided into three sections. In the first (15-7mm) the hardness of the alloy remains approximately constant. Then the curves (7-3 mm) show a monotonic decrease to ~120 $HV_{0.2}$ and a rapid drop to minimal values (3-0 mm). Usually, such trends in microhardness are typical for non-heat treatable alloys (see Figure 23 in Ref [4]). Obviously, changes in the transverse hardness are associated with various microstructure features (dislocation densities, grains sizes, phase composition).

The TEM technique was employed to interpret the observed drop in microhardness in nugget. The aim of these investigations was to characterize the grain structure, dislocation and precipitation state. The SAED patterns after cold deformation [2,5] contain only diffraction spots from the Al matrix, and no evidence of any precipitates in the images could be discerned. In the meantime, after peak-ageing

(Figure 2), TEM-images and SAED patterns provide robust evidence for the presence of two variants of θ''/θ' on orthogonal sets of the {001} planes. In addition, the SAED patterns exhibit intensity maxima at 1/3 and 2/3 $[220]_\alpha$ (red circles in the inset in Figure 2a) that is a feature (peculiarity) of the Ω-phase. A TEM-micrograph taken along the $<011>_\alpha$ directions, as well as streaking along $\{111\}_\alpha$, unequivocally confirm the presence of two variants of fine plate-shaped Ω-precipitates lying on the $\{111\}_\alpha$ planes of the matrix.

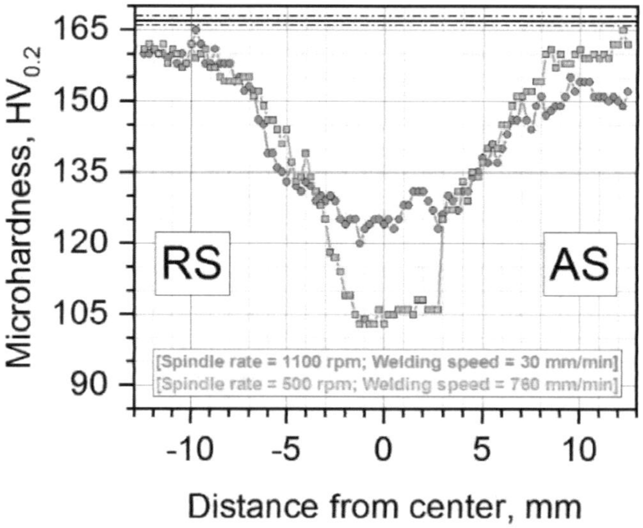

Figure 1. Microhardness profiles measured across transverse cross-sections of the welds. RS and AS denote retreating side and advancing sides, respectively.

Figure 2. Intragranular precipitation of secondary precipitates in the peak aged alloy. In (a) and (c) the electron beam was parallel to the $<001>_\alpha$ and $<011>_\alpha$ directions, respectively.

To compare the microstructural evolution as the function of the welding conditions, we introduce Figures 3 and 4, which represent a series of TEM-micrographs together with the corresponding SAED patterns obtained from the stir zone. Th diversity of the second phases in these conditions is fascinating. Apart from the non-homogeneous distribution of the coherent θ''-phase (Figures 3b, 4b), the analysis of SAED patterns, as well as EDX measurements (not shown) point out to the precipitation of the incoherent θ-phase. It should be noted that traditionally θ are nucleated in the interphase of θ'/matrix interfaces, grain boundaries (more than 9°) or directly from the matrix if the ageing temperature is high enough (i.e. >300°C) [7]. Consequently, it could be assumed that copper available for precipitation has to be shared between the strengthening $\theta''/\theta'/\Omega$ particles and the stable θ-phase. Also, it should be mentioned that presence of θ has a deleterious effect on the mechanical properties (tended to the enhanced corrosion along the rows of particles, reduction of the fatigue resistance of the material).

4. Conclusions

We have investigated the mechanical properties and the related microstructure of the friction stir welded and aged Al–Cu–Mg alloy. The major results are as follows:

1. Ageing after FSW of cold rolled AA2519 is an ineffective way to gain qualitative welds. This is due to the nugget does not contain sufficient solute capable of sustaining subsequent metastable precipitation and age hardening.
2. FSW makes θ formation easier by providing a great number of possible sites for heterogeneous nucleation.
3. The low hardness of the nugget zone is attributed to the low dislocation density and the non-homogeneous distribution of all precipitates (including coherent θ″ and stable θ), as revealed by transmission electron microscopy.

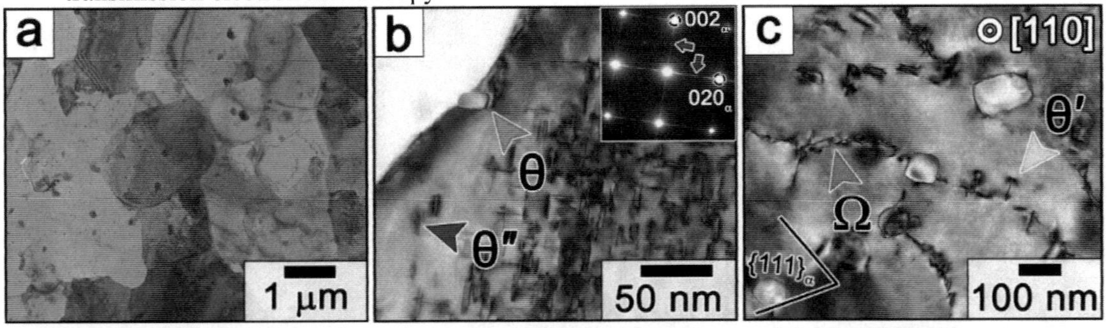

Figure 3. BF-TEM micrographs and corresponding SAED patterns of the "cold" weld.

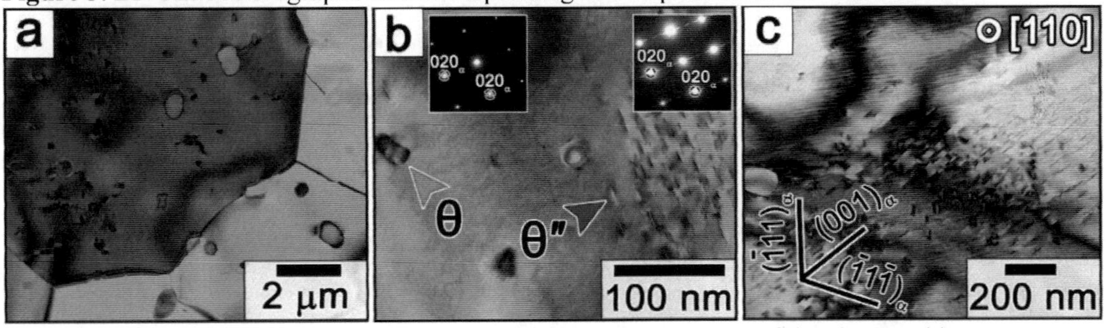

Figure 4. BF-TEM micrographs and corresponding SAED patterns of the "hot" weld.

References

[1] Palmer I, John D St, Nie J F, Qian M, Light Alloys: Metallurgy of the Light Metals, 5th ed., Butterworth-Heinemann, 2017
[2] Zuiko I, Gazizov M, Kaibyshev R, 2018 *Def. and Diff.* For. **385** 290
[3] Zuiko I, Gazizov M, Kaibyshev R, 2014 *Mater. Sci. For.*, **794–796** 888
[4] Threadgill P L, Leonard A J, Shercliff H R, Withers P J, 2009 *Int. Mat. Rev.* **54** 49
[5] Zuiko I, Kaibyshev R, 2018 *Mater. Sci. Eng.* A **737** 401
[6] Zuiko I S, Mironov S, Kaibyshev R, 2020 *Mater. Sci. Eng. A* **793** 139882
[7] Wang S C, Starink M J, 2005 *Int. Mat. Rev.* **50** 193
[8] Zuiko I, Kulitskii V, Kaibyshev R, 2018 *Def. and Diff. For.* **385** 364

AUTHOR INDEX

Institute of Physics
Dirac House, Temple Back
Bristol BS1 6BE UK

ISSN: 1757-8981
ISBN 978-1-7138-4548-5

CURRAN ASSOCIATES INC.
proceedings.com